Polypropylene Film for HVDC Capacitors

Boxue Du · Meng Xiao

Polypropylene Film for HVDC Capacitors

Springer

Boxue Du
Tianjin University
Tianjin, China

Meng Xiao
Tianjin University
Tianjin, China

ISBN 978-981-96-3028-8 ISBN 978-981-96-3029-5 (eBook)
https://doi.org/10.1007/978-981-96-3029-5

This Springer imprint is published by the registered company Springer Nature Singapore Pte Ltd.
The registered company address is: 152 Beach Road, #21-01/04 Gateway East, Singapore 189721, Singapore

Preface

Polypropylene film is the most widely used dielectric material in High Voltage Direct Current (HVDC) capacitors. Its dielectric properties affect the stable operation of power system. With the increase of power system voltage level and transmission capacity, higher requirements are put forward for the dielectric properties of polypropylene films. In addition, in actual operation, HVDC capacitors bear a variety of stresses, which will also affect the dielectric properties of polypropylene films. In recent years, researchers have studied various modification methods for polypropylene film to improve its breakdown field strength and energy storage density under operating conditions, while also reducing its dielectric loss.

This book comprehensively reports the relevant research of polypropylene film for HVDC capacitor. The book provides a detailed introduction to the application background, basic characteristics, and fundamental theory of polypropylene energy storage dielectrics. The relationship between the structure and properties of polypropylene film and the failure mechanism of polypropylene insulation under operating conditions are discussed. The modification methods of polypropylene film including ash washing, molecular chain structure regulation, aggregation structure regulation, nanoparticle doping, polyphenyl ring compound modification, cycloolefin copolymer blending, and interfacial molecular regulation are reviewed. The effects of high temperature and strong magnetic field on the dielectric properties of polypropylene films are discussed, and the polypropylene dielectrics resistant to high temperature and strong magnetic field are introduced. The book contains the dielectric constant, conductivity, breakdown field strength, energy storage density, and other relevant graphics and data of modified polypropylene film under different operating conditions, which clearly explains the improvement effect of various modification methods on the dielectric performance of polypropylene film.

This book provides a comprehensive reference for the research of polypropylene film for HVDC capacitor. It can help students, researchers, and technicians in related industries quickly understand the relevant theories of polypropylene film used in capacitors, as well as provide insights into the latest modification methods and improvements in the dielectric performance of polypropylene film.

Tianjin, China

Boxue Du
Meng Xiao

Contents

Part I
Fundamental Theory

Chapter 1
HVDC Capacitor

1.1 Introduction

In order to promote the revolution of energy production and consumption, and build a low-carbon, clean, safe and efficient energy system, a new power system with wind power generation, photovoltaic power generation and other new energy sources will be established vigorously in the world [1]. The high voltage direct current (HVDC) technology has rapidly developed due to its advantages of high power and long-distance transmission, which could efficiently deliver electricity generated from renewable energy sources to major load centers [2]. In equipment such as ultra-HVDC transmission valve hall and pulse energy storage system, metallized film capacitors assume important roles such as voltage support and energy conversion, whose safety and reliability mainly depend on dielectric insulation performance [3]. However, with the voltage level and capacity of new energy power equipment continuously improving, the deterioration of dielectric performance will occur during actual operation, which seriously threatens the safe and stable operation of the system [4]. Therefore, the working environment characteristics and structure of HVDC capacitors need to be explored to further improve its dielectric performance has received widespread attention from researchers.

This chapter will provide an overview of the application of metallized thin film capacitors in high-voltage direct current (HVDC) transmission projects. It will discuss their pivotal role in enhancing efficiency and reliability within HVDC systems, highlighting their specific advantages in terms of performance and longevity. Following this, the structure of metallized thin film capacitors will be detailed, focusing on their layered composition and the materials utilized to achieve optimal electrical properties. Subsequently, the manufacturing process of metallized thin film capacitors will be comprehensively explained. This section will delve into the industrial preparation of thin films, detailing the deposition techniques and material selection criteria critical for achieving uniformity and high performance. The process flow of capacitor assembly will also be outlined, covering each stage from initial

B. Du and M. Xiao, *Polypropylene Film for HVDC Capacitors*,
https://doi.org/10.1007/978-981-96-3029-5_1

substrate preparation to final encapsulation, emphasizing quality control measures implemented at each step to ensure consistency and reliability. Finally, the chapter will explore the detection methods employed for metallized thin film capacitors. It will discuss non-destructive testing techniques such as capacitance measurement, insulation resistance testing, and high-voltage breakdown testing. Specialized equipment and procedures used for these tests will be detailed, along with their significance in verifying capacitor performance and identifying potential defects or irregularities that could impact operational efficiency and reliability in HVDC applications.

1.2 Metalized Film Capacitor

1.2.1 Working Environment

With the rapid development of social economy, the demands for power resources have been showing an increasing trend. High voltage direct current (HVDC) system possesses the advantages of long transmission distance, large capacity, high efficiency, etc. [5]. Thus, it has become the best way to realize the optimal energy allocation and the new energy grid connection.

The first power transmission line in the world was a direct current line built in 1882. Alternating current (AC) technology has long dominated the field of power transmission since the famous 'battle of the currents' won by Westinghouse Electric Company and Nikola Tesla in the 1890s. However, since the middle of the twentieth century, DC transmission technology has begun to experience a renaissance. Since then, more than 200 HVDC projects have been established around the world, which have demonstrated significant efficiency advantages in transporting large amounts of power over long distances. In recent years, DC transmission has gained particular momentum owing to two key factors: firstly, advances in modern conversion technologies (in particular voltage source converters, VSCs), and secondly, the growing demand for offshore wind power transmission.

The reason why high-voltage direct current transmission can flourish in the world, in addition to the rectifier technology is attributed to the continuous progress and the emergence of high-power rectifier devices, more importantly, HVDC transmission has an incomparable advantage of AC transmission.

Low consumption. Traditional three-phase AC lines require three wires, and line corridor width. While DC transmission lines only need positive and negative poles, the structure of the tower is simple, and the line corridor is narrow. In addition, a DC transmission line with the same voltage transmission capacity of about two times the AC transmission line, its transmission efficiency is about two times or more than the AC line.

Long distance, large capacity. DC transmission does not exist in AC transmission power angle stability problems, so it will not be due to static stability or transient stability performance deterioration and reduce the transmission capacity. This is an important feature of DC transmission power transfer, but also one of its major advantages. DC transmission capacity is determined by the converter valve current allowable value, transmission capacity and distance are not subject to synchronous operation of the AC system at both ends of the limitations, in favour of long-distance high-capacity power transmission.

Easy interconnection. For the AC transmission system, the grid in the various systems must be synchronous operation, or may form a strong circulating current in the equipment and damage to the equipment, or blackout accidents. The AC systems at both ends of the DC transmission do not need to run synchronously, and non-simultaneous contact between AC systems of different rated frequencies or the same rated frequency can be realized. This improves the ability of AC systems on both sides to serve as a backup for each other as well as to provide emergency support in case of accidents, thus improving the stability of the AC system and the economy of power supply.

The imbalance of energy distribution in China poses a challenge to the rapid and sustainable development of the national economy and its derived flexible DC grid technology provide effective technical solutions to the problems of effective access to renewable energy sources, large-scale power transmission over long distances, and resource constraints in transmission corridors. With the significant progress of HVDC technology, the demand for power capacitors grows.

The DC transmission project includes converter stations, DC lines, power filters on the AC and DC sides, reactive power compensation devices, converter transformers, DC reactors, and protection and control devices, etc. In HVDC transmission system, the converter station relies on thyristors, capacitors, and reactors to realize the conversion of AC and DC power. Figure 1.1 shows the schematic diagram of the basic layout of the converter station. The AC and DC systems in the converter station are connected through the converter valve hall, in which metallized film capacitors (MFCs) are the core equipment. In actual projects, MFCs play a significant role in HVDC projects, such as filtering, voltage equalization, voltage support, auxiliary commutation, damping buffer and energy storage [6].

Figure 1.2 shows the topological structure diagram of modular multi-level converter valve. The large-capacity modular multilevel converter (MMC) is the main topological form adopted in the flexible DC transmission system based on a voltage source converter. In the topology structure of the sub-module, VT1 and VT2 represent insulated gate bipolar transistor (IGBT), and C0 represents a sub-module capacitor. D1 and D2 represent freewheeling diodes [7]. The input or removal of the capacitor is realized by the on–off of the IGBT and the capacitors play a role of voltage support. It can specifically play the following roles: (1) smoothing and filtering the output voltage of the rectifier, (2) absorbing the high-amplitude pulsating voltage to keep the voltage fluctuation on the DC bus within the allowable range, and (3) preventing the transient overvoltage of the "DC-link" from damaging the IGBT. While in conventional DC transmission projects, the arrangement of damping capacitors is shown in Fig. 1.3. The converter is divided into rectifier and inverse type, usually in the form of bipolar single or double 12-pulse. Each 12-pulse converter is composed of two 6-pulse converters containing 6 single valves. Each single valve consists 50–110

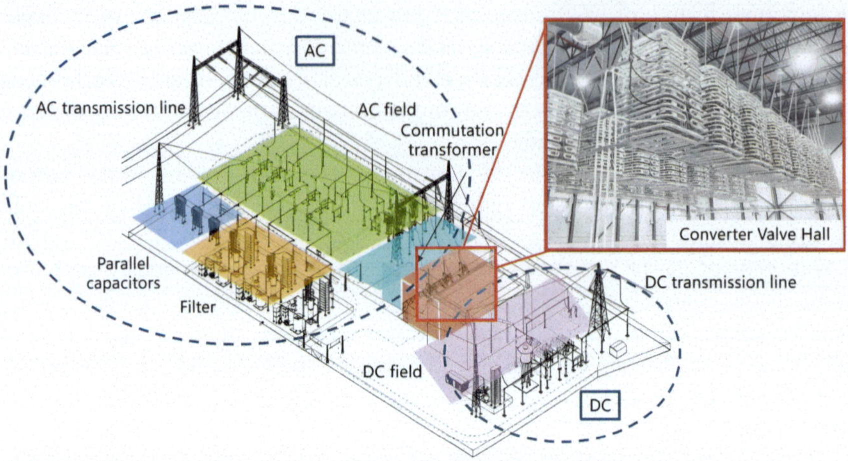

Fig. 1.1 Schematic layout of the converter station's basic configuration

thyristors in series and in parallel with a corresponding number of damping capacitors. The damping capacitors suppress the commutation overshoot voltage caused by switching on and off, which can prevent the damages of the thyristor caused by overvoltage [8]. In summary, the performance of HVDC capacitors is directly related to reliable operation of the converter valve. However, the converter valve failure caused by the performance degradation of the HVDC capacitor accounts for 20%. At present, with the continuously improving voltage level of the HVDC transmission system, the number of capacitor components and the operating voltage of the capacitor bank are further increased, thereby putting forward more stringent requirements for the operation reliability of the HVDC capacitor.

The new Lexicon Webster's Dictionary (1987) defines a capacitor as a: 'A device which gives capacitance, usually consisting of conducting plates or foils separated by layers of dielectric. A potential difference applied across the plates induces a separation of charge centers in the dielectric, thus storing electrical energy'. A capacitor consists of two such conducting plates that are separated by uniform insulating media of high quality, permitting the media to store electrical energy at field stress levels approaching the ultimate voltage withstand value of the material(s). The choice of insulating material between capacitor plates varies depending on the capacitor type, ranging from gases, air, organic films like polypropylene and Mylar, to inorganic films such as mica, ceramic, and oxides. The specific characteristics of the insulating medium greatly influence all capacitor performance parameters, except possibly peak current limits in extremely high-power applications. In such instances, the electrical conductivity of the capacitor plates and connections also affects achievable peak current limits and energy dissipation per pulse discharge cycle. It's crucial to highlight that certain insulating materials exhibit permittivity that vary with the applied voltage (V), leading to a nonlinear relationship between capacitance and voltage, particularly observed in certain types of ceramic capacitors. This nonlinearity can escalate

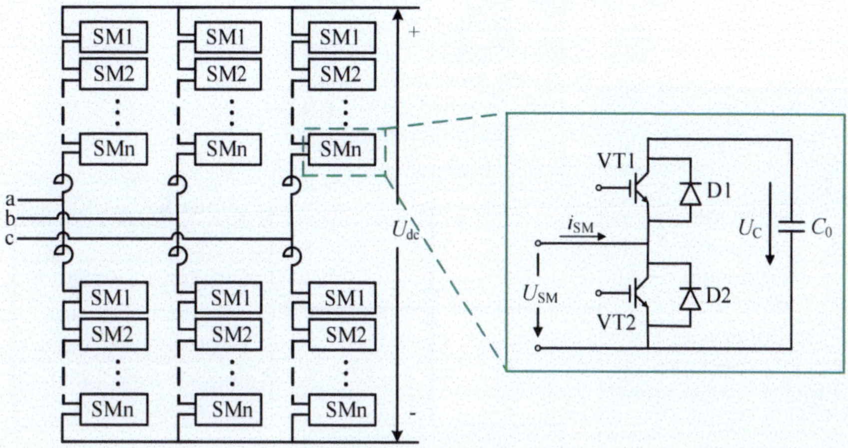

Fig. 1.2 Topological structure of a modular multi-level converter valve for flexible DC transmission system

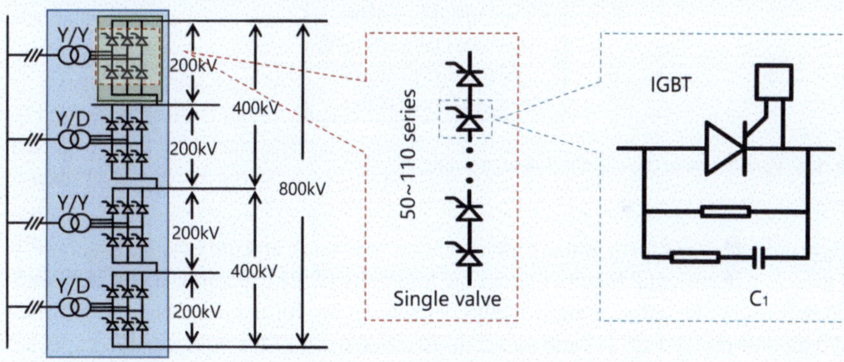

Fig. 1.3 Converter valve for a conventional HVDC transmission system

internal power dissipation within the capacitor and amplify electromechanical stress due to repeated charge and discharge cycles.

When choosing a capacitor for use in an electronic circuit, its properties are carefully tailored to ensure a predetermined level of reliability for the component over the circuit's operational lifespan. Figure 1.4 outlines several key parameters that impact the capacitor designer's decisions regarding geometry, connections, and materials. The process of capacitor design involves aligning the available capacitor characteristics and parameters with the specific requirements of the application. Apart from the fundamental capacitance value and voltage rating, specifying all the attributes detailed in Fig. 1.4 enables the supplier to offer the most economical capacitor suited to the application at hand. The primary parameters, as shown on the right side of Fig. 1.4, are predominantly influenced by environmental factors such as temperature

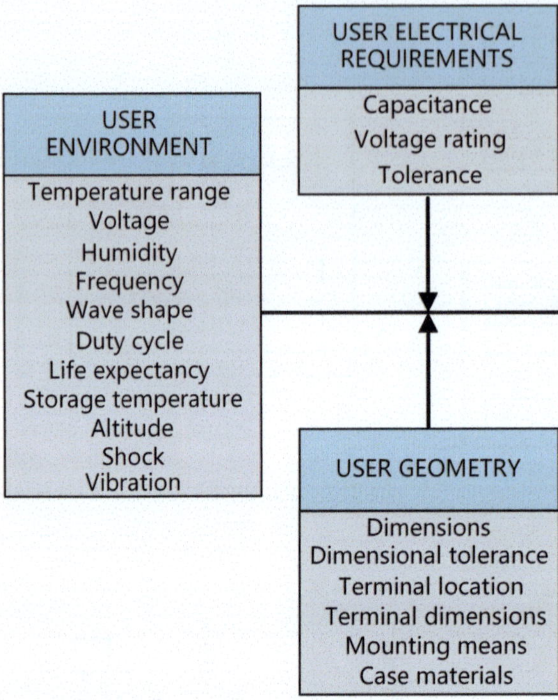

Fig. 1.4 Parameters in the capacitor design strategy

range, voltage, waveform, pulse repetition rate (rep-rate), and duty cycle. Essentially, all these environmental factors impact the longevity of the capacitor. It is at this juncture that users can achieve significant cost savings by furnishing designers with all the data outlined in Fig. 1.4 and an accurate assessment of the equipment's expected lifespan where the capacitor will be deployed.

The HVDC capacitors can be divided into two categories, foil film capacitors and metallised film capacitors, if they are classified according to the electrode form. Foil capacitors appeared earlier and are mainly composed of three parts: aluminium foil, dielectric film and impregnant. Metallised film capacitors, on the other hand, are made of two layers of metallised polymer film wound on top of each other, and are available in both oil-immersed and dry-type configurations. Compared with foil film capacitors, metallised film capacitors have self-healing properties and insulation can be restored after breakdown, making them an ideal replacement for high-voltage DC capacitors. At present, whether foil capacitors or metallised film capacitors, the dielectric material is polymer film, generally BOPP film.

Compared with traditional foil capacitors, MFCs show various superiority, including small size, light weight, low inductance, and self-healing [9]. The structure of the MFCs is simple, which is composed of multiple capacitor elements in series and parallel. The dielectric films deposited with metal electrode are wound to

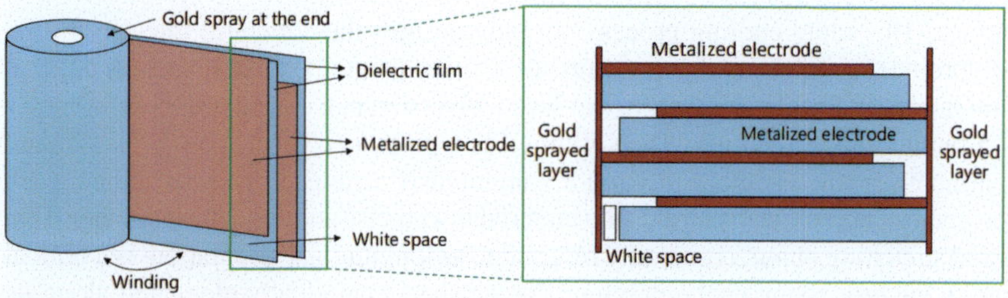

Fig. 1.5 Metallized film capacitor components

form a metalized film capacitor element [10], as shown in Fig. 1.5. After spraying gold at the end of the element, the welding wires lead to the electrodes at both ends. A certain width of non-evaporated electrode area is left at the other end of the metalized film, which is called white space, aiming to prevent short-circuiting of the electrodes at both ends. The metal electrode material is generally aluminum, zinc or zinc-aluminum composite.

1.2.2 Film Preparation

Biaxially Oriented Polypropylene (BOPP) film is produced through a meticulous process involving the fusion and extrusion of polypropylene melt from a dual-extruder system. This molten polypropylene subsequently traverses a slender extrusion die head, where it is then cooled on chilled rollers, transforming into a sheet. Following this initial cooling phase, the sheet undergoes a two-stage stretching process, first in the longitudinal direction and subsequently in the transverse direction, to achieve molecular orientation [11]. This critical orientation step significantly enhances the mechanical properties and dimensional stability of the film. The detailed steps of the production process are as follows [12]:

Melt Extrusion: Polypropylene resin is introduced into the sophisticated dual-extruder system, where it undergoes a meticulous heating process that elevates its temperature to the melting point. Within this system, the resin is thoroughly and homogenously mixed to ensure consistency and uniformity throughout. Subsequently, the melted and homogenized polypropylene is propelled under pressure through the narrow, precision-engineered gap of the extrusion die head. This process results in the formation of a continuous, smooth ribbon of molten polypropylene, ready for further shaping and processing into various products.

Cooling and Solidification: As the molten polypropylene ribbon gracefully emerges from the extrusion die head, it immediately encounters a series of chilled rollers, strategically positioned to facilitate rapid cooling. The contact with these cold surfaces initiates a swift thermal exchange, effectively dissipating the heat and solidifying the molten polymer into a flat, continuous sheet with a smooth and even surface. This rapid cooling process not only ensures the structural integrity of the polypropylene sheet but also prepares it for subsequent operations, such as cutting, laminating, or further shaping, to produce a diverse range of end products.

Longitudinal Stretching: The solidified polypropylene sheet, now in its flat and continuous form, advances through a meticulously arranged series of rollers, each designed to incrementally stretch the material in a precise and controlled manner. This initial stretching phase, commonly referred to as longitudinal orientation, is a critical step in the processing of the polymer. As the sheet is gradually elongated along its lengthwise direction, the polymer chains within are aligned parallel to the direction of stretching. This alignment reinforces the material's structural integrity, resulting in a notable enhancement of its strength properties along the stretched axis. Additionally, by orienting the polymer chains in this manner, the sheet exhibits reduced shrinkage tendencies in the longitudinal dimension, enhancing its dimensional stability and suitability for various applications requiring precision and durability.

Transverse Stretching: Subsequently, the polypropylene film progresses to the transverse stretching phase, a pivotal step in the bi-axial orientation process. This stage is typically executed utilizing a sophisticated tenter frame, a specialized piece of equipment designed to precisely manipulate the film's dimensions. Within the tenter frame, the film is securely gripped at its edges and then subjected to a controlled lateral stretching force. This action, coupled with the longitudinal stretching previously achieved, completes the bi-axial orientation process, resulting in a film that is optimized in both directions.

Heat Setting: Following the transverse stretching process, which further enhances the mechanical properties and dimensional stability of the polypropylene film, the next crucial step is heat-setting. This stage serves to stabilize the molecular structure of the film, ensuring that the desirable properties acquired during stretching are retained and fixed. The film is carefully exposed to a precisely controlled environment of temperature, where it is held for a specified duration. This heat treatment allows the polymer chains to relax into their new, oriented configuration, strengthening the film's internal structure and preventing it from reverting to its pre-stretched state. By undergoing heat-setting, the film becomes more resistant to shrinkage, distortion, and stress relaxation, making it an ideal material for applications requiring high dimensional stability and durability.

Corona Treatment: The principle of corona treatment involves the application of a high-frequency voltage to electrodes, resulting in the ionization and discharge of air adjacent to the film's treatment surface. This ionization process generates high-energy particles that, under the influence of a strong electric field, rapidly impact the moving film. This impact causes the chemical bonds within the surface molecular chains to break, leading to degradation. Concurrently, during the discharge process, ozone is formed, which oxidizes the film's surface molecules and introduces polarity. When air undergoes ionization, it gives rise to a plasma that comprises electrons, oxygen ions, photons, and ozone. This plasma penetrates the film, disrupting molecular bonds, exciting free radicals, and interacting with ozone ions to generate oxidized polar groups. This comprehensive process enhances the surface properties of the film, preparing it for better adhesion and compatibility with subsequent processing or coating layers.

The process of turning dielectric films into capacitor devices involves several steps, including metallization, film slitting, core winding, heat treatment, metal spraying, energizing, and encapsulation [13]. Metallization involves a thin metal electrode layer on the film, creating a self-healing film. Film slitting is used for winding, and core winding forms the film capacitor's core. Heat treatment removes impurities and stabilizes the core's shape. Metal spraying applies a metallic layer to both ends, facilitating terminal connection. Energizing applies voltage to the core, enhancing its electrical properties. Finally, encapsulation encapsulates the core within a protective housing [14].

1.2.3 Device Fabrication

For metallized film capacitors with high accuracy requirements and low rated capacitance, to ensure stable capacitance and prevent decay, the assembly manufacturing process of the capacitor must be strictly controlled. The wound core is subjected to a thermal pressing process to expel residual air, and after thermal pressing, the ends of the core are sprayed with gold to ensure good electrical conductivity between the metal layer and the dielectric material. Lead wires are soldered to both ends of the core to serve as conductors, and then the core is filled with epoxy resin, which is cured at a certain temperature and for a certain period of time. The welding process of polypropylene film capacitors is directly related to the electrical performance, stability, and service life of the capacitors. Traditional welding processes often have problems such as significant damage to the core end surfaces and poor welding consistency, which makes it difficult for film capacitors to meet the application requirements for high precision and high stability. Film capacitors can be divided into foil capacitors and metallized capacitors according to the structure. Metallized film capacitors use a vacuum deposition process to coat a thin layer of metal on the film as electrodes, which have the advantages of large capacitance and strong self-healing ability. However, due to the extremely thin electrode layer, any carelessness during the welding process may cause damage, affecting the performance of the capacitor [15].

The main indicators reflecting the quality of capacitors include the dielectric loss tangent, capacitance, insulation resistance, and dielectric strength of the capacitor. The dielectric loss of the film capacitor is an important indicator to measure the quality of the capacitor. It can be defined as the energy consumed by the capacitor due to heat generation per unit time under the action of the applied voltage, known as the loss of the capacitor. The metal loss of the film capacitor varies with its capacity and contact resistance, and the higher the frequency, the greater the loss.

Poor welding can reduce the bonding strength between the metal layer on the and the sprayed gold layer, or between the lead wire and the sprayed gold layer, leading to changes in contact resistance. When a capacitor is charged and then short-circuited for discharge, the large current passing through the sprayed gold layer generates heat due to contact resistance. The amount of heat generated is directly proportional to the contact resistance and the square of the instantaneous current. Transient high currents can cause local high temperature. Prolonged exposure to such conditions can cause the film to shrink due to heat, thereby increasing the loss value at high frequencies. In severe cases, it can lead to the detachment of the metal layer from the sputtered gold layer, causing the capacitor to failure.

In addition, the welding power is also a significant factor that affects the quality of the welding, with the main process parameter being the welding current. If the welding power is too low, it is prone to result in a cold solder joint, which lacks the necessary firmness. If the welding power is too high, the instantaneous temperature can be excessively high, leading to an increase in the dielectric loss tangent, which is common knowledge in the quality control of the welding process during the production of capacitors. The traditional welding process for film capacitors mainly includes steps such as lead wire pre-treatment, welding, and inspection. However, traditional welding methods often require high temperatures to ensure a good bond between the lead wires and the electrode layer, and high temperatures can easily cause the electrode layer to melt or peel off. Due to the imprecise temperature control and inconsistent welding time during the welding process, it is difficult to ensure the quality of the welding. High temperatures and mechanical stresses during the welding process are prone to damage the end faces of the core, affecting the overall performance of the capacitor.

To resolve the aforementioned challenges, the automatic laser welding technology for metallized film capacitors has garnered widespread attention. From a long-term perspective, compared to traditional electric welding, laser welding offers uniform weld spots, higher welding efficiency, greater utilization of solder wire, and reduced labor requirements. In the rapidly developing present, the energy-saving and high-efficiency characteristics of laser soldering are more in line with the production requirements of contemporary enterprises.

Packaging process is a critical step to ensure the stable performance of the capacitor, neatness of appearance, and ease of use. The quality of the packaging process directly affects the service life and reliability of the capacitor. The packaging materials for film capacitors mainly include the casing and epoxy resin. The casing material should possess good insulation, mechanical strength, and corrosion resistance to ensure that the capacitor can still function normally under harsh environmental conditions. Epoxy resin, as a potting material, has excellent electrical insulation, heat resistance, and chemical corrosion resistance, which can effectively protect the internal structure of the capacitor and enhance the stability and reliability of the capacitor. The disadvantages of epoxy resin encapsulation include the fact that the viscosity of the epoxy resin fluctuates considerably with seasonal temperature changes, which complicates the control over the size of the encapsulated capacitor. Furthermore, the solidification process of liquid epoxy resin is slow, leading to a prolonged production cycle.

The packaging process of thin film capacitors usually includes the following steps. First, put the welded capacitor core into the shell to ensure the insulation and fixation between the core and the shell. The assembly accuracy of the shell directly affects the sealing and stability of the capacitor. Secondly, epoxy resin is injected into the shell to form a protective layer through the curing process to isolate the external environment from the corrosion of the internal structure of the capacitor. The pouring amount, temperature and curing time of resin should be controlled in the potting process to ensure the potting effect. Finally, after packaging, appearance inspection, electrical performance test and reliability test are needed to ensure the quality and performance of the capacitor meet the standard requirements.

The casing and potting technologies of film capacitors directly affect the sealing and stability of the encapsulation. Traditional manual casing methods have low efficiency and it is difficult to ensure quality. To ensure the potting effect, it is necessary to control the amount of resin injected, temperature, and curing time. In response to the low level of automation and unstable product quality in the current assembly of film capacitors, the automatic assembly process of capacitors has received widespread attention. To improve the level of automation and production efficiency, automatic casing equipment can be used, which achieves rapid and accurate casing operations by precisely controlling the casing position and assembly accuracy. Similarly, the use of automatic potting equipment, combined with an accurate control system, can achieve uniform resin injection and precise curing, thereby improving the quality of encapsulation and production efficiency. The encapsulation process of film capacitors is a key link to ensure their stable performance and reliable quality. By optimizing key technical links such as casing and potting, and improving automation.

1.2.4 Testing Method

Limited by the quality of the insulating medium, production process conditions and the diversity of production environments, metallized film capacitor insulation structure will inevitably exist within the uneven distribution of electric field and dielectric defects, which in turn triggered the film medium partial discharge phenomenon. With the metallized film capacitor working field strength increases, the probability of film breakdown also increases. The breakdown characteristics of such capacitors are determined by their unique metallized electrode structure, and the extremely thin metal vapor-deposited layer forms the basis for reliable self-healing of metallized film capacitors. Typically, metallized film capacitors use organic film as the insulating medium, however, during the production and operation of the capacitor, many defects are introduced into the medium, resulting in localized areas where the breakdown threshold is significantly lower than the surrounding area, which are referred to as "electrically weak points".

Self-healing discharge phenomenon refers to when the applied voltage reaches a certain amplitude, the dielectric weakness of the short-term breakdown and then restore the insulation dynamic process, metalized film capacitors. When the applied voltage rises, the first in the weakness of the film where the dielectric breakdown occurs, and the formation of the discharge path, accompanied by the flow of large currents generated by the Joule thermal effect, the local temperature rises abruptly. The high temperature prompts the evaporation of the surrounding thin metal layer and spread around, while further ionization occurs, the formation of plasma.

As the evaporation expands, the conductive path is cut off, the plasma discharge is terminated, and the metallized film capacitor is restored to its insulating state, a self-healing process known as "self-healing". This property enhances the stable operation and voltage resistance of the storage dielectrics in high-intensity electric fields, which significantly improves the operating field strength and cycle life of the energy storage components.

(1) Capacitance

The measurement of capacitance in metallized film capacitors utilizes specialized equipment such as LCR meters or dedicated capacitance measuring instruments. These tools are essential for accurately assessing the capacitor's capacitance to ensure it aligns with specified nominal values. During the measurement process, the capacitor is connected to the LCR meter or capacitance meter, which applies a controlled test signal. This signal allows the instrument to calculate capacitance by analyzing the resulting impedance. This meticulous method is critical for confirming that the capacitor meets the exacting requirements necessary for its designated application, particularly in electronic circuits where precise capacitance values are fundamental for optimal functionality and performance. Regular capacitance testing plays a vital role in maintaining the reliability and uniformity of capacitor performance across various production runs and environmental conditions.

(2) Equivalent Series Resistance (ESR)

To measure ESR, specialized equipment is utilized. This equipment applies a known AC voltage across the capacitor and then measures the resulting voltage drop. By analyzing the relationship between the applied voltage, the current flowing through the capacitor, and the observed voltage drop, the ESR value can be precisely calculated. Accurate ESR measurement is essential for ensuring that capacitors meet stringent performance standards, especially in demanding electronic circuits where stability and reliability are crucial. Regular ESR testing is key to identifying potential issues early on, thereby ensuring optimal performance and extending the lifespan of metallized film capacitors across various operational conditions.

(3) DC Voltage Stability

DC voltage stability testing of metallized film capacitors is crucial to assess their reliability and performance over time. This test involves subjecting the capacitors to a constant DC voltage for a specified duration under controlled conditions. By monitoring voltage variations and ensuring they remain within acceptable limits, this test evaluates the capacitor's ability to maintain consistent performance and electrical characteristics under operational conditions. It helps ensure that capacitors meet stringent reliability standards, providing assurance of their suitability for various electronic applications where stable voltage performance is essential.

(4) High Voltage Endurance Capability

High-voltage testing of metallized film capacitors represents a pivotal stage in their evaluation, characterized by stringent protocols to gauge their resilience and reliability under extreme electrical stress. This testing regimen is meticulously conducted within specialized laboratory environments equipped with state-of-the-art equipment tailored for high-voltage applications. The procedure commences with the careful installation of capacitors in controlled settings conducive to precise voltage manipulation. This environment ensures safety while allowing capacitors to be subjected to voltages exceeding their nominal operational thresholds. Such controlled overvoltage conditions simulate real-world scenarios where capacitors may encounter exceptional electrical demands. Throughout the testing phase, a structured approach is employed to incrementally apply increasing voltages while monitoring critical performance metrics. These metrics encompass parameters such as leakage current, insulation resistance, and capacitance stability. Each parameter is scrutinized meticulously at various voltage intervals, with ample time allocated for capacitors to stabilize at each level before precise measurements are taken. The culmination of this testing regimen lies in the comprehensive analysis of gathered data. This analysis delves into intricate details regarding the capacitor's response to high-voltage stressors, identifying any signs of electrical breakdown or performance degradation. Such insights are pivotal for assessing the capacitor's suitability for applications demanding steadfast reliability under elevated electrical potentials. Ultimately, the findings from high-voltage testing are encapsulated in meticulously crafted reports that outline the experimental methodology, observed results, detailed analyses, and consequential implications. These reports serve as authoritative resources within engineering circles, guiding decisions on the capacitors' deployment in sectors necessitating steadfast electrical performance and enduring reliability under challenging voltage conditions.

(5) Temperature Characteristic

Temperature characteristic testing of metallized film capacitors is crucial for assessing their operational stability across varying temperatures, which is essential for their reliable performance in diverse applications. This testing involves subjecting the capacitors to controlled temperature environments to evaluate how their electrical properties respond to temperature changes. The procedure begins with setting up the capacitor in a controlled laboratory setting equipped with a temperature chamber capable of precise temperature control. The capacitor is then exposed to a defined temperature range that typically includes ambient temperature and extends to extreme temperatures relevant to its intended applications. During the testing process, specific temperatures within this range are selected for measurement. Key electrical parameters such as capacitance, dissipation factor (tanδ), insulation resistance, and leakage current are meticulously measured at each temperature point. This data collection phase is crucial for understanding how these parameters fluctuate with temperature variations. To ensure accuracy, sufficient stabilization time is allotted at each temperature point before measurements are taken. This ensures that the capacitor reaches thermal equilibrium, thereby providing reliable and consistent test results. Following the data collection phase, thorough analysis is conducted to interpret the capacitor's temperature coefficients. This analysis provides insights into how the capacitor's electrical characteristics, particularly its capacitance stability, are influenced by temperature changes. These insights are crucial for determining the capacitor's suitability for specific temperature-sensitive applications. Ultimately, the findings from temperature characteristic testing are compiled into a comprehensive test report. This report details the experimental setup, environmental conditions, measurement results, and detailed analysis. It serves as a critical resource for engineers and researchers evaluating the capacitor's performance under various temperature conditions and facilitates informed decision-making regarding its application in practical settings.

(6) Mechanical Stability

Vibration and shock testing of metallized film capacitors represents a crucial phase in their evaluation, designed to ensure robust performance and durability in demanding operational environments. This testing regimen is conducted meticulously in specialized laboratories equipped with sophisticated equipment tailored for assessing the capacitors' resilience against mechanical stresses. The process begins with capacitors securely mounted in test fixtures that replicate conditions akin to those encountered during transportation, handling, or operational use. These fixtures subject the capacitors to controlled vibrations across a spectrum of frequencies and amplitudes, simulating real-world scenarios where mechanical oscillations could impact performance. In parallel, shock testing involves subjecting capacitors to sudden, controlled impacts that replicate the effects of drops or abrupt mechanical jolts. This phase assesses the capacitors' ability to withstand sudden mechanical forces without compromising electrical integrity or physical integrity, such as lead wire connections. Throughout both vibration and shock testing phases, stringent monitoring protocols are employed to measure and analyze critical parameters. These include resonant frequencies, acceleration levels, and the capacitors' response to mechanical stressors. Comprehensive data analysis focuses on identifying any deviations from expected performance, potential weak points, or structural vulnerabilities. The culmination of these tests is the synthesis of detailed reports that document the experimental procedures, observed results, and analytical insights. These reports serve as authoritative resources, guiding decisions on the capacitors' suitability for applications requiring resilience against mechanical vibrations and shocks, ensuring reliable performance and longevity in challenging operational environments.

1.3 Summary

The development of high-voltage direct-current (HVDC) transmission has progressed dramatically because of its advantages, such as a long conveying distance, large capacity and high efficiency. As the core component in converter stations, metallized film capacitors play a significant role in suppressing the commutation overshoot voltage and avoiding severe damage to thyristors. As a result of an increase in the voltage level and transmission capacity, the requirements for the performance of metalized film capacitors have become more stringent. Following this, the chapter will provide a detailed examination of metallized thin film capacitors, focusing on their layered composition and the specific materials chosen to optimize their electrical properties. The manufacturing process will then be thoroughly explored, encompassing the industrial preparation of thin films, deposition techniques, and the critical criteria for material selection to ensure both uniformity and high performance. The assembly process of capacitors will also be outlined, starting from initial substrate preparation through to final encapsulation, highlighting the implementation of rigorous quality control measures at each stage to uphold consistency and reliability. Finally, the chapter will delve into various detection methods, such as capacitance measurement, insulation resistance testing, and high-voltage breakdown testing. This section will elaborate on the specialized equipment and procedures employed to validate capacitor performance and detect any potential defects that could impact the operational efficiency and reliability of HVDC applications.

References

1. C. Buttay et al, "State of the art of high temperature power electronics," *Mater. Sci. Eng: B*, vol. 176, no. 4, pp. 283–288, 2011.
2. M. Xiao, Y. S. Zhao, Y. N Song and B. X. Du, "Polypropylene films for HVDC capacitors with enhanced dielectric performance by irradiation crosslinking modification," *IEEE Trans. Dielect Elect. Insul.*, vol. 31, no. 2, pp. 666–673, 2024.
3. M. Xiao, Y. N. Song, H. L. Liu, M. D. Zhang, B. X. Du and Z. L. Xing, "Dielectric properties of cross-linked polypropylene by Gamma-ray irradiation for film capacitor," *IEEE Trans. Dielect Elect. Insul.*, vol. 31, no. 2, pp. 625–632, 2024.
4. M. Xiao Z. Y. Zhang and B. X. Du, "Dielectric performance improvement of polypropylene film by hierarchical structure design for metallized film capacitors," *J. Phys. D: Appl. Phys.*, vol. 57, no. 38, pp. 5501–5509, 2024.
5. S. H. Kung and G. J. Kish, "Multiport modular multilevel converter for DC systems," *IEEE Trans. Power Deliv.*, vol. 34, no. 1, pp. 73–83, 2019.
6. M. Xiao et al, "Breakdown strength improvement of polypropylene by introducing long-chain branched structures for film capacitors," *J. Phys. D: Appl. Phys.*, vol. 55, no. 24, pp. 5501–5508, 2022.
7. L. B. Han, L. Liang, Y. Kang and Y. F. Qiu, "A review of SiC IGBT: models, fabrications, characteristics, and applications", *IEEE Trans. Power Electr.*, vol. 36, no. 2, pp. 2080–2093, 2020.
8. L. Wang, H. R. Liang, A. V. Prokhorov, H. Mokhlis and C. K. Huat, "Modal control design of damping controllers for thyristor-controlled series capacitor to stabilize common-mode torsional oscillations of a series-capacitor compensated power system," *IEEE Trans. Ind. Appl.*, vol. 55, no. 3, pp. 2327–2336, 2019.
9. A. Schneuwly, P. Gröning and L. Schlapbach, "Uncoupling behavior of current gates in self-healing capacitors," *Mater. Sci. Eng. B Solid State Mater. Adv. Technol.*, vol. 55, no. 3, pp. 210–220,1998.
10. H. Li, Y. Chen, F. Lin, B. Peng, F. Lv, M. Zhang and Z. Li, "The capacitance loss mechanism of metallized film capacitor under pulsed discharge condition," *IEEE Trans. Dielect Elect. Insul.*, vol. 18, no. 6, pp. 2089–2094, 2011.

11. M. Rabuffi and G. Picci, "Status quo and future prospects for metallized polypropylene energy storage capacitors," *IEEE Trans. Plasma Sci.*, vol. 30, no. 5, pp. 1939–1942, 2002.
12. M. H. El-Husseini, P. Venet, A. Al-Majid, M. Fathallah, G. Rojat and J. A. Ferreira, "Manufacturing technology effect on current pulse handling performance of metallized polypropylene film capacitors," *J. Phys. D: Appl. Phys.*, vol. 36, no. 18, 2003, Art. no. 2295.
13. Y. Shizukuni, O. Eiji, S. Hisashi, K. Hidekazu, F. Masanori, T. Tomohiro and H. Kinoshita, "Polyacene capacitors," *J. Power Sources*, vol. 60, no. 2, pp. 207–212, 1996.
14. J. L. Nash, "Biaxially oriented polypropylene film in power capacitors," *Polym. Eng. Sci.*, vol. 28, no. 13, pp. 862–870, 1988.
15. S. Francis, "Techniques, equipment and procedures for production welding of electronics," *IEEE Trans. Compon. Packag. Manuf. Technol.*, vol. 7, no. 1, pp. 13–17, 1963.

Chapter 2
Polypropylene Film

2.1 Introduction

At present, HVDC capacitors widely use polymer film as the solid dielectric material, and the cost of polymer film accounts for about 70% of the production cost of capacitors. The quality of the polymer film directly affects the capacitor's capacity stability, voltage resistance, safety and its life cycle and other performance indicators. The history of the development of power capacitors is essentially the history of the development of power capacitors with dielectric materials. It can be said that the development of dielectric materials is the basis for technological progress of power capacitors. With the development of capacitors to the direction of low loss, miniaturization, capacitor solid dielectric materials experienced cable paper, capacitor paper, polymer film and capacitor paper composite, and finally to all the polymer film development process. 1920s, to kraft paper as a solid dielectric foil capacitors in the United States was born. In the 1970s, polypropylene (PP) film capacitor dielectric changes, followed by the emergence of paper-film composite dielectric, to the 1980s, the use of polypropylene film is much more than the use of paper media, in the 1990s, polypropylene film has become a power capacitor's main dielectric material.

At present, in practical applications, due to the biaxially oriented polypropylene (Biaxially Oriented Polypropylene, BOPP) film has the advantages of stable dielectric properties, high resistance to electric fields and easy to scale up the preparation. Its relative dielectric constant of about 2.2, dielectric loss angle tangent value is generally 2×10^{-4} (0–30 °C), BOPP film has become the most important power capacitors preferred dielectric material, and is currently the most widely used dielectric material in power capacitors. If the dielectric performance of the capacitor film is weak, it is more prone to breakdown and self-healing under high electric field, thereby leading to a faster decrease in the overall capacitance of the capacitor. Therefore, improving the dielectric properties of capacitor films can contribute to extending the service life of capacitors.

B. Du and M. Xiao, *Polypropylene Film for HVDC Capacitors*,
https://doi.org/10.1007/978-981-96-3029-5_2

This chapter will introduce the basic theory of dielectrics during operation is presented, including dielectric polarization, dielectric loss, dielectric breakdown, charge transport behaviour and energy storage of dielectric. Finally, the dielectric properties of different types of dielectric materials were compared, and the specific reasons why polypropylene stood out among them were further explained.

2.2 Fundamental Theory

2.2.1 *Dielectric Polarization*

When voltage is applied to two parallel conductors, positive and negative charges will accumulate on the two conductors respectively. The charge stored by applying a unit voltage is called a capacitor, which is shown in Formula (2.1). Such two conductors make up the simplest. The two conductors make up the simplest vacuum plate capacitor.

$$C = \frac{Q}{V} \tag{2.1}$$

where C is the capacitance value, Q is the charge stored on the conductor, and V is the applied voltage.

The capacitance of the vacuum capacitor is affected by the spacing d and area A of the two conductors, and the capacitance value C_0 is calculated as follows.

$$C_0 = \frac{\varepsilon_0 A}{d} \tag{2.2}$$

where ε_0 is the vacuum dielectric constant, $\varepsilon_0 = 8.854187817 \times 10^{-12}$ F/m.

The capacitance value increases after inserting an insulating material between the above two plates, as shown in Fig. 2.1. The increased capacitance is calculated as follows.

$$C_0 = \varepsilon_r C_0 = \frac{\varepsilon_r \varepsilon_0 A}{d} \tag{2.3}$$

where ε_r is the relative dielectric constant, $\varepsilon = \varepsilon_r \varepsilon_0$ is the dielectric constant of the insulating material. The above insulation material is actually the dielectric material. The dielectric material is embedded in the vacuum capacitor. When the voltage is applied, the surface of the dielectric material near the positive electrode plate induces negative charges, and an equal amount of positive charges are induced on the dielectric material near the negative electrode plate, thereby increasing the amount of charge on the positive and negative plates. The induced charges are also called as bound charges. The phenomenon in which the dielectric induces bound charges

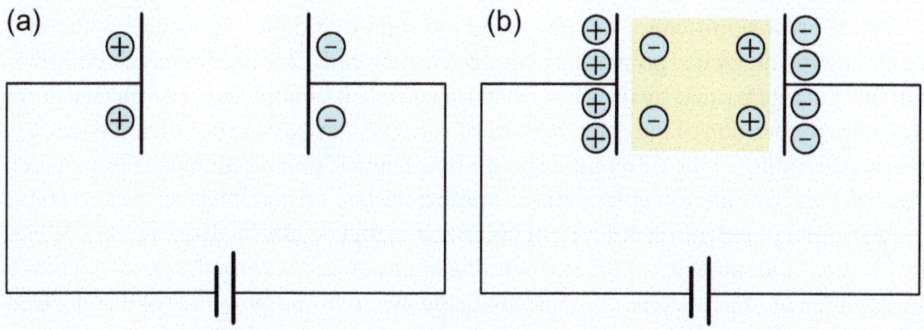

Fig. 2.1 Structure diagram of **a** the vacuum capacitor and **b** the capacitor inserted into dielectrics

under the action of an electric field is called dielectric polarization. The greater the polarization, the stronger the ability of the flat capacitor to store charges.

Under the influence of an external electric field, the free movement of charges within a molecule or in local space is restricted, resulting in local migration of macroscopic bound charges (also known as polarization charges). This physical phenomenon is called polarization. The substances that could produce electric polarization are collectively referred to as dielectrics. In order to quantitatively describe the polarization phenomenon of dielectrics, the concepts of electric moment and electric dipole moment were introduced. The electric moment of a point charge relative to a reference point is expressed as follows.

$$\mu = qr \tag{2.4}$$

where μ is the electric moment, q is the charge amount of point charge, and r is the radial vector from the reference point to the point charge. The multiple charge systems that exhibit electrical neutrality are the most common, which could be simplified as a charge system composed of two positive charge centers(+ Q) and negative charge centers(− Q) at a distance. The system is usually called dipole, and the electric moment μ is called electric dipole moment or dipole moment. The electric dipole moment per unit volume is the polarization intensity P, which characterizes the polarization degree of the dielectric.

The experiment shows that the polarization intensity is influenced by the external electric field and the built-in electric field generated by the bound charge. For isotropic dielectrics, the polarization strength could be expressed as follows.

$$P = (\varepsilon_r - 1)\varepsilon_0 E \tag{2.5}$$

where P is the polarization intensity, ε_0 is the vacuum dielectric constant (8.85×10^{-12} F/m), ε_r is the relative dielectric constant of the material, and E is the applied electric field strength.

It is of great significance to understand the source of dielectric constant, thereby regulating the dielectric properties of dielectric materials. The macroscopic polarization intensity generated by dielectric materials under the influence of an electric field is essentially the sum of the contributions of various microscopic polarization mechanisms. According to the different response frequencies, polarization usually includes electron polarization, ion polarization, dipole polarization and interface polarization.

Electronic polarization is a type of polarization that occurs in all dielectrics. When the electric field intensity of the surrounding environment is zero, the positive charge center of the atomic nucleus coincides with the negative charge center of the electron orbital. However, under the influence of an external electric field, the electron orbitals shift relative to the atomic nucleus, and the positive and negative charge centers no longer coincide, forming a dipole moment parallel to the direction of the electric field. The electronic mass is very small and responds quickly to the electric field (DC to optical frequency). However, compared to other types of microscopic polarization mechanisms, the polarization generated by electron displacement is smaller. Similar to electron displacement polarization, ions could also undergo displacement when an electric field is applied, specifically referring to the elongation of ionic bonds in ionic crystals. Compared to electrons, ions have a larger mass, resulting in longer polarization time (10^{-12} to 10^{-13} s).

The electron polarization is related to the electric displacement of the electron cloud relative to the nucleus under the applied electric field, and responds in the visible light frequency range. Free electrons can enhance electron polarization, while the enhancement of electron polarization is usually accompanied by a decrease in energy band. Therefore, there is an upper limit for the enhancement of electron polarization. Ion polarization usually occurs in ion materials and responds in the infrared frequency region. The addition of other elements in ion materials can effectively enhance ion polarization. Usually, the ionic polarization is 10–50% of the electronic polarization. Since the above two polarization types are determined by the internal structure of the molecule, they occur instantaneously in the resonance state at high frequencies. Therefore, the above two polarization types are not affected by temperature and contribute less to the dielectric constant of the polymer. The related dielectric loss mechanism occurs above the infrared frequency, and there is no dielectric loss in the lower frequency range [1].

Dipole polarization, also known as directional polarization, occurs in polar substances carrying permanent dipoles and contributes greatly to the dielectric constant of polar polymers. The polarization intensity is affected by the polar groups and the geometry of the polymer chain. The dipole polarization is related to frequency. When the frequency increases to a certain range, the orientation speed of the dipole is difficult to keep up with the change frequency of the external electric field, showing a decrease in polarization. The interface polarization response frequency is low and the duration is long, which exists in almost all polymer dielectrics.

Space charge polarization occurs in non-uniform dielectric. The space charges distributed chaotically at grain boundaries and the second phase move outward along the negative and positive electrodes, resulting in macroscopic polarization. Therefore, it is a common method to improve the dielectric constant of composites by filling high dielectric fillers to enhance interfacial polarization. Dielectric materials store and release energy by means of polarization and the energy density is related to the dielectric constant and the electric field strength. The increase of both is beneficial to the improvement of energy density. For the method of dielectric material modification, it is usually possible to start from these two points to stimulate stronger polarization behavior under higher electric field.

2.2.2 Dielectric Loss

The macroscopic characterization parameter of dielectric polarization is the dielectric constant of the medium, which is one of the basic parameters of dielectric materials and is usually represented by ε. It reflects the ability of the material to store potential energy under the influence of an applied electric field. The ratio of the dielectric constant ε to the vacuum dielectric constant ε_0 is the relative dielectric constant ε_r. It reflects the rate at which the spatial electric field intensity decreases after the medium is filled. That is, owing to the polarization of the dielectric, the electric field strength decreases from the vacuum electric field strength E_0 to E. It is defined as the ratio of the capacitance C of the capacitor to the capacitance C_0 under vacuum condition at the same size. Under the alternating electric field, the dielectric constant of the material changes with the frequency of the applied electric field. The dielectric constant is expressed in the complex form. ε' and ε'' are the real and imaginary parts of the dielectric constant, respectively, which are related to the dielectric constant and dielectric loss. The ratio of the imaginary part and the real part is the loss tangent $\tan\delta$.

For the ideal capacitor, ε'' is equal to 0, that is, there is no dielectric loss. However, there must be two loss forms in real capacitors. The long-range charge transport process causes conduction loss, which is independent of frequency. The dielectric relaxation loss caused by the energy absorbed by the charge motion occurs in the high frequency range. For polymer dielectrics, relaxation can be divided into two types. The segment relaxation-α relaxation caused by the micro-Brownian motion of the segment, and the group relaxation-β relaxation and γ relaxation caused by the molecule or small segment unit. The β-relaxation is related to the side group containing the C–C structure. The conformational change of the repeating unit leads to the change of the orientation of the polar group or the local motion of the dipole group on the main chain causes β-relaxation. The small unit movement of the main chain or side chain causes γ relaxation, which usually occurs at low temperatures. The activation energy required for α relaxation is higher than that for β relaxation and γ relaxation.

In a weak electric field, the phenomenon in which carriers (electrons, ions, holes, polarisers, etc.) produce a macroscopic mean migration along the direction of the external electric field to directly transport current is known as conductance, and the expression is given below:

$$\begin{cases} j_c = nq\overline{\mu_E}E = nq\overline{v_E} = \gamma E \\ \overline{\mu_E} = \overline{v_E}/E \end{cases} \tag{2.6}$$

where n is the carrier number density per unit volume; $\overline{\mu_\mathrm{E}}$ is the average mobility of the electric field equation; and q is the charge with conductivity $\lambda = nq\overline{\mu_\mathrm{E}}$.

Conductivity loss is the main cause of performance degradation of polymer dielectric materials at high temperatures and high fields. High temperatures will lead to an increase in carrier mobility and carrier number within the dielectric material. Charge is injected from the electrodes into the conduction or valence band of the insulating material through the Fowler–Nordheim effect or the Schottky effect. It has been shown that an increase in temperature causes a sharp increase in the leakage current within the dielectric, which tends to rise exponentially. A study of the conductive current in polypropylene films at elevated temperatures shows that the conductive activation energy of polypropylene films increases from 0.2–0.52 eV to 1.05 eV at temperatures above 80 °C, and the conductive current increases.

The contribution of various conductivity mechanisms to the conductivity loss of dielectric films varies in different temperature ranges and field strengths. In polymer dielectrics, at low temperatures and low fields, the various conductivity mechanisms involving hot electrons contribute less to the leakage current, and the dominant conductivity mechanism is ohmic conductivity; at medium temperatures and medium fields, hopping conductivity is the dominant conductivity mechanism in most polymer dielectrics; and at high temperatures and high fields, leakage currents due to Schottky injection and P-F emission are the main sources of conductivity loss. This is because for the same material, the jump conductance activation energy is smaller than the Schottky injection barrier, and both are smaller than the P-F emission activation energy. Therefore, the suppression of Schottky injection and P-F emission is the key to improve the performance of polymer dielectric materials at high temperature and high field.

2.2.3 *Dielectric Breakdown*

The breakdown phenomenon refers to the process in which the material is subjected to electrical failure in a short period of time under an applied electric field, excluding aging under high electric field [2]. The breakdown strength is the value of the maximum applied electric field that the dielectric material could withstand. After exceeding this value, the dielectric will change from an insulating state to a conductive state.

Although the breakdown process of dielectrics has not been thoroughly studied so far, intrinsic breakdown, thermal breakdown, partial discharge breakdown and electromechanical breakdown have been widely recognized.

(1) Intrinsic Breakdown

The intrinsic breakdown strength refers to the highest electric strength that a pure and uniform material can withstand, which is only related to the properties of the material itself. Intrinsic breakdown includes electronic breakdown and avalanche breakdown. A large number of electrons with sufficient energy in the dielectric will transition from the valence band to the conduction band, leading to breakdown. For the ideal dielectric, the bandgap is large and the electron transition is difficult. However, the dielectric material contains defects such as vacancies and impurities, which could adsorb free electrons to form defect energy levels.

An electron migrates from the negative electrode to the positive electrode under the action of an electric field, and a portion of the energy obtained from the electric field is used to accelerate motion, while a portion is lost by collision with lattice atoms. If the energy obtained from the electric field is greater than the ionization potential, an additional electron will be generated during the collision between electrons and lattice atoms, which will continue to be accelerated by the electric field and collide with other atoms to produce more and more electrons, leading to avalanche breakdown.

(2) Electromechanical Breakdown

Under the influence of an electric field, a large amount of bound charges are generated at both ends of the dielectric, leading to a certain amount of pressure due to mutual attraction. If the pressure generated exceeds the compressive strength of the dielectric, the dielectric breakdown failure will occur.

Mechanical breakdown is related to the Young's modulus. In addition, the mechanical breakdown and the electrostriction effect are closely related. The strain caused by electrostriction is directly proportional to the square of the polarization intensity, that is, directly proportional to the square of the dielectric constant. Therefore, the larger the dielectric constant, the more obvious the electrostriction effect, which in turn leads to a decrease in the breakdown strength. Therefore, the dielectric constant and breakdown field strength cannot be increased simultaneously.

(3) Thermal Breakdown

When an electric field is applied to a dielectric, the heat could be generated due to the factors such as dielectric loss and leakage current, which is then transferred to the environment through conduction and radiation. When the rate of heat generation exceeds the rate of heat dissipation, the dielectric breakdown failure will occur. Therefore, it is of great significance to reduce the heat generation. On the one hand, the heat generation rate could be reduced by reducing conductivity and dielectric loss. On the other hand, the heat dissipation rate could be improved by increasing the thermal conductivity of material.

At present, the generally accepted breakdown principle holds that the injection and accumulation of carriers into the polymer is the main reason for the destruction of the polymer [3]. Under the action of an applied electric field, carriers accumulate in the polymer and collide with atoms, resulting in atomic-scale damage [4]. Moreover, new carriers are generated, thereby forming a vicious cycle. When the breakdown electric field strength is reached, the carrier density increases sharply instantaneously, and the number of electrical weaknesses accumulates rapidly, resulting in the deterioration of dielectric insulation performance. A large number of experimental and simulation results show that the wide energy band gap is helpful to improve the breakdown strength of dielectrics. The band gap is the energy difference between the valence band and the conduction band of the dielectric. The valence electrons in the high band gap polymer need higher energy to activate, which means that the carrier surge is slower and the corresponding breakdown strength is higher.

For the two-component dielectric materials containing fillers, the relationship between the breakdown strength and the dielectric constant of the polymer matrix (P) and the filler (F) and the volume fraction of the filler is as follows.

$$E_P = E_0[f_P(1 - \frac{\varepsilon_P}{\varepsilon_F}) + \frac{\varepsilon_P}{\varepsilon_F}]^{-1} \tag{2.7}$$

$$E_F = E_0[f_P(\frac{\varepsilon_P}{\varepsilon_F} - 1) + 1]^{-1} \tag{2.8}$$

where E_0 is the external electric field strength, E_P is the actual electric field strength of the polymer matrix, E_F is the actual electric field strength of the filler, f_P is the volume fraction of the polymer, ε_P and ε_F are the dielectric constants of the polymer and the filler, respectively.

When the dielectric constant of the filler is much higher than that of the polymer matrix, the electric field strength of the polymer will increase sharply. For example, when the metal filler is added to the polymer, the E_F tends to 0, while the E_P will reach a very high value, which is easy to cause the breakdown of the composite material.

2.2.4 *Charge Transport Behaviour*

(1) Charge Injection

When moving the charge inside the metal electrode to the outside of the electrode, it is necessary to overcome the mirror charge Coulomb gravity of the electrode. According to the Coulomb's law, the gravitational force (F0) between the charge and the mirror charge is as follows.

$$F_0 = \frac{e^2}{16\pi \varepsilon d^2} \tag{2.9}$$

where e is the element charge, ε is the dielectric constant of the medium outside the electrode, and d is the distance between the two charges. The potential distribution (φ_0) generated by the mirror charge can be obtained by integrating the gravity.

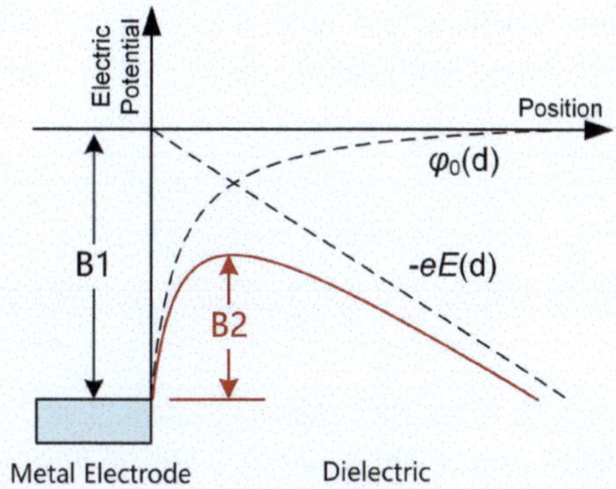

Fig. 2.2 Charge injection model

$$\phi_0(d) = \int_d^{\infty} F dx = -\frac{e}{16\pi \varepsilon d} \tag{2.10}$$

The distribution of the potential φ_0 is shown in Fig. 2.2. If the charge in the metal electrode is moved to infinity, the energy that needs to be absorbed from the outside is represented by B1, which is also called the interface barrier where the charge is injected from the metal electrode into the dielectric. When an external electric field is applied, the potential curve (φ_0) is superimposed with the electric field potential curve ($-eE$), and the interface barrier of charge injection is reduced from B1 to B2. When the external electric field is enhanced, the slope of the potential line formed increases, the interface barrier decreases more, and the interface barrier is further reduced.

Due to the unique parallel plate insulation structure of the capacitor, the insulation area (specific surface area) per unit volume is significantly higher than other insulation forms. For example, the insulation area of a capacitor cylindrical core with a diameter of 5 cm and a height of 10 cm can reach tens of thousands of square meters, while the insulation area of high-voltage cables of the same size is only a few square meters. The large specific surface area determines that the capacitor faces a serious charge injection problem for a long time. Specially, the difference between the conduction band of the dielectric film and the Fermi level of the metal electrode is the electron from the electrode. The electron injection barrier φ_B in the injected dielectric. Correspondingly, the difference between the valence band of the dielectric film and the Fermi level of the metal electrode is the hole injection barrier.

As the ambient temperature increases, the energy of the carriers inside the metal electrode and the dielectric increases. The charge in the metal electrode is easier to cross the interface barrier and inject into the dielectric, and the charge in the valence band and the local state in the dielectric is easier to transition to the high energy level to participate in charge transport. A larger number of carriers are frequently trapped and detrapped, resulting in more energy being dissipated in the form of heat energy, which aggravates the increase of temperature. Considering the temperature factor, the injection current density (j_{sc}) under the Schottky model and the excitation current density (j_{pf}) under the P-F model could be expressed by Formulas 2.11 and 2.12, respectively [5].

$$j_{sc} = AT^2 \exp(-\frac{ew_{metal}}{kT}) \exp(\frac{e}{kT}\sqrt{\frac{eE}{4\pi\varepsilon}}) \tag{2.11}$$

$$j_{pf} = AT^2 \exp(-\frac{ew_{trap}}{kT}) \exp(\frac{e}{kT}\sqrt{\frac{eE}{4\pi\varepsilon}}) \tag{2.12}$$

where A is the Richardson constant, T is the ambient temperature, e is the element charge, w_{metal} and w_{trap} are the metal injection barrier and the dielectric trap depth respectively, k is the Boltzmann constant, E is the electric field strength, and ε is the dielectric constant of the dielectric. The Schottky injection current increases exponentially with $E^{0.5}$ and increases super-exponentially with temperature T, so the Schottky injection is more significant than the jump conductance at high temperature.

When the barrier width is less than the De Broglie wavelength of the electron, the electron still has the probability to pass through the interface barrier even if it does not have enough energy to cross the barrier, thus forming charge transfer. This is caused by the volatility of electrons. When the external electric field is enhanced, the barrier height decreases and the barrier width decreases, as shown in Fig. 2.3. Therefore, more electrons are injected into the dielectric in the form of tunneling under high electric field. This process is also known as Fowler Nordheim tunneling. If the dielectric thickness is thin enough, the barrier width is small enough when the electric field is low, and the electron can tunnel directly. When the electron energy is between the Fermi level of the dielectric conduction band metal electrode, the electron may also undergo hot electron-field emission [6].

(2) Charge Transport

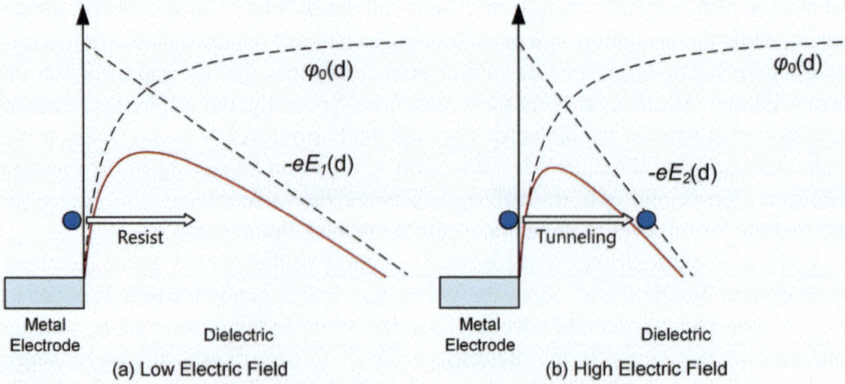

Fig. 2.3 Charge interface injection under **a** low and **b** high electric field

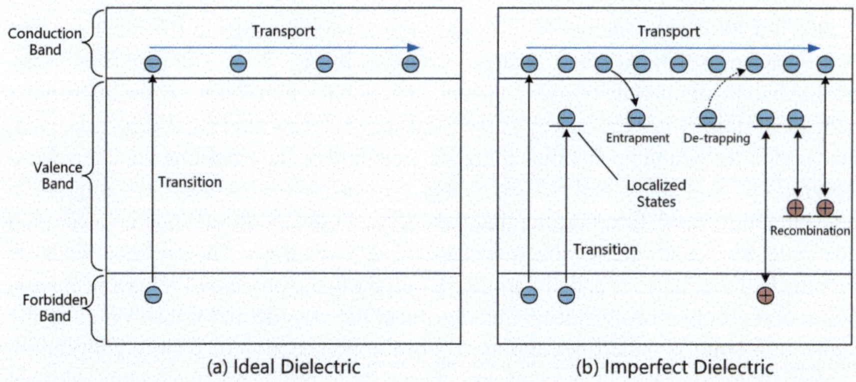

Fig. 2.4 Carrier transport behavior

In an ideal polymer insulating material, the molecular chains are periodically arranged closely, and there is no other energy state between the valence band and the conduction band, as shown in Fig. 2.4a. Meanwhile, owing to the wide band gap of the polymer, only a very small amounts of electrons could jump to the conduction band due to the probability fluctuation, and transport along the conduction band under the action of the electric field. However, the conductivity of the polymer is generally in the order of 10^{-45} S/m, which is obviously much lower than the conductivity values measured in experiments and engineering.

There are complex structures in the non-ideal polymers, so that the molecular orbitals of the materials are no longer only distributed in the theoretical conduction band and valence band. On the molecular chain scale, some catalyst groups may be attached to the molecular chain during the polymerization of the material and not removed. Free radicals may be formed during the crosslinking reaction of the material. Moreover, additional chemical compounds resulting from reactions could potentially be generated during the post-treatment of the material, and there is a possibility of introducing impurity particles, such as dust, into the manufacturing process of the equipment.

On the aggregation scale, the molecular weight of the polymer molecular chain is very large, indicating that the polymer chain could not be arranged as highly ordered as the atoms in the ceramic crystal. Therefore, almost all polymer dielectric materials are semi-crystalline or amorphous. The irregularity of the molecular chain could lead to the change of the crystal morphology and the spherulite size, thereby inhibiting the crystallization. Furthermore, the interface transition zone between two or more phases could occur in the composite film [7]. The aforementioned factors disrupt the periodicity and uniformity of the energy level arrangement within the polymer's local region, resulting in the formation of localized energy states, commonly referred to as local states.

When the energy level of the local state is located in the valence band or the conduction band, the transport process of the positive and negative charges will not be affected. When the energy level of the local state is located in the band gap, the carrier transport behavior will change, as shown in Fig. 2.5b. The electrons in the valence band will not only transit to the conduction band for migrating, but also transit to the local state. Since the energy of the local state is lower than the conduction band, the electron transition probability is higher. In addition, the electrons are likely to be captured by local states and release energy outward when the electrons migrate in the conduction band. Considering that the energy of the localized state is lower than the conduction band, the localized state could serve as traps. The process of charge trapping by local states is called trapping. Contrary to the process of releasing energy, the process of electrons absorbing energy from the outside and transitioning to the conduction band is called detrapping. From the perspective of energy conversion, the detrapping of electrons is an inverse process of trapping, and the energy released by during trapping process may be fully or partially applied to the another electron detrapped. In addition, the electrons and holes between different energy levels will also recombine. Considering that the energy levels of the electron and hole traps are closer, the probability of charge recombination in the traps is higher.

(3) Charge Excitation

Compared with the conduction band, the energy of the local state is lower, thus attracting and capturing the charge. The closer to the local state, the lower the carrier potential energy, as shown in Fig. 2.5a. If the influence between charge traps is ignored, the carriers in the trap must absorb enough energy to cross the trap depth to continue to migrate in the conduction band. When the dielectric is placed in the electric field, the trap potential and the external electric field potential are superimposed to form a new potential distribution, as shown in Fig. 2.5b. At the same time, the barrier height and width in the right direction of the trap in the figure decrease,

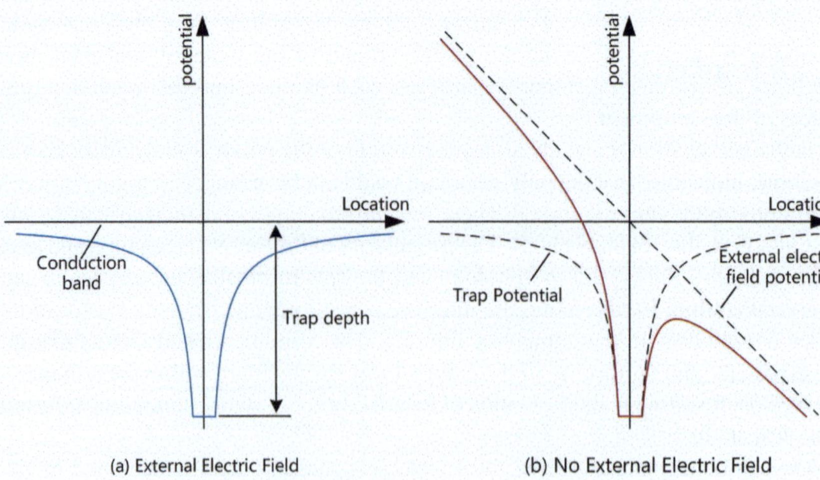

Fig. 2.5 Charge trap potential distribution under strong electric field

and the carriers in the trap are more likely to absorb energy and excite from the local state. This model is called Poole–Frenkel (P-F) model [8].

The model shown in Fig. 2.5 is not limited to the phenomenon of charge excitation in the local state. Charges at any level below the conduction band may be excited from the low level to the high level due to the action of an external strong electric field. For example, with the increase of the electric field, the inclination of the external electric field potential increases gradually from the valence band excitation to the trap or from the valence band excitation to the conduction band, and the charge excitation phenomenon becomes more and more obvious.

(4) Charge Acceleration

The charge injected at the interface or excited by the bulk will migrate in the conduction band of the polymer under the strong electric field. However, there are a large number of unoccupied three-dimensional cavities in the polymer in addition to the space occupied by the molecular chain. Due to the loose molecular chain structure, the molecular chain could move freely. This part of the space is called free volume. Under the action of the strong electric field, electrons could accelerate in the free volume. When the free volume is large enough, the electrons could accumulate enough energy to cross the intermolecular barrier, so that the leakage current of the insulating material will increase sharply. Furthermore, high-energy electrons may even directly impact the molecular chain of the polymer, causing macromolecular cracking and insulation degradation [9]. Therefore, the free volume is further increased.

(5) Partial Discharge

In the production process of polypropylene film, micro-defects such as bubbles and impurities are inevitably introduced into it. Due to the difference in dielectric parameters between the micro-defects and the matrix, the strong electric field concentration and electric field distortion could occur at the weak point. When the local electric field strength exceeds the breakdown field strength at the defect, the electrical weakness will be broken down, thereby resulting in partial discharge. Partial discharge refers to the phenomenon that may occur at any position of the insulating material without forming a breakdown channel through the positive and negative electrodes.

In addition to ionizing the material in the defect to form ions, the partial discharge will also produce high-energy photons and phonons. High-energy photons impact the molecular chain near the defect, so that the long-chain degradation of the polymer produces small molecular free radicals, which further catalyze the degradation of the polymer to form a chain reaction. Furthermore, the generated phonons aggravate the lattice vibration near the discharge position, promote the thermal motion of the molecular chain and reduce the stability of the chemical bond. Under the combined action of sound and light, the material near the defect gradually undergoes thermal oxygen aging and deterioration, the size of the defect gradually increases, and gradually forms a visually visible insulation degradation channel, eventually leading to penetrating insulation damage.

2.2.5 Energy Storage

The charging and discharging of dielectrics are based on the polarization and depolarization processes when the electric field is applied and then removed. During the charging process, polarization gradually increases with the increase of electric strength. During the discharge process, the polarization intensity gradually decreases with the decrease of the electric field intensity until the electric field intensity is zero. Due to dielectric loss, the stored energy cannot be fully released. The energy storage density U of the dielectric is affected by the electric field E and potential shift D. The energy storage density formula is as follows.

$$U_e = \int_{D_r}^{D_m} EdD \tag{2.13}$$

where Ue is the discharge energy density, Dm is the maximum electric displacement, Dr is the residual electric displacement, and E is the applied electric field strength.

For linear dielectrics, the dielectric constant does not change with the applied electric field, and the remanent polarization is basically zero. The discharge energy density formula can be further simplified as follows.

$$U_e = \frac{1}{2}\varepsilon_0\varepsilon_r E^2 \tag{2.14}$$

In order to increase the released energy storage density of dielectric materials, it is of great significance to increase the electric field strength and potential shift. Therefore, developing materials with high dielectric constant and high breakdown field strength is the key to achieving high releasable energy storage density.

In addition to releasing energy storage density, energy storage efficiency is also an important parameter, which could reflect the energy utilization efficiency of energy storage devices. Most importantly, the energy lost will be converted into heat, thereby leading to the deterioration of dielectric energy storage performance and even thermal breakdown. In order to improve energy storage efficiency, it is essential to reduce energy loss. Therefore, it is necessary for dielectrics to choose materials with low dielectric loss, low residual polarization, and thin and elongated electroconductive loops, thereby promoting the service life and operational stability of HVDC capacitors.

2.3 Polymer Dielectrics

2.3.1 Types of Polymer Dielectric

The development process of the capacitors is accompanied by the renewal process of its dielectric films. In different historical periods, the dielectric films of capacitors have experienced the development from cable paper, capacitor paper, polymer/paper composite, and finally to all polymer films [10]. In the 1940s, the insulation of power capacitors was mainly paper. With the rapid development of the petrochemical industry, petroleum is no longer only used for refining gasoline, diesel, asphalt, etc., and many organic polymer materials have appeared. In 1951, a kind of semi-crystalline polymer, polypropylene (PP) was discovered. Then the application of polyethene (PE), polystyrene (PS), poly tetra fluoroethylene (PTFE), polyethylene terephthalate (PET), polycarbonate (PC), PP and other materials were discussed by researchers. Traditional dielectric materials are ceramics with high dielectric constant and high thermal conductivity. However, ceramic materials are difficult to meet the application requirements in advanced power electronics due to their brittleness, difficulty in processing, low breakdown strength and poor stability. Polymer dielectrics usually have the advantages of convenient processing, high breakdown voltage, long life, low energy loss, high frequency current resistance and strong self-cleaning ability. They occupy an important position in the capacitor material market and have been more and more widely studied. It is proved that PP was very suitable for power system applications compared with other materials and the detailed introduction is as follows.

The performance of dielectric thin film is the core factor to determine the key parameters such as energy storage density and service life of the HVDC capacitor. According to the different molecular structure and dielectric properties, the common capacitor energy storage media is mainly divided into linear dielectric, ring dielectric, dipole dielectric and ferroelectric dielectric. The four types of dielectrics are introduced below.

(1) Linear Dielectrics

Linear dielectric refers to the dielectric whose electric displacement intensity (D) changes linearly with the electric field intensity (E) in a certain range of electric field intensity, as shown in Fig. 2.6. Such dielectrics are generally composed of non-polar molecules with high symmetry of molecular chain structure, and the positive and negative charge centers are almost coincident. Under the ideal condition of pure and uniform dielectric, the polarization behavior of the dielectric under strong electric field is mainly manifested as the separation of positive and negative charge centers in the molecule, that is, the electron polarization process [11]. Considering the existence of impurities, the ion polarization process and dipole polarization process may also occur under normal operating conditions.

The typical linear dielectrics are polyethylene (PE) and polypropylene (PP). At present, the biaxially oriented polypropylene film (BOPP) has already been the

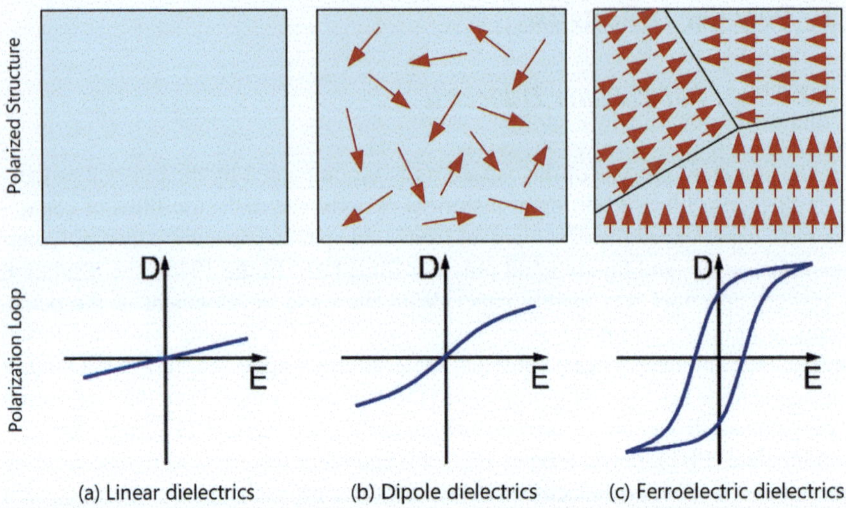

Fig. 2.6 Polarization structure and polarization D-E loop of different dielectrics

most mature commercial thin film material. Compared with other dielectric medias commonly applied in HVDC capacitors, the BOPP film has the highest breakdown field strength (> 800 kV/mm), the highest charge and discharge efficiency (> 99% at room temperature), the most excellent self-healing ability and the most excellent processing performance [12].

Compared with other materials, the dielectric constant of polypropylene dielectric is low, resulting in limited energy storage density. At present, the energy storage density of commercial BOPP film is about 2–3 J/cm^3. In addition, the melting point of polypropylene dielectric is low (about 170 °C), and the dissociation of polypropylene molecules is more serious at high temperature. When the temperature exceeds 85 °C, the insulation performance of polypropylene film decreases rapidly. At present, finding modification methods to maintain the advantages of polypropylene while improving its energy storage density and dielectric properties at high temperature has been widely investigated.

(2) Ring Dielectrics

The ring dielectric is mainly composed of benzene ring or other ring structure, which is a typical stable planar structure. The attraction of the electron cloud could be formed between the ring structures to produce π–π stacking interaction, so that the dielectric properties of material could remain stable at high temperature. The molecular asymmetry of the ring dielectrics is slightly higher than that of linear dielectrics, and the dielectric constant is between 2.5 and 3.5 [13]. However, the polarity of the overall dielectric is still weak owing to the ring structure, indicating that the polarization form is similar to that of the linear dielectric. The advantages of the ring dielectrics are mainly reflected in the strong stability at high temperature.

Polyimide (PI), Polyetherimide (PEI), Polycarbonate (PC) and Poly (ether-ether-ketone) (PEEK) are currently the best temperature-resistant capacitor film materials. At present, a large number of modification methods have been proposed to improve the dielectric properties of ring dielectrics. By constructing the PI/Al_2O_3 and PI/HfO_2 composite systems, the breakdown field strength of pure PI film at 150 °C could reach 422 kV/mm, and the energy storage density at 400 kV/mm could be increased to 3.8 J/cm^3. By blending PEI with Al_2O_3@ZrO_2 core–shell nanoparticles, the energy storage density of pure PEI could be increased to 5.19 J/cm^3 at 150 °C and 520 kV/mm. In addition, by multi-layer compounding PI and PEI, the energy storage density of the modified sample could reach 5.3 J/cm^3 at 200 °C and 300 kV/mm. The mechanism analysis shows that the establishment of the sandwich structure effectively inhibits the migration of carriers, thereby reducing the degree of damage to the molecular structure [14–16].

However, the polymer stability of the ring structure has brought other drawbacks. The higher glass transition temperature (Tg) of dipole dielectrics leads to much worse ductility than that of linear dielectrics, which is difficult to prepare micron-sized capacitor films in batches. Furthermore, the conjugated structure of the cyclic molecule greatly reduces the band gap of the material. Once the material is applied as the substrate of the capacitor film, the barrier width of the carrier tunneling and hopping is significantly shortened and the leakage loss of the capacitor film is greatly increased, which accelerates the energy loss rate of the capacitor and aggravates the heat diffusion. In addition, the ring structure determines the high C/H ratio (C/H wt%), which will increase the possibility of persistent short circuit (self-healing failure) formed by the accumulation of carbon at self-healing sites.

(3) Dipolar Dielectrics

The large group with strong polarity is introduced into the dielectric matrix to improve the polarizability of the dielectric material. The positive and negative charge centers of the polar groups are far away and the charge amount is large, indicating that the contribution of the electric dipole moment to the dielectric constant is more significant. Therefore, the dipole dielectric could exhibit better energy storage density. As the polarization of the dipole tends to be complete, the dielectric constant growth rate of the material gradually decreases with the electric field strength increasing and eventually remains stable. The D-E loop of the dipole dielectric is no longer completely linear, as shown in Fig. 2.6.

At present, a large number of preparation methods have been proposed to improve the dielectric properties of dipolar dielectrics. Polymethylsulfonyl ethyl methacrylate (PMSEMA) was synthesized by using methacrylate monomer. The dielectric constant of PMSEMA could reach 11.4 at 1 Hz, and the energy storage density could reach 4.54 J/cm^3 [17]. Sulfonated poly (2,6-dimethyl-1,4-phenylene oxide) (SO_2-PPO) was prepared by grafting method, and the discharge energy density of the modified sample could reach 16 J/cm^3 [18]. Polystyrene methyl methacrylate (PMMA) was obtained by copolymerization of polystyrene (PST) and polymethyl methacrylate (PMMA) [19], and its energy storage density was still greater than 12 J/cm^3 at 90 °C.

However, although the large dipole brings greater polarization potential to the dielectric, it will have obvious polarization relaxation hysteresis at higher frequencies, resulting in higher polarization loss, making the capacitor film prematurely reach the service life. In addition, the speed of energy release by dipole relaxation is much slower than that by electron polarization, and further slows down with the deepening of discharge degree, which is not conducive to the precise control of the AC/DC power conversion.

(4) Ferroelectric Dielectrics

The ferroelectric material has unique spontaneous polarization properties. It undergoes a paraelectric-ferroelectric phase transition in a specific temperature range, thereby forming a large-sized polarization domain and increasing the polarization strength of the material, as shown in Fig. 2.6. Therefore, the dielectric constant of the ferroelectric dielectrics is often significantly higher than that of other types of dielectric materials.

Polyvinylidene fluoride (PVDF) and its derivatives are the most widely applied ferroelectric polymer materials. At present, a large number of modification methods have been proposed to improve the dielectric properties of ferroelectric dielectrics.

By doping $BaTiO_3$ nanoparticles into PVDF, a composite film with an energy storage density of 18.8 J/cm^3 was prepared. By blending PVDF with P (VDF-CTFE) at a ratio of 7:3, a capacitor film with an energy storage density of 7 J/cm^3 was obtained. Furthermore, by doping P (VDF-TrFE-CFE) into the PVDF matrix, the energy storage density of the film could be further increased to 20 J/cm^3.

The polarization relaxation of ferroelectric polymers is microscopically reflected in the movement of their internal domain walls, which leads to the slower polarization of ferroelectric materials and their derivatives than all other polymer materials. In the process of polarization relaxation, the dielectric loss of ferroelectric polymers could reach 40%. A large amount of energy is lost in the form of heat energy, which accelerates the thermal aging of materials. In addition, under high electric field, PVDF molecules are prone to eliminate the reaction to release HF and corrode the metal electrode [20–22].

Figure 2.7 summarizes the advantages and disadvantages of the four dielectrics in the reported research on various key performance indicators. Although the synthesis of new high-performance energy storage materials has a higher upper limit of performance, no new materials with better performance than or close to polypropylene have been found from the existing research progress. Taking polypropylene as the matrix, maintaining its advantages and modifying its performance short board is still the most promising research idea at present.

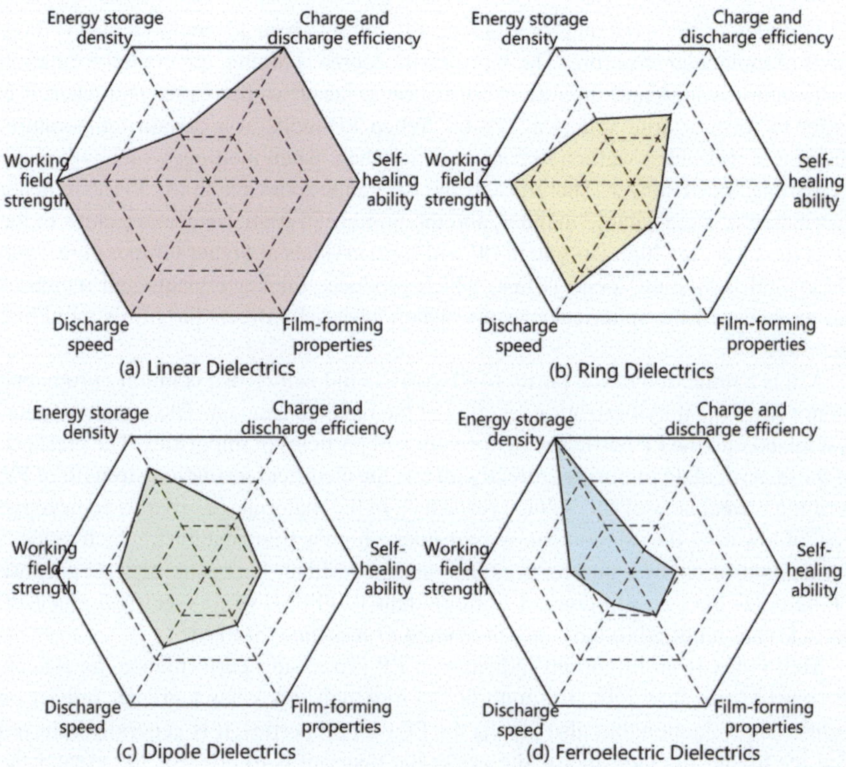

Fig. 2.7 Comparison of key performance indicators of different dielectrics

2.3.2 *Polypropylene*

(1) Raw Material

PP is a crystallizable polymer material, whose isotropy has a large influence on the crystallization ability of PP. The degree of crystallinity is an indicator of the proportion of crystalline regions in a polymer. It is generally accepted that a high degree of crystallinity results in a high melting point. The process of crystallization is the orderly arrangement of molecular chains, whereas melting destroys the molecular structure. The higher the crystallinity, the more regular the arrangement of the molecular chains, the higher the temperature required to break them down, and therefore the higher the melting point. A higher melting point means for capacitors that they can operate at higher temperatures, so a higher degree of crystallinity is desired. Electrical grade PP particles can have a crystallinity of 45% or more, general purpose PP particles have a lower crystallinity.

Whether the distribution of cubic defects or isotropic sequences is reasonable or not is one of the main reasons affecting the stability of film forming. Isotropy refers to the percentage of all isomeric polymers in the total polymer, isotropy is an important indicator to reflect the molecular structure of polymer materials. In the

PP particles in the existence of isotactic PP, intermediate PP and atactic PP three kinds of molecular structure. The fat isotactic degree is higher, the crystal formation temperature is high and the rate of formation is faster, so the wafer is thicker, it is easier to form a large spherical crystal. When PP melts at a certain temperature, because of its thicker wafers, to consume more heat when melting, which will make its melting speed slower. Therefore, if the iso-gauge sequence in the PP resin is not distributed in a reasonable manner, and the number of high iso-gauge grades in the PP is too large, the high iso-gauge PP will form crystals at higher temperatures, and the crystal formation speed is fast, which produces thicker crystals and increases the thickness of the spherical crystals, which adversely affects the film's moulding stability.

Ash is a reflection of the purity of PP particle indicators, ash is mainly propylene monomer in the polymerization reaction of the residual catalyst. The metal salt catalyst mixed into the PP particles, the formation of particles of impurities. The presence of these particulate impurities directly affects the electrical insulation strength of PP, forming a weakness of the voltage resistance in the material. In order to achieve the goal of lowering the ash content, we consider a high activity catalyst, which reduces the amount of catalyst and reduces the catalyst residue. Domestic general purpose PP particles have an ash content of more than 100 PPM, while electrical grade PP particles are required to have an ash content of less than 20 PPM.

Melt index is an important indicator of PP processing performance, he reflects the plastic melt flow rate. Too high or too low melt index has a greater impact on industrial production, but also affects the PP film properties. It is generally believed that the higher the melt index, the easier the material is to process, the greater the yield, the better the barrier properties, but its impact resistance, resistance to environmental stress capacity deterioration. With a low melt index, the material is difficult to process, but the fracture strength, toughness and aging resistance are improved. Melt index is mainly related to molecular weight, molecular weight distribution and molecular structure, so controlling the appropriate molecular weight of PP to obtain the appropriate melt index is a crucial factor.

(2) Stretching Process

Nationalized PP film stretching utilizes a more advanced processing technology. Biaxially oriented stretching includes the tube film method and the flat film method. Tube film is produced through longitudinal stretching and blowing, while flat film is manufactured through longitudinal and transverse stretching. The flat film process can be further divided into simultaneous biaxial stretching and progressive biaxial stretching.

The manufacturing process of bi-directional stretching flat film is as follows: the raw material is melted in the extruder, extruded into a sheet through the die, and then shaped into a sheet using the cold press rollers; the sheet goes through the longitudinal stretching machine, where it is stretched longitudinally by the difference in the rotational speeds of the rollers, and then transversely stretched through the transverse stretching machine. Meanwhile, the manufacturing process of the bi-directional stretching flat-film method is as follows: the extruded plastic sheet is simultaneously stretched longitudinally and transversely in a stretching machine, and the film is produced through hot and cold treatment. In this method, the extruded sheet is clamped by a fixture at the inlet of the stretching machine. Subsequently, the sheet is fully preheated in the preheating zone with the assistance of the fixture moving in the same direction and synchronously. In the stretching zone, the fixture expansion helps gradually increase the distance between the clip movements to achieve longitudinal and transverse stretching orientation of the film simultaneously. Finally, after heat treatment and cooling, the entire film is completed. The bi-directional stretching method involves one-time bi-directional stretching orientation, where the film does not come into contact with any rollers during stretching. This method is not affected by the quality of the roll surface and can produce scratch-free film. When the film is stretched using this method, the sheet is constrained by the fixture, preventing any necking of the sheet. This is highly advantageous for enhancing the uniformity of film thickness. It is easy to produce films with different stretch ratios in the longitudinal and transverse directions, and, of course, it is also possible to produce films with the same stretch ratio in both directions. The main process parameters controlled during the stretching process are line speed, temperature, and stretch ratio.

The stretching temperature primarily affects the orientation of the PP molecular chain. In order to enhance the orientation of the PP chain, it may be suitable to increase the temperature. However, high temperatures can easily lead to the deorientation of the PP chain, causing the PP molecular chain or microcrystalline structure to fracture during stretching. This results in a decrease in the strength and stability of the BOPP film. Therefore, the temperature range in the stretching process is generally limited to the interval between the PP glass transition temperature and the viscous flow temperature. Such as longitudinal stretching operating temperature can generally be set to 80–110 °C, transverse stretching operating temperature can generally be set to 120–150 °C.

When the stretching temperature in the β crystals and α crystals is between the melting point, containing α crystals and β crystals of PP through bidirectional stretching, lower density β crystals will be converted to higher density α crystals. In the β crystals, in the presence of the place of the formation of "craters", induced the surface of the film will form a large number of interlinked ring grooves. This process is called film Surface Roughening. The control of surface roughness of BOPP film for capacitors is crucial, and the surface roughness must be maintained within the ideal range. Excessive roughness can impact the electrical properties of the film. When the film is stretched in two directions, it may form micropores or excessive pores, leading to a decrease in dielectric strength. Conversely, if the film surface is too smooth, with a small surface area, it may not facilitate subsequent processes like impregnation, vapor deposition, winding, etc. For BOPP capacitor film, it is generally necessary to control the surface roughness to around 0.25–0.65 Ra μm^{-1}.

During the manufacturing process of BOPP films, the degree of orientation of the PP molecular chains increases with the rising tensile ratio at a constant tensile temperature. In addition, as the stretching rate increases, the duration of tensile stress action decreases, which ultimately affects the orientation effect of the film. Excessive stretching rates can lead to the breakage of PP molecular chains during stretching. This occurs because the rate of tensile deformation exceeds the deformation caused by the relaxation of the PP molecular chains. At this point, the molecular orientation of BOPP film and the degree of stretching will continue to increase with the stretching speed and reduce. Therefore, in the bidirectional stretching process, it is essential to control the stretching speed reasonably to ensure that the material's crystallization rate is relatively slow. This is conducive to the smooth progression of the stretching orientation. If the stretching process produces crystallization too early or too quickly, or results in the emergence of larger PP crystal particles, there may be an issue of film breakage during stretching.

The sequence of longitudinal and transverse stretching, as well as the ratio between the two, must be controlled within the appropriate range to guarantee balanced film properties. Good stretching orientation is conducive to the regularity of the arrangement of PP crystals or molecular chains, making it easier to produce a tandem crystal interlocking structure, which positively affects the mechanical properties of the film. The bidirectional stretching effect can also lead to the breakage of larger crystals in the polymer, thereby reducing the fogging caused by crystallization, and increasing the surface gloss and transparency of the PP film. PP film subjected to bidirectional stretching can reduce haze by 50%. Capacitors with BOPP film also requires the thickness deviation to be as minimal as possible to ensure there is no local breakdown. When the film thickness deviation is small enough, the electric field of the prepared capacitor becomes more uniform, making the capacitance easier to control. It has been shown that the surface roughness of the film varies with the stretching ratio in bi-directional stretching. The surface roughness parameter of the film reaches a maximum value at a stretching ratio of 2. In general, BOPP film production involves a longitudinal stretch ratio of 4 to 5.5 and a transverse stretch ratio of 7 to 9.

(3) Coating Techniques

The evaporated material is evaporated in a vacuum chamber with the aid of a heated evaporation source. The evaporated material moves freely through the vacuum; when the evaporated material comes into contact with the surface of the substrate, it cools down and forms the coating.

To enhance the self-healing capabilities of film capacitors, the safety film design ensures the safe operation of the capacitor by controlling the amount of injected energy. The safety film utilizes a split electrode design, where the block electrodes are separated by a non-metallized insulating tape and connected solely by very fine metallized filaments. The safety film block construction reduces the self-healing energy and enables the capacitor to last up to six times longer than non-safety film capacitors. However, the safety film design has some drawbacks. The metallized filaments lead to slightly higher capacitor losses compared to normal metallized film. Additionally, the insulation gap takes up nearly 5% of the effective electrode area, reducing the usability of the metallized film. Copper Peak Electronics proposes a grid metallized film with a gradual change in fuse size, introducing a fuse gradient structure to address the variation in self-healing energy based on the gradual change in square resistance. This design enables the metallized film to be utilized with enhanced resistance to peak and inrush currents, while preserving its self-healing capability.

The thicker square resistor design effectively suppresses the problem of electric field distortion at the edge of the metal electrode. When the capacitor is operated under AC voltage, the electric field will be distorted and increased at the edge of the metal layer, making it more susceptible to corona discharge or breakdown compared to other parts of the capacitor. If air and moisture are present at the edge of the electrode plate, corona discharge at the edge is more likely to occur, which, in turn, causes corrosion and retreat of the metal layer at the edge of the metallized film. The area of electric field distortion decreases with increasing electrode thickness. In other words, the degree of electric field distortion is more pronounced when the square resistance is high. Adjusting the thickness of the coating at the edge position can decrease the edge field strength and minimize the corrosion recession of the metallized film coating.

The asymptotic square resistance significantly reduces the heat generation of the capacitor in engineering, provided that the average value of the square resistance remains constant. This is due to the fact that the asymptotic square resistance is designed in a way that the position with the highest current on the metallized film has the smallest square resistance, while the position with the lowest current has the largest square resistance. The gradual increase in resistance facilitates the balanced design of properties such as electric field strength, current carrying capacity, and self-healing energy. Setting the anti-oxidation film on the metallized film can protect the metal film and effectively prevent oxidation. Additionally, placing the wave-shaped elastic sheet between the anti-oxidation film and the metal film can act as a buffer, enhancing the compressive strength of the entire metallized film. Metallized PP film capacitors are typically vapor-coated with aluminum, zinc, or a combination of aluminum and zinc on the surface, providing them with excellent self-healing properties. However, in practice, the metallized film may still exhibit issues because the self-healing energy cannot be released in time to facilitate proper bonding within the film, potentially resulting in capacitor breakdown. Therefore, improving the temperature resistance of the film and the self-healing environment is key to enhancing the performance of metallized film capacitors.

Alumina exhibits excellent electrical insulation and thermal conductivity, with a relative dielectric constant ranging from 9.5 to 11.2 and a thermal conductivity of 30 W/(m K). Based on coated diaphragm technology, Tongfeng Electronic Company evaporates alumina onto the surface of PP film to create a composite insulating medium. On one hand, alumina, as a component of the capacitor's insulating medium, can increase the relative dielectric constant of the composite film. On the other hand, the presence of alumina can enhance the thermal conductivity of the capacitor core. This allows the heat generated by heat loss or self-healing of the dielectric to be quickly dissipated from the core. It also helps prevent the heat generated by self-healing from causing bonding between the layers of PP film and adjacent self-healing points. PP film is placed next to the self-healing point to prevent interlayer bonding, control the extent of self-healing, and act as a barrier for "isolated self-healing".

2.4 Challenge

With the development of HVDC transmission technology, HVDC capacitors are developing towards high voltage and large capacity, it is very important to improve their ability to handle high voltage and high current, reduce loss, improve energy storage density, and reduce capacitor volume and weight under the premise of meeting voltage output quality. There are many factors affecting the performance of high-voltage capacitors, such as the performance of dielectric materials, the structure of the capacitor and so on. After nearly one hundred years of development, the structural design of power capacitors has become more and more perfect, while the performance of dielectric materials has become a fundamental factor affecting the development of HVDC capacitors. Capacitor energy storage density depends on the dielectric constant and electric field tolerance. Therefore, for power capacitors, it is very important to improve the voltage withstand capability of dielectric polymer film. In addition, the internal temperature rise of power capacitors is also a factor that cannot be ignored during long-term operation. Capacitor internal temperature rise for polymer dielectric aging, breakdown performance and dielectric properties have a very important impact. High-voltage DC capacitor in operation, will be subject to a large number of harmonic currents, the electrode resistance of the heat, the presence of polymer dielectric loss, will lead to the capacitor internal temperature rise, so that the aging of the insulating medium is accelerated, and may even cause the breakdown phenomenon occurs. Electric field and thermal field together for a long time is the main cause of dielectric aging, temperature and over-voltage of the joint role of the capacitor dielectric material aging effects cannot be ignored.

The DC support capacitor operates inside the converter valve, and the dielectric loss caused by the high temperature environment inside the converter valve poses a challenge to the insulation performance of metalized thin film capacitors. In general, the operating temperature of the power equipment inside the converter valve is less than 70 °C, but under abnormal operating conditions, its temperature can reach 90–100 °C. Meanwhile, the self-heating of metallized thin film capacitors operating in high-temperature environments cannot be ignored. The electrode of the metallized thin film capacitor is an evaporated electrode with a thickness of nm. A thinner electrode thickness will result in a higher electrode resistance and the resistance loss will be generated when current flows through the deposited electrode, thereby leading to an increase in the internal temperature of the metallized thin film capacitor. In addition, the heat generated during self-healing of metalized thin film capacitors cannot be dissipated in a timely manner, which will also cause a local temperature rise inside the equipment. The high temperature of the environment and the temperature rise inside the metallized film capacitor will cause a sharp increase in the dielectric loss of the polypropylene film in the equipment. The metallized thin film capacitor applied in flexible DC converter valves bears the effect of DC superimposed harmonic voltage, and its polarization loss is very small. The dielectric loss of polypropylene film is mainly caused by conductivity loss. High temperature will lead to an increase in carrier mobility and concentration within polypropylene films. Charge is injected into the conduction or valence band of the insulating material through the Fowler Nordheim effect or Schottky effect from the electrode. Research has shown that an increase in temperature can cause a sharp increase in the leakage current inside polypropylene films, showing an exponential upward trend. It was found by studying the conductivity current of polypropylene film at high temperatures that the activation energy of conductivity of polypropylene film increased from 0.2–0.52 eV to 1.05 eV when the temperature was above 80 °C, thereby resulting in an increase in conductivity current [23]. The conductivity loss caused by leakage current of polypropylene film will generate a large amount of Joule heat inside the metalized film capacitor in long-term operation, causing further temperature rise inside the metalized film capacitor. There is a mutual promotion effect between the high-temperature environment and the dielectric loss of polypropylene film. The high-temperature environment leads to an increase in the dielectric loss of polypropylene film, and the increase in dielectric loss of polypropylene film promotes the temperature rise.

In high-temperature environments, the vicious cycle between the internal temperature rise of metallized film capacitors and the conductivity loss of polypropylene film will lead to a decrease in the dielectric performance of polypropylene film. Under the combined effect of long-term electric field and temperature field, polypropylene film will undergo electrothermal aging, and the crystallinity and morphology of polypropylene film will change, leading to increased dielectric loss of polypropylene film. Research has shown that after a certain period of thermal aging of polypropylene films, it is found that the crystalline morphology of polypropylene films changes, leading to a decrease in their breakdown field strength. After conducting electric thermal aging on two polypropylene films with different crystallinity at 65 °C and 19.2 kV/mm [24], it was found that the polypropylene film with higher crystallinity had higher breakdown field strength before and after aging, and the breakdown field strength of the two polypropylene films with different crystallinity after electric thermal aging was lower than that before aging.

As a key component of metallized film capacitors, the breakdown field strength of capacitor films directly determines the working field strength and safety reliability of metallized film capacitors. Studies have shown that the breakdown field strength of polypropylene films is very high, up to 700 V/μm [25]. However, in practical engineering, the actual field strength of metallized thin film capacitors applied in DC applications is generally only 200 V/μm [26]. The reason for this is that the polypropylene film has defects, so the probability of breakdown at the defect site is the highest. The generation of defects in polypropylene films comes from multiple aspects, including the irregularity of the polypropylene structure, the inevitable introduction of defects such as bubbles and impurities during the production process. Polypropylene film is a typical semi crystalline polymer, with both crystalline and amorphous regions present inside the film. Molecular chains in the crystalline region are arranged in an orderly manner, while those in the amorphous region are arranged in a disorderly manner. There are inevitably defects such as micropores, chain bending, dislocations, and interfaces between crystalline and amorphous regions within the membrane. Meanwhile, there may also be impurities and defects caused by the aggregation of small amounts of residual catalysts, antioxidants, etc.

inside the polypropylene film. In addition, under the combined effect of electric field and temperature field for a long time, polypropylene films will also produce new defects due to electrical aging, thermal aging, and other reasons. The various defects present in polymer media could lead to band breaks, forming charge capture centers (trap energy levels) that could affect charge transport and subsequently affect the macroscopic dielectric properties of polymers, including conductivity and breakdown field strength. What's more, polypropylene film capacitors could store energy. The energy is stored within the insulating material separating the capacitor plates and is released into the load upon closing the discharge switch. It's crucial to recognize that discharging this stored energy generates internal mechanical forces within the capacitor, potentially leading to material deterioration. During repetitive use, the energy discharge may trigger mechanical resonances within the capacitor's structure, ultimately leading to severe damage and electrical failure.

The polypropylene film in the metallized film capacitor is subjected to a complex electric field effect caused by the superposition of DC voltage and harmonic voltage, and the breakdown characteristics of polypropylene film under complex electric field stress are not yet clear. Research has shown that the DC breakdown voltage of polymer dielectric materials is 1.7–1.8 times that of 50 Hz AC breakdown voltage [27]. The study on the distribution of space charge and electric field under DC voltage and AC voltage shows that under DC voltage, the electrode injects same polarity space charge into the sample. The built-in electric field formed by the same polarity space charge weakens the electric field strength near the interface between the sample surface and the electrode, while enhancing the electric field strength inside the sample. Under DC voltage, breakdown begins inside the medium. Under AC voltage, the polarity change of the voltage causes the positive and negative charges injected into the sample by the electrode to mainly concentrate on the surface of the sample. At the peak of AC voltage, the electric field distortion on the surface of the sample is strongest, and under AC voltage, the breakdown of the medium begins at the surface of the medium. The waveform of applied voltage has a significant impact on the breakdown performance of polypropylene film, leading to differences in the lifespan of capacitors. In practical engineering, the service life of capacitors applied in DC applications is only one-third of that used in AC applications. At present, the influence of DC superimposed harmonic electric field on the breakdown characteristics of polypropylene films is not clear, and the correlation between charge transport and breakdown characteristics of polypropylene films under DC superimposed harmonic voltage is still lacking in systematic and in-depth research.

The IGBT heating inside the flexible DC converter valve and the self heating of the metallized film capacitor cause the polypropylene film to work in high-temperature environments. High temperature will also affect charge transport and the microstructure of polypropylene film, thereby affecting the breakdown characteristics of polypropylene film. Research has shown that as the temperature increases, the breakdown field strength of polypropylene films significantly decreases. At the same time, high temperature promotes the movement of polymer molecular chains, leading to an increase in the free volume inside the polymer. According to the theory of free volume breakdown, as the free volume increases, electron energy will increase, resulting in a decrease in breakdown field strength. From the structure of metallized film capacitors, it can be concluded that the local breakdown field strength of polypropylene film directly determines the actual service life of metallized film capacitors. When the temperature rise inside the DC support capacitor is uneven, it will cause uneven expansion and contraction of the polypropylene film, resulting in an air gap. The existence of the air gap will exacerbate partial discharge and even cause dielectric breakdown. Research has shown that high temperatures will cause a decrease in the lifespan of capacitors. For every 8 °C increase in temperature, the lifespan of capacitors will decrease by about 50%. The impact of high temperature on the breakdown characteristics of polypropylene film and the lifespan of capacitors cannot be ignored [28].

At present, there is relatively little research on the breakdown characteristics of polypropylene film under high temperature and DC superimposed harmonic voltage operating conditions for metallized film capacitors. Conducting research on the breakdown characteristics of polypropylene films under the coupling effect of high temperature and DC superimposed harmonic voltage and analyzing the breakdown mechanism of polypropylene films under the coupling effect of high temperature and DC superimposed harmonic voltage could provide a research basis for exploring methods to improve the breakdown performance of polypropylene films. It is of great significance for improving the working field strength and actual service life of the DC support capacitor inside the flexible DC converter valve.

Direct current support capacitors are prone to surface discharge issues inside, posing a threat to the safe and stable operation of the capacitors. Due to the use of a evaporated electrode structure in the metallized film capacitors, as shown in Fig. 2.8, there is a margin (A) between the two electrodes. At the same time, after self-healing, the electrodes around the breakdown point evaporate to form surface insulation (B). Under the action of external voltage, the electric fields at A and B in Fig. 2.8 are relatively concentrated and prone to surface discharge. The charges generated by surface discharge migrate and accumulate on the surface of the polypropylene film, ultimately developing into a penetrating discharge channel, namely surface flashover, thereby leading to capacitor failure.

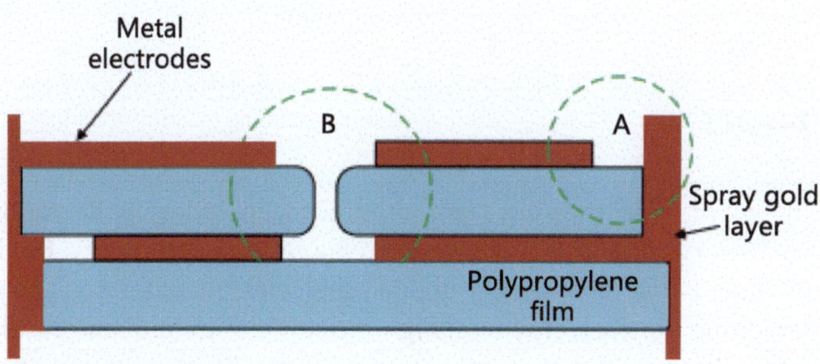

Fig. 2.8 Surface discharge process

At present, research on capacitor discharge mainly focuses on the causes, measurements and effects on insulation media of partial discharge. Research has shown that excessive concentration of electric field in areas with insulation defects at the edge of electrodes can cause partial discharge, leading to breakdown of the insulation medium. The interlayer air and uneven mechanical stress during the film winding process could lead to the degradation of partial discharge performance of polypropylene films. The heat generated by partial discharge cannot be dissipated in a timely manner, and the gradual accumulation of heat causes the local temperature to be too high, leading to the aging of the polypropylene film. At the same time, it also leads to an increase in defects in the dielectric material, resulting in a decrease in the dielectric properties of the polypropylene film. The external application conditions have a significant impact on the surface discharge and flashover voltage of polymer materials. Relevant studies have shown that temperature has a significant impact on surface discharge, and an increase in temperature will lead to an increase in discharge frequency and discharge capacity. Under the superimposed harmonic voltage of DC, the DC ripple coefficient affects the flashover voltage value, and the influence of DC ripple coefficient will also increase with the increase of electrode spacing. However, current research on surface discharge of capacitors is not in-depth, especially when surface discharge develops into flashover. The mechanism of surface flashover under high temperature and DC superimposed harmonic voltage needs to be explored, and there is a lack of research on methods to suppress surface discharge from developing into flashover.

In summary, due to the tendency of surface discharge to occur inside DC support capacitors, the development of surface discharge into continuous flashover poses the most serious threat to capacitors, and external conditions have a significant impact on surface discharge and flashover voltage of polymer materials. Therefore, conducting research on the surface flashover characteristics of polypropylene film under high temperature and DC superimposed harmonic voltage provides a research basis for exploring methods to improve the flashover voltage of polypropylene film in metallized film capacitors. It is of great significance to improve the surface discharge tolerance and actual service life of metallized film capacitors in flexible DC converter valves.

2.5 Summary

The basic theory of dielectrics has been introduced, including polarization theory, dielectric loss theory, breakdown theory, and charge transport theory. Furthermore, the polypropylene is the most widely applied dielectric material due to its notable attributes, including low dielectric loss, high breakdown strength and excellent self-healing. In this chapter, the preparation process of PP films is introduced, and the preparation parameters and design of BOPP films are described in detail from the perspectives of raw materials, stretching process and metal electrode vapour deposition process.

References

1. J. Wei and L. Zhu, "Intrinsic polymer dielectrics for high energy density and low loss electric energy storage," *Prog. Polym. Sci.*, vol. 106, 2020, Art. no. 101254.
2. M. Xiao, K. X. Wang, Y. N. Song and B. X. Du, "Dielectric performance of polypropylene films modified by UV irradiation for HVDC capacitors," *IEEE Trans. Dielect Elect. Insul.*, vol. 31, no. 1, pp. 176–183, 2024.
3. H. L Liu, B. X. Du, M. Xiao and Y. W. Ma, "Carrier deflection behavior in polypropylene under strong electromagnetic field," *IEEE Trans. Dielect Elect. Insul.*, vol. 30, no. 6, pp. 2625–2632, 2023.
4. H. L Liu, B. X. Du, M. Xiao and Y. W. Ma, "Effects of electromagnetic field on partial discharge behavior in BOPP film capacitors," *IEEE Trans. Dielect Elect. Insul.*, vol. 30, no. 6, pp. 2617–2624, 2023.
5. P. R. Emtage and W. Tantrapom, "Schottky emission through thin insulating films," *Phys. Rev. Lett.*, vol. 8, no. 7, 1962, Art. no. 267.
6. J. G. Simmons, "Richardson-Schottky effect in solids," *Phys. Rev. Lett.*, vol. 15, no. 25, 1965, Art. no. 267.
7. S. T. Li, D. M. Min, W. W. Wang and C. Chen, "Modelling of dielectric breakdown through charge dynamics for polymer nanocomposites," *IEEE Trans. Dielect Elect. Insul.*, vol. 23, no. 6, pp. 3476–3485, 2016.
8. C. M. Vodenicharov and M. B. Vodenicharova, "Schottky and Poole-Frenkel mechanisms of conductivity in thin-film Me-GeS-Me systems," *Phys. Status Solidi A.*, vol. 57, no. 2, 1980, Art. no. 483.
9. M. Xiao, Y. S. Zhao and B. X. Du, "Dielectric performance of furan-functionalized polypropylene film for HVDC capacitors," *IEEE Trans. Dielect Elect. Insul.*, vol. 31, no. 1, pp. 160–167, 2024.
10. J. Ho, T. R. Jow and S. Boggs, "Historical introduction to capacitor technology," *IEEE Electr. Insul. Mag.*, vol. 26, no. 1, pp. 20–25, 2010.
11. H. L. Liu, B. X. Du, M. Xiao, Z. Y. Ran and Y. W. Ma, "Dielectric properties of polymer films in strong electromagnetic field for energy storage capacitor," *IEEE Trans. Dielect Elect. Insul.*, vol. 29, no. 5, pp. 1745–1753, 2022.
12. M. Ritamaki, I. Rytoluoto and K. Lahti, "Performance metrics for a modern BOPP capacitor film," *IEEE Trans. Dielect Elect. Insul.*, vol. 26, no. 4, pp. 1229–1237, 2019.
13. X. J. Liu, M. S. Zheng, G. Chen, Z. M. Dang and J. W. Zha, "High-temperature polyimide dielectric materials for energy storage: theory, design, preparation and properties," *Energy Environ. Sci.*, vol. 15, no. 1, pp. 56–81, 2022.
14. H. L. Liu, B. X. Du, M. Xiao, and Z. Y. Ran, "Improved conductivity and breakdown performance of polypropylene film by parylene blending for power capacitor," *IEEE Trans. Dielect Elect. Insul.*, vol. 29, no. 3, pp. 997–1004, 2022.
15. L. L. Ren et al, "High-temperature high-energy-density dielectric polymer nanocomposites utilizing inorganic core-shell nanostructured nanofillers," *Adv. Energy Mater.*, vol. 11, no. 28, 2020, Art. no. 2101297.
16. Y. J. Niu et al, "Significantly enhancing the discharge efficiency of sandwich-structured polymer dielectrics at elevated temperature by building carrier blocking interface," *Nano Energy*, vol. 97, no. 15, 2022, Art. no. 107215.
17. J. J. Wei et al, "Achieving high dielectric constant and low loss property in a dipolar glass polymer containing strongly dipolar and small-sized sulfone groups," *ACS Appl. Mater.*, vol. 7, no. 9, pp. 5248–5257, 2015.
18. Y. F. Zhu, Z. B. Zhang, M. H. Litt and L. Zhu, "High dielectric constant sulfonyl-containing dipolar glass polymers with enhanced orientational polarization," *Macromolecules*, vol. 51, no. 16, pp. 6257–6266, 2018.
19. Q. Z. Li, J. J. Liu, X. Zhang, S. B. Tang, J. Y. Lu and Z. C. Zhang, "Tuning the dielectric and energy storage properties of polystyrene-based polymer dielectric by manipulating dipoles and their polarizing behavior," *Phys. Chem. Chem. Phys.*, vol. 21, no. 28, pp. 15712–15724, 2019.

20. Prateek, V. K. Thakur and R. K. Gupta, "Recent progress on ferroelectric polymer-based nanocomposites for high energy density capacitors: synthesis, dielectric properties, and future aspects," *Chem. Rev.*, vol. 116, no. 7, pp. 4260–4317, 2016.
21. Z. H. Yi, Z. Wang, Y. X. Li, D. Wu and Y. Xue, "Improving the energy storage performance of all-polymer composites by blending PVDF and P(VDF-CTFE)," *Macromol. Rapid Commun.*, vol. 44, no. 4, 2022, Art. no. 2200728.
22. X. Jiang et al, "Significantly improved dielectric breakdown strength and energy density in P(VDF-TrFE-CTFE) polymer via a facileuniaxial drawing process," *ACS Appl. Polym. Mater.*, vol. 4, no. 7, pp. 4741–4750, 2022.
23. A. Kahouli, O. Gallot-Lavallée, P. Rain et al, "Dielectric features of two grades of bioriented isotactic polypropylene," *J. Appl. Polym. Sci.*, vol. 132, no. 28, 2015, Art. no. 42224.
24. J. Kurimsky, M. Kosterec and B. Vargova, "Breakdown voltage of polypropylene film during DC and thermal ageing," 2017 18th International Scientific Conference on Electric Power Engineering (EPE), Kouty nad Desnou, 2017, pp. 1–4.
25. L. H. S. Duarte and M. F. Alves, "The degradation of power capacitors under the influence of harmonics," 10th IEEE International Conference on Harmonics and Quality of Power, Rio de Janeiro, Brazil, 2002, pp. 334–339.
26. L. Cheng, Z. Li, J. Wang, Z. Xu, W. Liu and S. Li, "Degradation behavior and mechanism of metalized film capacitor under ultrahigh field," *IEEE Trans. Dielect Elect. Insul.*, vol. 30, no. 2, pp. 509–517, 2019.
27. H. L. Liu, B. X. Du, M. Xiao and T. Tanaka, "High-Temperature performance of dielectric breakdown in BOPP capacitor film based on surface grafting," *IEEE Trans. Dielect Elect. Insul.*, vol. 28, no. 4, pp. 1264–1272, 2021.
28. P. Kreczanik, P. Venet, A. Hijazi et al, "Study of supercapacitor aging and lifetime estimation according to voltage, temperature, and RMS current," *IEEE Trans. Ind. Electron.*, vol. 61, no. 9, pp. 4895–4902, 2014.

Chapter 3
Polypropylene Microstructure

3.1 Introduction

The electrical properties of polymers are closely related to their microstructure, which includes molecular structure and aggregation structure. The molecular structure of polypropylene is mainly related to the arrangement of methyl groups on the main chain of the molecular chain. In the molecular chain of polypropylene, the methyl groups connected to the tertiary carbon atoms have different arrangement modes. According to the different positions of the methyl groups, polypropylene can be divided into isotactic polypropylene (iPP), syndiotactic polypropylene (sPP) and atactic polypropylene (aPP). The methyl group of the iPP molecular chain is distributed on one side of the main chain, the methyl group of the sPP molecular chain and the hydrogen atom are regularly spaced on both sides of the main chain, and the methyl group of the aPP molecular chain is irregularly distributed on both sides of the main chain. In iPP, methyl groups are distributed on the same side of the main chain, which is easier to crystallize than aPP and sPP, and has high melting point, excellent processing performance, better electrical and mechanical properties. In addition to the molecular structure of polypropylene affecting its electrical properties, the aggregation structure of the molecular chain can also affect its electrical properties. As a typical semi-crystalline polymer, there are crystal zone, amorphous zone, crystal zone and amorphous zone interface in polypropylene. In the crystal region, the molecular chains are neat and closely arranged; in the amorphous region, the interface between the crystalline region and the amorphous region, the molecular chains are arranged disorderly and loosely. The disorder and loose arrangement of the molecular chain will destroy the periodic potential field generated by the periodic arrangement of the crystal region, affect the distribution of energy levels, and then affect the transport of carriers. Different crystal structures lead to different amorphous regions, crystal regions and amorphous region interfaces. Studies have shown that the amorphous region and the interface between the crystalline region and the amorphous region have an important influence on the breakdown strength

B. Du and M. Xiao, *Polypropylene Film for HVDC Capacitors*,
https://doi.org/10.1007/978-981-96-3029-5_3

and conduction loss of the polymer. The ash content is also a parameter that characterizes the microscopic properties of the material. Ash content is a parameter used to characterize the impurity content in polypropylene matrix. Studies have shown that some chemical structures in impurities will form a 'trap center' in polypropylene, causing space charge accumulation and causing local breakdown of the material. At the same time, agglomerated impurities cause distortion of the internal electric field of the material, resulting in a decrease in breakdown field strength. The analysis and study of the influence of these microstructures on the dielectric properties of polypropylene and the mechanism will help us to improve the understanding of polypropylene and optimize the properties of polypropylene films [1].

3.2 Ash Content

The ash of polypropylene refers to the residual substance that cannot be volatilized after burning at a high temperature of (850 ± 50)°C. It generally exists in the form of oxide during detection. The ash in polypropylene mainly comes from the catalyst system used in the polymerization process. The ash in polypropylene also contains various additives and system impurities. In addition, in the production and processing process, mechanical equipment and its processing environment may also cause the ash content of polypropylene increased, for example, due to the poor cleaning effect of the mechanical device, resulting in the introduction of rust and iron chips in the product.

The dielectric properties of polypropylene films depend largely on the chemical impurities and other factors. In this paper, the influence of polypropylene ash on dielectric properties was analyzed from macro to micro perspectives.

When the ash content of polypropylene is overmuch, the film may have structural defects, resulting in a decrease in dielectric properties. The residual catalyst will lead to the occurrence of crystal points in polypropylene film products. This is due to the complete crystallization and high melting point of polypropylene aggregates formed by excessive polymerization during processing, which cannot be uniformly dispersed with the surrounding structure, thus forming local defects in the film [2]. When the crystal point size in the polypropylene film is large, the local thickness of the film increases, which will produce ' fish eye ' defects, resulting in a decrease in the dielectric properties of the polypropylene film and a decrease in the long-term thermal aging resistance, which is easy to cause insulation breakdown failure. In general, the formation of ' fisheye ' is attributed to the residual catalyst in polypropylene, which is mainly composed of metal elements such as Mg and Al [3].

The ash in polypropylene film is mostly metal element, which has a very significant effect on the dielectric properties of polypropylene. It usually leads to the increase of conductivity and permittivity, and the increase of dielectric loss [4, 5]. In addition, the melting and flow of materials will occur during the processing of polypropylene, which will cause the migration and aggregation of ash in the matrix, and form granular impurities, which will become the electrical weakness in the polypropylene film.

Under the electric field, it will lead to local charge accumulation and cause local electric field distortion [6, 7]. Through simulation calculation in Ref. [8], it is found that the degree of local electric field distortion caused by metal impurities increases with the increase of impurity particle size. For the polypropylene film for capacitors, the DC electric field distortion caused by charge accumulation can reach up to 80%, which is easy to induce partial discharge, resulting in the overall insulation breakdown failure of the film [9].

There are a large number of chemical impurities or defects in the ash of polypropylene, such as carbonyl groups, which usually lead to changes in the electronic structure of the matrix, resulting in interference of electron density between macromolecular chains. The electronic band gap is an important index to measure the insulation performance. Because the band gap of polypropylene material is as high as 7 eV, it has good dielectric properties in theory. However, most of the chemical impurities in polypropylene will introduce localized states in the band gap, forming a carrier capture center, resulting in the cycle of carrier transport, charge capture, bond cleavage and new defect state generation, until the final breakdown of polypropylene [2, 5, 10]. Based on density functional theory, Ref. [10] found through simulation calculations that chemical defects in polypropylene introduce occupied and non-occupied states in its band gap, forming holes and electron traps, resulting in a significant decrease in dielectric breakdown performance. Reference [11] found that the impurity state introduced in the band gap of polypropylene will increase the mobility of holes, resulting in an increase in the conductivity of polypropylene and an increase in the conductivity loss (Fig. 3.1).

Ash will affect the long-term stability of polypropylene. Generally speaking, high-voltage DC capacitors will inevitably be exposed to complex environmental stresses such as high temperature and oxygen during operation. The polypropylene film that constitutes the capacitor medium is the most affected component, and its sensitivity to stress such as temperature is greatly affected by internal impurities and defects [9]. Due to the introduction of ash during manufacturing and processing, oxidative degradation of polypropylene films occurs during practical applications, usually accompanied by the breakage of macromolecular chains and the formation of oxidation products, resulting in a significant decrease in long-term stability. Although the ash content in polypropylene is often at the ppm level, its promotion effect on the thermal oxidation degradation reaction is extremely obvious. It is proposed in Ref. [12] that among various impurities and structural defects, the effect of catalyst residues on the degradation of polypropylene is often the most significant. The order of activity of different impurities or defects on the degradation of polypropylene is: metal ions $\gg$ hydroperoxides $\gg$ polycyclicaromatic hydrocarbons $\gg$ carbonyl groups [12]. In the presence of high temperature and oxygen, the tertiary carbon atoms of polypropylene are prone to dehydrogenation and produce free radicals. The free radicals react with oxygen to form peroxy radicals, which extract another unstable hydrogen atom from the macromolecular chain to form hydroperoxides. Hydroperoxides will undergo a cracking reaction to generate two free radicals, which promotes the spontaneous reaction. This is a key process that affects the rate of thermal oxidative degradation

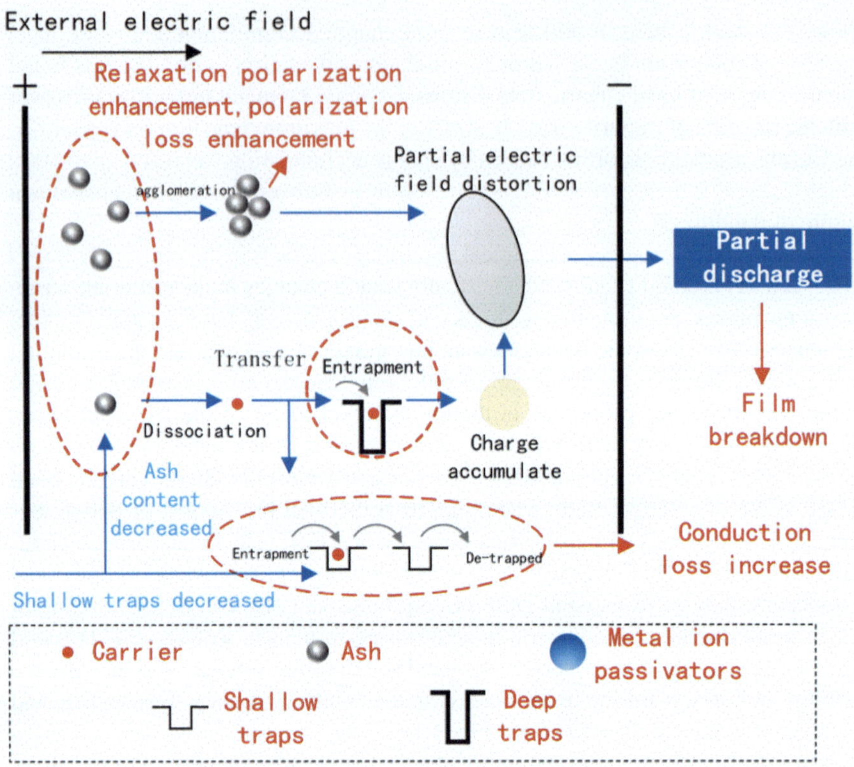

Fig. 3.1 Dielectric property improvement process of deashing PP

of polypropylene, resulting in a typical induction stage in the oxidative degradation process [13]. The residual metal impurities in the catalyst of polypropylene can catalyze the cracking of hydroperoxide, promote the ' automatic acceleration' of oxidative degradation reaction, accelerate the breakage of macromolecular chains, and lead to a significant decrease in the dielectric properties and long-term stability of polypropylene (Fig. 3.2).

It was found in Ref. [12] that the presence of metal impurities Fe, Ti, Al and Mg could accelerate the oxidative degradation process of polypropylene, which reduced the oxidation induction period of polypropylene from 125 min to 10, 25, 40 and 70 min, respectively. In Ref. [14], it was found by thermogravimetric analysis that the presence of metal impurity Fe significantly reduced the temperature and activation energy when the thermogravimetric loss of polypropylene was 10 wt%, and the thermal decomposition rate was increased to 2.1 times of the original. In order to improve long-term stability, various additives such as antioxidants are usually added to commercial polypropylene products to inhibit the occurrence of oxidative degradation reactions [14]. Nevertheless, it has been demonstrated that the presence of residual catalyst within polypropylene can expedite the degradation of phenolic antioxidants. This acceleration ultimately results in the deterioration of

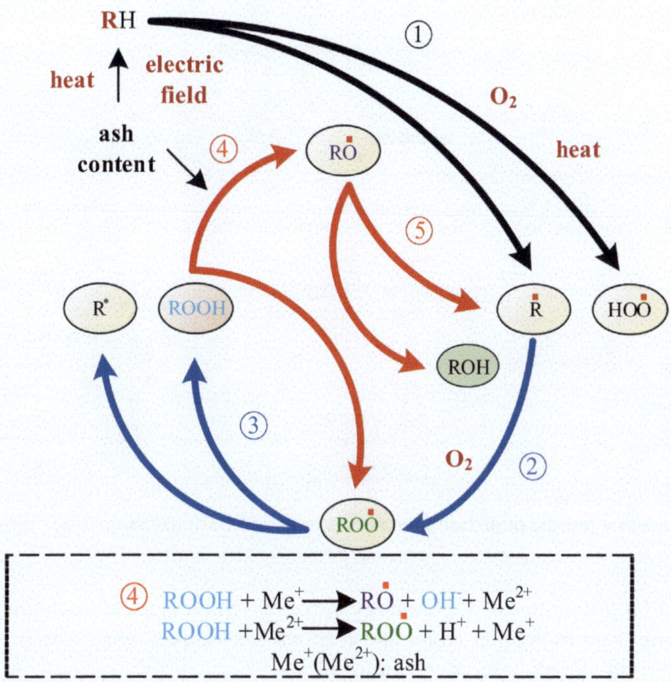

Fig. 3.2 Thermal oxidation degradation process of PP

antioxidant effectiveness, consequently compromising the thermo-oxidative stability of polypropylene materials.

Based on the characterization results of the microscopic physical and chemical properties of polypropylene films with different ash contents, combined with the evolution law of the macroscopic dielectric parameters of the film, the mechanism of ash leading to the insulation failure of polypropylene film under the action of multiple physical fields is proposed, as shown in Fig. 3.3.

During the crystallization process of polypropylene, part of the ash will act as a nucleating agent, increasing the number of nucleation centers to improve the crystallinity, while the remaining ash will be expelled from the crystal region and exist in the amorphous region. The residual ash in the amorphous region is ionized under the continuous action of the electric field to generate carriers, and the carriers generated by the self-oxidation dehydrogenation of the polypropylene molecular chain are obtained under the action of the electric field. These carriers obtain energy for migration and are captured by some local defects in the material to form charge accumulation, resulting in local electric field distortion inside the dielectric. In addition, the agglomerated ash forms an electric weakness to form an interfacial charge agglomeration under the action of an electric field, which aggravates the electric field distortion. When the local electric field threshold is close to the threshold of the material, the partial discharge breakdown occurs in this area, and the partial discharge

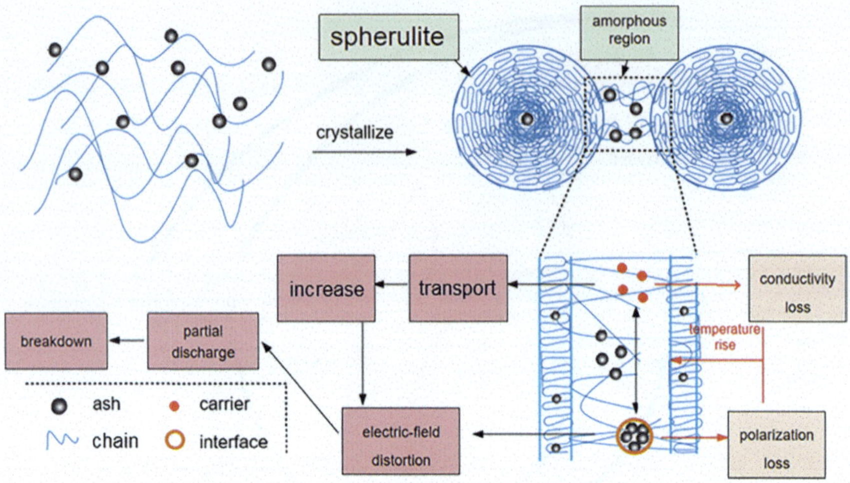

Fig. 3.3 Evolution process of dielectric properties induced by ash inducing

produces a large number of high-energy carriers, which causes the degradation of the polypropylene molecular chain and forms a low-density area, which gradually develops into the breakdown of the entire film.

The migration of carriers generated by the ionization of residual ash under the action of electric field leads to an increase in ion conductivity, which is manifested as a significant increase in DC conductivity, which will lead to an increase in conductivity loss. In addition, the polarization of the interface between the agglomerated ash particles and the matrix will also lead to an increase in polarization loss. The continuous loss will produce a large amount of waste heat in the polypropylene film, resulting in a higher ambient temperature for the capacitor polypropylene film that works at a higher temperature. The high temperature will further aggravate the generation of carriers and the dehydrogenation behavior of the molecular chain, forming a vicious cycle, and finally the insulation failure of the polypropylene film occurs under the action of the electrothermal composite field. It can be seen that although the high ash content will increase the crystallinity, the ash in the amorphous region determines the dielectric properties of the film.

3.3 Isotacticity

The molecular chain structure of polypropylene resin includes the size and shape of the molecular chain, which determines the stacking arrangement characteristics of the molecular chain, and then affects the migration characteristics of the carrier in the polypropylene film. Therefore, the analysis of the molecular chain structure and aggregation structure of polypropylene resin is very important for the research of polypropylene resin for high performance capacitors. Isotacticity is used to characterize the degree of regularity of the arrangement of-CH_3 groups on the side of the polypropylene molecular chain, specifically referring to the proportion of isotactic polypropylene molecules in the total polymer. In the polypropylene molecular chain, the methyl groups connected to the tertiary carbon atoms have different arrangements. According to the different methyl position, polypropylene can be divided into isotactic polypropylene (iPP), syndiotactic polypropylene (sPP) and atactic polypropylene (aPP). Studies have shown that compared with aPP and sPP, iPP is easier to crystallize, has high melting point, excellent processing performance, and good electrical and mechanical properties due to the spatial regularity of CH_3 groups [15]. Some scholars have also compared the molecular structure and electrical properties of sPP and iPP, and found that the spherical grain size formed by sPP at high temperature is 20–30 times smaller than that of iPP. Studies have shown that the structure of the molecular chain will affect the polymerization state, resulting in changes in macroscopic properties. High isotacticity helps to improve the crystallinity of polypropylene materials and reduce the proportion of amorphous regions of polypropylene materials [16]. Studies have shown that isotactic polypropylene has a high Young 's modulus due to the spatial regularity of the CH_3 group, and the pressure resistance is higher than that of the unoriented film [17]. When the molecular chain orientation is small, it will cause molecular chain winding and bending, which directly affects the insulation breakdown field strength. The main chain skeleton carbon atom connected to the side methyl group is the tertiary carbon atom. A large number of tertiary carbon atoms in the amorphous region molecular chain are easily dehydrogenated into free radicals under the action of a strong electric field. The combination of highly active free radicals with the polypropylene molecular chain will lead to long chain rupture and reduce the insulation breakdown strength [18, 19].

3.4 Molecular Weight Distribution

Whether it is isotacticity or molecular weight, the effect on the properties of polypropylene films is to affect the migration characteristics of carriers in the films by producing different aggregated structures, which in turn leads to differences in dielectric properties and breakdown properties. Therefore, it is very important to study the crystal structure and characteristic parameters of the film for establishing the relationship between the microstructure and macroscopic properties of polypropylene. In the case of no significant difference in molecular chain structure and isotacticity, the crystallization characteristics of the polymer are significantly affected by the size and distribution of molecular weight, thus affecting the dielectric properties. The increase of weight average molecular weight leads to the decrease of crystallinity and melting temperature at the same time. For macromolecular polymers, the larger the molecular weight, the lower the crystallinity, and the crystallinity tends to be stable after the molecular weight reaches a certain value. The crystallization process mainly depends on the speed of diffusion and regular accumulation of the chain segments to the crystal nucleus. The larger the molecular weight, the slower the diffusion and migration of the chain segments in the crystallization process, which leads to the decrease of crystallinity to a certain extent, thus reducing the breakdown field strength of the medium and increasing the dielectric loss [20].

The molecular weight of polymers has two characteristics: one is much larger than that of small molecules; second, the molecular weight has polydispersity, so the molecular weight and its distribution of the polymer only have statistical significance, and the molecular weight measured by the experimental method is only the average value with statistical significance. At present, the commonly used statistics describing the molecular weight and its distribution of polymers are: numerical mean molecular weight M_n, weight average molecular weight M_w, Molecular weight distribution width coefficient MWD. Their respective definitions are as follows:

$$M_n = \frac{\sum N_i M_i}{\sum N_i} \tag{3.1}$$

$$M_w = \frac{\sum N_i M_i^2}{\sum N_i M_i} \tag{3.3}$$

$$MWD = \frac{M_w}{M_n} \tag{3.3}$$

In the formula: M_i—Molecular weight of each component, N_i—number of molecules.

The crystal morphology of polypropylene changes under different molecular weights and their distributions. The molecular weight distribution width and crystallinity, spherulite size and its distribution have good regularity, while the influence of molecular weight is poor, and the distribution of data points is more dispersed. Therefore, it is more appropriate to use the parameter of molecular weight distribution width MWD to describe the variation of crystal morphology of materials under different molecular weights and their distributions [21].

The polydispersity of molecular weight means that there is an obvious interaction between high and low molecular weight components. When the molecular weight distribution becomes wider, the interaction between high and low molecular weight components increases. Due to the different crystallization rates of high and low molecular weight components, the crystallization rate of low molecular weight components is fast and crystallizes first, resulting in the pre-crystallized polymer chain ‘freezing’ the uncrystallized polymer chain and reducing the crystallinity. At the same time, the rapid crystallization of low molecular weight components limits the growth of spherulites, resulting in the decrease of spherulite size and the broadening of spherulite size distribution. As the crystallinity of polypropylene decreases, the crystalline region decreases, resulting in more free volume in the polymer, which weakens the pinning effect on carrier migration. The limitation of carrier mobility is weakened, resulting in an increase in carrier mobility and conductivity. The low crystallinity of PP leads to an increase in electron transfer in the amorphous region, resulting in an increase in impact ionization. In addition, the low density of the spherical polymer leads to a decrease in the dielectric breakdown strength of the PP material. If the molecular weight distribution becomes narrower, the interaction between the high and low molecular weight components decreases, and they crystallize at a similar rate to each other, which is beneficial to the growth of spherulites and makes the spherulite distribution tend to be uniform. The increase in crystallinity and crystal density reduces the free volume and limits the charge transfer, resulting in a decrease in conductivity and an increase in breakdown strength [22].

Polymer materials are usually molded in a fluid state, such as injection molding, blow molding or spinning. The polymer fluid will inevitably be affected by the shear flow field before cooling and curing. Under the action of the external flow field, the molecular chain in the fluid can undergo dynamic mechanical behaviors such as conformational transition, orientation, and relaxation, thereby affecting the crystallization kinetics, crystal morphology, and crystal structure of the crystalline polymer. Isotactic polypropylene (iPP) is a kind of homogeneous polycrystalline crystalline polymer. In general, iPP crystals form stable α-type crystals, and β-type crystals with slightly poor stability can be formed under the action of shear or tensile flow fields [23].

Polymer crystallization is essentially the performance of molecular chain behavior. Therefore, the structure of the molecular chain is the internal reason that affects the crystal structure and crystallization behavior. Studies have found that higher molecular weight and molecular weight distribution are conducive to the formation of β-iPP. The higher the molecular weight, the larger the average length of the molecular chain in the melt, the more the number of the same chiral helical segments formed by a single molecular chain, resulting in the increase of β crystal content with the increase of molecular weight, thereby increasing the breakdown field strength and reducing the conductivity [24].

3.5 Crystallinity

The macroscopic properties of polypropylene are largely affected by the microstructure such as crystallinity. The crystallization process of PP is a complex physical process. Its crystallization behavior depends on its own structure, and it is also closely related to the thermal experience in the crystallization process, the presence or absence of heterogeneous nuclei, and the stress state. There are two main ways of PP nucleation: homogeneous nucleation and heterogeneous nucleation. Homogeneous nucleation refers to the process of spontaneous nucleation of polymer melt due to the orderly arrangement of its molecular segments due to the decrease of temperature. The formation of its crystals is generally divided into two stages: As the melt temperature of PP decreases, some ordered polymer chains begin to form very small crystal billets. As the crystal billets gradually grow to a certain size, thermodynamically stable crystal nuclei are formed; under the influence of thermal motion, the surrounding random macromolecular chains are continuously adsorbed by the crystal nucleus, and finally enter the stable growth stage. It can be seen that the two stages of nucleation formation and grain growth rate determine the total rate of PP crystallization process. This way of spontaneous nucleation by thermal motion alone has defects such as few crystal nuclei, large spherulite size, low crystallization rate and slow crystallization speed, which has a negative effect on the processing and application performance of materials. Heterogeneous nucleation refers to the process in which some solid ' impurities ' in the polymer melt are not destroyed, and the polymer adsorbs on its surface to form a crystal nucleus. Because the crystal nucleus growth rate of PP is medium in crystalline resin, PP has a good response to heterogeneous nucleation and non-thermal nucleation with the help of foreign crystal nucleus, which makes the crystallization speed faster, and can significantly improve the crystallinity, crystallization temperature and even change the crystal morphology of the product, which is of great benefit to improve the application performance of the product [25].

The microstructure of polypropylene determines the high temperature dielectric properties of the material. The molecular chain of polypropylene has a linear structure. Under high temperature environment, the film deformation caused by the thermal motion of molecular chain is an important reason for the increase of conductivity loss and the decrease of breakdown strength of PP. In addition, the movement of PP molecular segments and the enhancement of intermolecular slip at high temperature leads to the increase of carrier free volume and average free path. PP is a semi-crystalline polymer, which mainly includes crystalline region, amorphous region and interface region between crystalline region and non-crystalline region. Compared with the crystalline region with ordered molecular chains and compact structure, the intermolecular interference in the amorphous region is smaller. The thermal motion of the molecular chain in the amorphous region is significantly enhanced at high temperature, so the high temperature dielectric properties of PP are obviously affected by the structure of the amorphous region [26]. By increasing the crystallinity and limiting the movement of carriers, the DC conductivity of the material is reduced. Due to the regular structure of the film with high crystallinity, the conduction loss is reduced, and the dielectric loss is limited at low frequencies [27]. At the same time, due to the increase of crystal region, the migration of carriers is limited. The carrier transport process is closely related to the free volume of the polymer. The free volume of the amorphous region is large, so the binding force of the molecular chain to the carrier is weak. The carriers move more freely in the amorphous region than in the crystalline region, resulting in greater conduction losses. High crystallinity and low spherulite size affect the motion path of carriers under DC electric field. The reduction of amorphous region in the film forces carriers to transport inside or at the interface of spherulites, prevents the molecular chain from being impacted and broken, and improves the breakdown strength of the film. The increase of crystallinity and crystal density reduces the free volume. The denser crystal structure hinders the transport of carriers, limits the charge transfer, reduces the conduction loss of polypropylene film, and improves the dielectric properties of polypropylene [28, 29].

As the dielectric material of capacitors, the dielectric loss factor of polypropylene film is the most important electrical performance parameter. Although there is no inherent dipole moment in polypropylene as a non-polar medium. Various additives, anti-aging agents, modifiers and adsorbed molecules (such as water molecules) will be added in the production process of polypropylene, so it is inevitable to contain a small amount of impurities to produce relaxation polarization. Inside the polymer, the turning process of dipoles is difficult to occur in the crystalline region where the molecular chains are arranged regularly and closely. A large number of dipole turning relaxation polarizations occur in the amorphous region and the region where the molecular chains on the surface of the wafer are arranged irregularly. Therefore, the greater the crystallinity, the thicker the wafer thickness, the higher the activation energy of the turning polarization, the longer the relaxation process is established, the lower the corresponding dielectric loss peak frequency, and the lower the dielectric loss.

Polypropylene is a semi-crystalline polymer. Because the molecular arrangement is directly affected by temperature, the breakdown process and mechanism can be divided into three stages according to the temperature: below the glass transition temperature, it is dominated by electron avalanche breakdown; when the temperature is between the glass transition temperature and the melting temperature, it is dominated by the collective breakdown of amorphous dielectric; when the temperature is close to the melting temperature, it is dominated by electro-mechanical breakdown. For electron avalanche breakdown, the increase of crystallinity will lead to the increase of the average free path of electrons, which is helpful for electrons to accelerate to obtain sufficient kinetic energy under the action of electric field for collision ionization. Therefore, the increase of crystallinity will lead to the decrease of breakdown field strength. For the breakdown test at room temperature, this breakdown process is the collective breakdown of amorphous dielectric. Compared with the dense and orderly structure inside the spherulites, there are a large number of defects formed by irregularly arranged small molecules and impurities in the grain boundary. From the perspective of traps, there are usually a large number of shallow traps at the amorphous region and the interface between the crystalline region and the amorphous region. The existence of shallow traps can promote the electron transport process to a certain extent, so that the breakdown path generally develops along the grain boundary. Therefore, higher crystallinity helps to improve the breakdown field strength [30].

3.6 Crystal Type

Firstly, analyze how electrons transfer in polypropylene films. Figure 3.4 show the model of electron transfer in PP film. From region 1, it can be seen that the spherulites are composed of radially grown lamellae and epitaxially tangential lamellae. Some molecular chains in the amorphous region are bonded between the layers. Therefore, this arrangement is relatively close. Region 2 is the interface between spherulites. Due

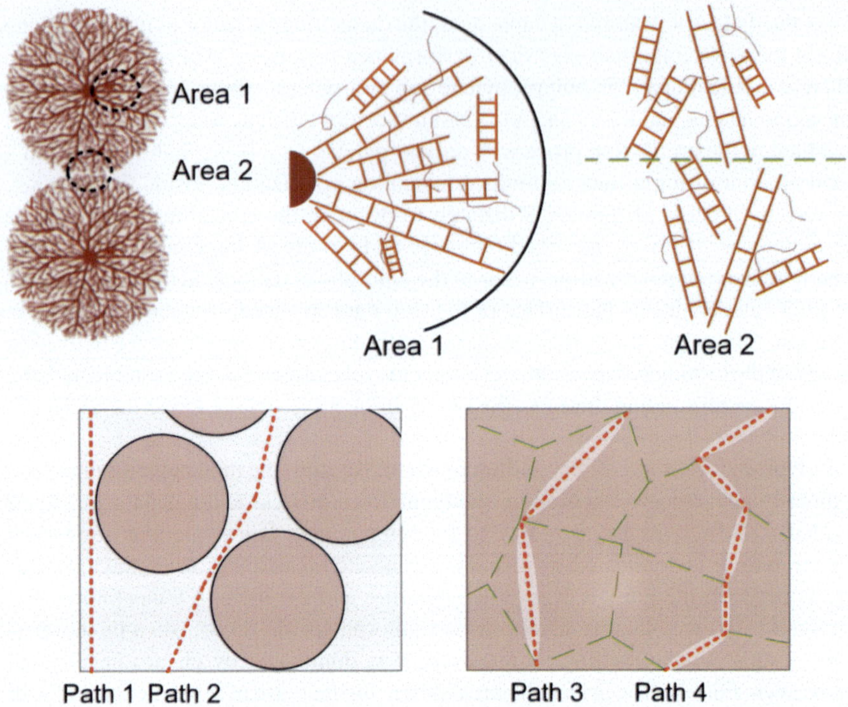

Fig. 3.4 Electron transport process under different aggregation states [25]. Copyright Clearance Center's RightsLink© License Number: 5811440993277

to the high density of lamellae in a single spherulite, the connection between lamellae is not close. The molecular chains at the spherulite boundary are tightly entangled, which increases the difficulty of electron transfer. The closer the amorphous region is to the crystalline region, the greater the molecular chain is constrained by the lamellae, and the more orderly the distribution is.

When the DC electric field is applied to the pure PP film sample, the electrons are preferentially transmitted in the amorphous region, where the molecular chains are loosely arranged in a disordered manner, such as path 1 and path 2. After improving the crystallization by doping nucleating agent and slow cooling, the microstructure of B-1 was improved, the density of spherulites increased, and the shape changed slightly. The decrease in the number of amorphous regions in the film may force electrons to move around the spherulites. Spherulite is a system in which crystalline phase and amorphous phase coexist. There are few amorphous parts inside the spherulites. If the electrons are selected to transport in the amorphous region of the crystal, the path is simplified to path 3 composed of region 1. If electrons try to avoid moving in spherulites, they can only choose to transport along the grain boundary as path 4 composed of region 2. Whether the electrons are transferred from the inside of the spherulites or from the interface between the spherulites, their movement is greatly limited, which not only reduces the conduction loss, but also suppresses the damage of the molecular chain and improves the breakdown strength of the PP film. In summary, the crystal structure determines the transport of electrons, which affects the electrical properties of PP films. The control of the electron transport process links the microstructure with the macroscopic electrical properties, making it possible to improve the breakdown strength by adjusting the crystallization characteristics.

In the crystal region of the polymer, due to the change of the molecular chain conformation and the stacking mode of the molecular chain, different crystal morphologies will be produced, which will affect the dielectric properties of the polymer. As a typical semi-crystalline polymer, polypropylene has three common crystal forms:α, β and γ. Studies have shown that the space charge density of polypropylene containing β crystal form is smaller than that of polypropylene containing only α crystal form. The crystal form also has an important influence on the crystal growth mode, crystal morphology, thermal and electrical properties of polypropylene films. The interface between α crystals is obvious, and the carriers are easy to migrate along the spherulite interface or the gap between spherulites. More molecular segments are introduced at the interface of β crystal, and the trap energy level is deepened. The test proves that at high temperature, compared with α-type polypropylene, the DC breakdown field strength of the film containing β crystal increases by more than 11% [11]. As the temperature increases, the deep trap energy level and shallow trap energy level of all samples increase, and the trap energy level density decreases. As the temperature increases, the mobility of the polypropylene molecular chain gradually increases, and the activation energy of the carrier increases, and the charge is easy to fall out of the trap with deeper energy levels. Therefore, the temperature rise increases the trap energy level of polypropylene. At the same time, because the increase of temperature does not change the crystal structure of polypropylene film, the deep trap density of polypropylene containing β crystal is still higher than that of pure polypropylene. The increase of the deep trap energy level, on the one hand, when the electrode injects charge into the medium, the injected charge is captured by the deep trap at the interface between the electrode and the medium and is not easy to detrap. The captured charge will form an electric field opposite to the direction of the applied electric field, which weakens the charge injected into the sample by the electrode. On the other hand, the deep trap energy level can make the trapped charge difficult to detrap, inhibit the carrier migration process, and thus reduce the carrier mobility. Therefore, the increase of β crystal form leads to the increase of deep trap energy level in polypropylene, which inhibits the injection and migration of carriers, resulting in the decrease of conductivity of polypropylene. The dielectric loss of polypropylene film includes two parts, polarization loss and conduction loss. The reason for the decrease of conduction loss is that the crystal regulation of polypropylene film inhibits the injection and migration of carriers, resulting in the decrease of conduction current. The change of crystal form leads to the decrease of conductivity loss and the decrease of dielectric loss of polypropylene film.

The molecular dynamics simulation software Materials Studio was utilized for modeling and conducting calculations on both the α and β crystal structures of polypropylene. By conducting calculations on the energy bands and density of states of both the α and β crystal structures, the study analyzed how regulating the crystal form influences the insulation properties of polypropylene. Firstly, polypropylene α crystal and β crystal models were constructed. According to the crystal parameters of α-PP and β-PP crystals proposed by Immirzi [31] and Ferro [32], the α-PP and β-PP crystal models were constructed by MS Visualizer module, respectively. The obtained α-PP and β-PP crystal models are shown in Fig. 3.5. The band diagram of α-PP and β-PP is shown in Fig. 3.6. From Fig. 3.6, it can be seen that the band gap of α-PP is 6.235 eV, and the band gap of β-PP is 5.975 eV [33].

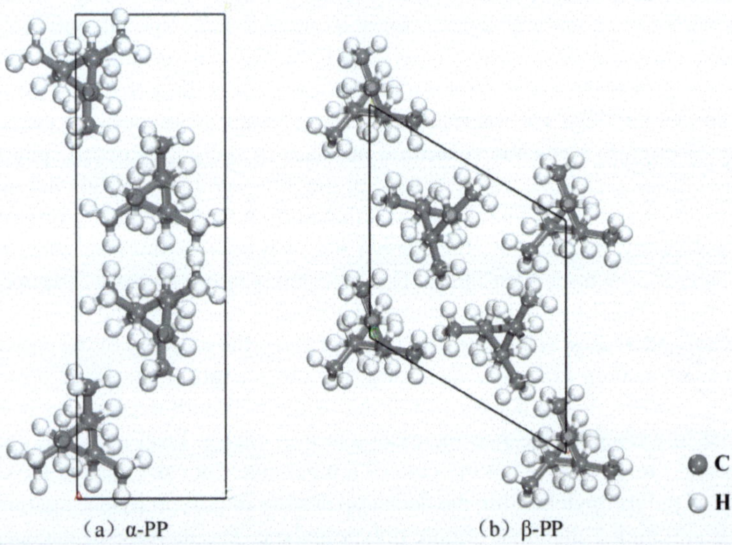

Fig. 3.5 Crystal models of α-PP and β-PP

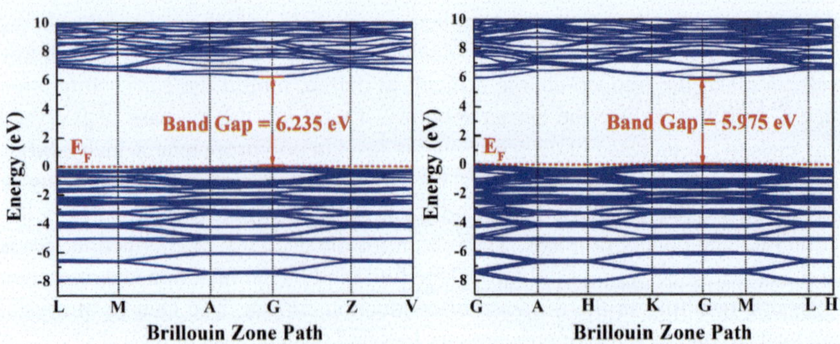

(a) Band diagram of α-PP crystal (b) Band diagram of β-PP crystal

Fig. 3.6 Band diagram of α-PP and β-PP crystals

Compared with the band gap of the polypropylene molecular chain simulated in Sect. 6.4.3 of Ref. [1], which is 9.72 eV, the band gap of α-PP and β-PP cell is slightly lower. This is because the α-PP and β-PP unit cell contains the ordered arrangement of multiple molecular chains. Because of the energy level splitting resulting from electron delocalization, the band gaps observed in the unit cell simulations of α-PP and β-PP are smaller compared to those of individual molecular chains. Simultaneously, considering the band gap of polymer dielectrics typically ranges from 6 to 9 electron volts (eV), the crystal simulation outcomes presented in this study fall within a rational spectrum. As illustrated in Fig. 3.6, the simulation suggests that the band gap of the β crystal is marginally less than that of the α crystal, implying that the β crystal exhibits a higher propensity for electron charge capture compared to the α crystal.

The energy band structure of α-PP and β-PP cell provides a theoretical reference for analyzing the change of trap energy level caused by different crystal forms. Because the trap energy level distribution is related to many factors such as crystal structure, crystal region and crystal region interface, crystal region and amorphous region interface, the simulation results provide support for analyzing the trap energy level distribution measured by the experimental method.

Further, the cell state density distribution of α-PP and β-PP is obtained by simulation calculation, as shown in Fig. 3.7. It can be seen from Fig. 3.7a, b that the density of states of β-PP is significantly higher than that of α-PP, indicating that the β crystal form will lead to an increase in the trap energy level density. This simulation result is consistent with the trap energy level distribution measured by isothermal discharge current method, which can well verify the experimental results. The β crystal introduces more trap energy levels in the film.

Polypropylene is a representative semi-crystalline polymer, and the interface between its crystalline structures plays a significant role in determining its insulation performance. Only analyzing the difference of crystals could not fully explain

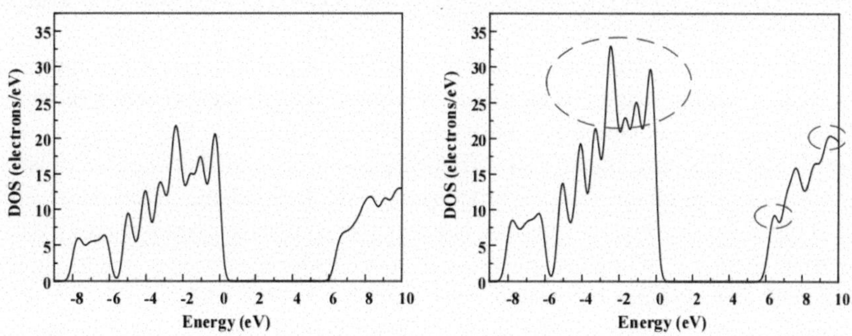

(a)State density diagram of α-PP crystal (b)State density diagram of β-PP crystal

Fig. 3.7 State density diagram of α-PP and β-PP crystals [1]. Copyright Clearance Center's RightsLink© License Number: 5811411218932

the mechanism of the influence of crystal form on insulation performance. Therefore, according to the structural characteristics of α crystal and β crystal, this section analyzes the mechanism of crystal regulation from the perspective of the influence of crystal form and crystal interface on charge transport. The dielectric loss and breakdown properties of polypropylene films are closely related to the distribution of trap energy levels, and there is a correlation between the distribution of trap energy levels and the aggregation state of polypropylene films. According to the analysis in Ref. [1], a part of β crystal form was induced inside the polypropylene film after adding β nucleating agent. With the increase of β nucleating agent content, the α crystal form in the sample gradually increased and decreased, and the β crystal form gradually increased. It is the change of crystal morphology in polypropylene film that causes the change of dielectric loss and breakdown characteristics of polypropylene film. According to the POM diagram of polypropylene film Ref. [1], the crystal characteristics of α crystal form and β crystal form can be summarized: The lamellae of the α crystal form grow radially from the central nucleus to the periphery, and the tangential lamellae are attached to the radially grown lamellae. The radial lamellae and the tangential lamellae are 100° or 80°, and the radial and tangential lamellae are interlaced to form a network structure. The α crystal form and its lamellae structure are shown in Fig. 3.8 (a1) and (a2). The β crystal form is formed by the bending growth of the bundle-like parallel lamellae to both ends, accompanied by the bifurcation growth of the lamellae, and finally the lamellae contact with each other to form a complete spherulite. The β crystal form and its lamellar structure are shown in Fig. 3.8 (b1) and (b2). Compared with α crystal form, there are more molecular chains between parallel lamellae and lamellae in β crystal form. Studies have shown that carriers are easy to migrate between irregularly arranged molecular chains (i.e., amorphous regions), free volume, micropores, etc. [22]. In Fig. 3.9a, the obvious interface between α crystal and α crystal makes it easy for carriers to migrate along the spherulite interface under the action of electric field. In Fig. 3.9b, more molecular segment entanglement between β crystal and β crystal introduces more deep traps.

The difference of morphology between α and β crystal forms will also lead to the difference of interface characteristics between different crystal forms. There is an obvious interface between α spherulites and α spherulites, while the interface between β spherulites and β spherulites is blurred due to the existence of more molecular segment entanglement. Different spherulite interface models are shown in Fig. 3.9. Studies have shown that carriers are easy to migrate between irregularly arranged molecular chains (i.e., amorphous regions), free volume, micropores, etc. [22]. In Fig. 3.9a, the obvious interface between α crystal and α crystal makes it easy for carriers to migrate along the spherulite interface under the action of electric field. In Fig. 3.9c, the interface between α crystal and β grain boundary is more blurred than that between α–α crystal. The reason is that β crystal connects more molecular segments at the interface. From Fig. 3.9b, c, it can be seen that the existence of β crystal can introduce more molecular chain segments at the interface, which leads to the deepening of trap energy level, the increase of trap energy level density, the obstruction of carrier migration and the decrease of carrier mobility, thus reducing the conductivity of polypropylene film. At the same time, the crystallization increases

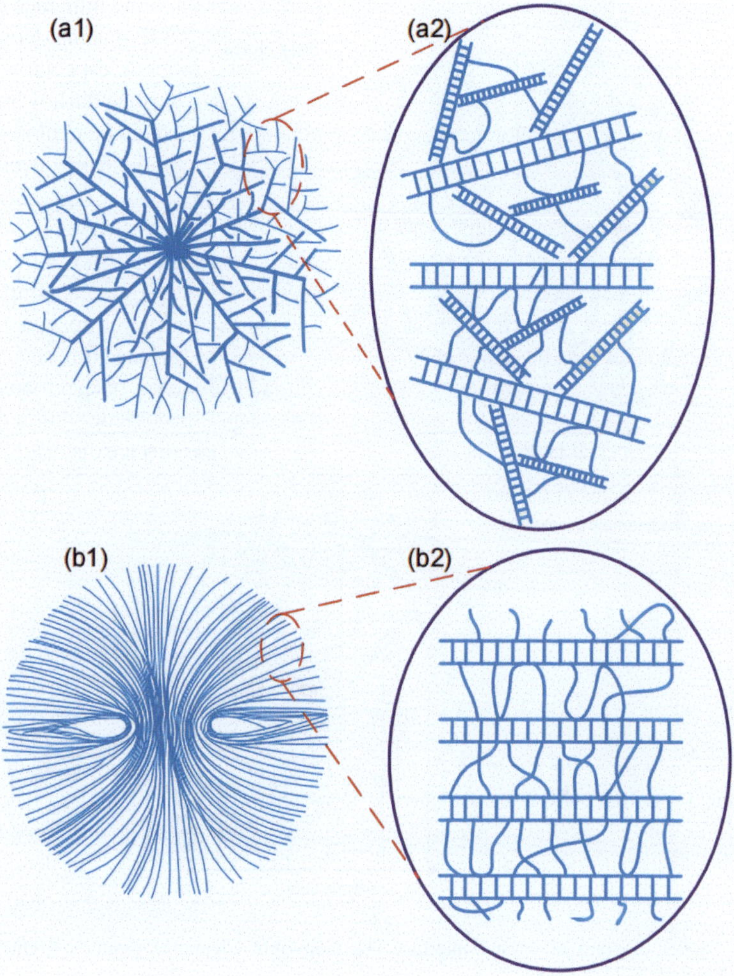

Fig. 3.8 Model diagram of α crystal form and β crystal form. Copyright Clearance Center's RightsLink© License Number: 5811411218932

after the addition of β nucleating agent, and there are more molecular segments between the β crystal lamellae, which lead to the decrease of the free volume inside the medium, thus reducing the carrier migration energy. On the other hand, due to the increase of deep trap energy level and deep trap density introduced by β crystal form on the surface of the sample, the homopolar charge injected by the electrode into the sample is limited to the surface of the sample. The reverse electric field formed by the homopolar charge reduces the charge injected by the electrode into the sample. The increase of deep trap energy level and deep trap density inside the sample limits the migration of carriers inside the sample and reduces the occurrence

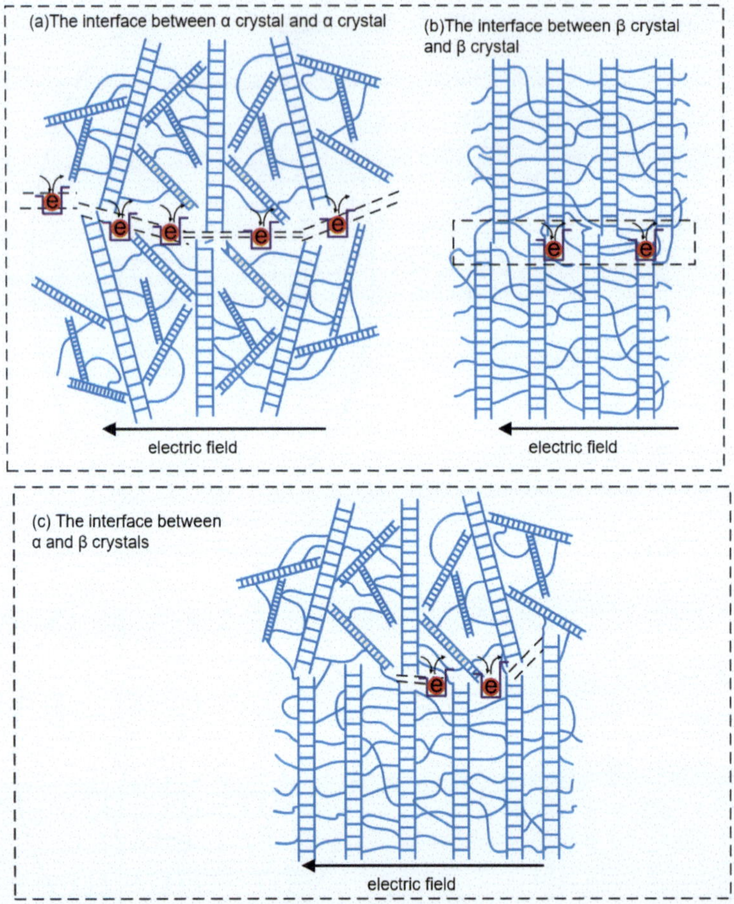

Fig. 3.9 Model diagram of crystal interface [1]. Copyright Clearance Center's RightsLink© License Number: 5811411218932

of collision electron avalanche. Therefore, the existence of β crystal form improves the breakdown field strength of polypropylene film.

3.7 Summary

In this paper, the effect of ash content on the dielectric properties of thin films was studied. It was found that the ash element became dense with the increase of content; the spherulite size of polypropylene showed a downward trend with the increase of ash content. At the same time, the number of nucleation centers increased, the area of amorphous region expanded, and the conductivity of polypropylene increased rapidly with the increase of ash content. The conductivity of polypropylene increases rapidly with the increase of ash content. Under the action of electric field, polypropylene ash ionization produces a large number of carriers, and the continuous migration of carriers increases the conductivity of polypropylene, resulting in conductivity loss. At high temperature, the ionization of polypropylene ash produces more carriers, and the conduction loss heat generation is more serious. The carriers generated by the ionization of polypropylene ash migrate under the action of electric field and are captured by some local defects to form charge accumulation. At the same time, the agglomerated ash forms an electric weakness, which together leads to the distortion of the internal electric field of the polypropylene film and the decrease of the breakdown field strength. At high temperature, the generation, migration and aggregation behavior of carriers in polypropylene are intensified, and the electric field distortion is more complicated. The breakdown field strength of polypropylene samples decreases especially seriously. It was found that polypropylene ash would lead to a slight increase in the dielectric constant of the film. The dipole polarization and interfacial polarization of ash lead to a significant increase in the dielectric loss factor of the sample. Since ash can cause a significant decrease in breakdown strength, the maximum discharge energy density of polypropylene decreases significantly with the increase of ash content. Under the action of electric field, the agglomerated ash and the residual energy at the interface of polypropylene lead to the decrease of discharge efficiency.

In this paper, the relationship between crystal morphology and trap energy level distribution was discussed. The dielectric loss and breakdown characteristics of crystal-regulated polypropylene films were analyzed. The mechanism of crystal regulation to improve the dielectric properties of polypropylene films was discussed. The coexistence of α spherulites and β spherulites leads to the change of trap energy levels in polypropylene films. The β spherulites generated in polypropylene can introduce more molecular segments at the spherulite interface, resulting in a decrease in the density of shallow traps and an increase in the density of deep traps in polypropylene films. The β crystal form formed in polypropylene films reduces the dielectric loss and conductivity loss of polypropylene films. The inhibition of crystal band structure, crystal form and crystal interface on charge transport behavior leads to the increase of breakdown field strength of crystal-regulated polypropylene films.

In this paper, the effect of crystallinity on the dielectric properties of polypropylene films was discussed. The increase of crystallinity will increase the crystal region, thereby reducing the free volume, limiting the migration of carriers, and improving the dielectric properties of polypropylene. The isotacticity and molecular weight of polypropylene will affect its crystallinity, thus affecting the dielectric properties of polypropylene films.

References

1. R. R. Xu, B. X. Du, M. Xiao, J. Li, H. L. Liu, Z. Y. Ran, J. W. Xing, "Dielectric properties dependent on crystalline morphology of PP film for HVDC capacitors application," Polymer., vol. 213, pp. 123204, 2021.
2. H. M. Umran, F. Wang and Y. He, "Ageing: causes and effects on the reliability of polypropylene film used for HVDC capacitor," *IEEE Access*, vol. 8, pp. 40413–40430, 2020.
3. J. X. Chen, Z. X. Zhong, Z. H. Su, Z. P. Li, Z. F. Weng and S. Y. Luo, "Analysis of defects in biaxially oriented polypropylene films by micro-fourier transform infrared and raman spectroscopies," *Chin. J. Anal. Chem.*, vol 48, no. 10, pp. e20134–e20138, 2020.
4. B. X. Du, J. W. Xing, Z. Y. Ran, H. L. Liu, M. Xiao and J. Li, "Dielectric and energy storage properties of polypropylene by deashing method for DC polymer film capacitors," *IEEE Trans. Dielectr. Electr. Insul.*, vol. 28, no. 6, pp. 1917–1924, 2021.
5. B. X. Du, J. H. Zhang, Meng Xiao, H. L. Liu and Z. Y. Ran, "Dielectric breakdown performance of deashed polypropylene for HVDC film capacitors," *IEEE Trans. Dielectr. Electr. Insul.*, vol. 29, no. 6, pp. 2209–2217, 2022.
6. H. L. Liu, B. X. Du, M. Xiao, Z. Y. Ran and Y. W. Ma, "Dielectric properties of polymer films in strong electromagnetic field for energy storage capacitor," *IEEE Trans. Dielect Elect. Insul.*, vol. 29, no. 5, pp. 1745–1753, 2022.
7. H. L. Liu, B. X. Du, M. Xiao, and Z. Y. Ran, "Improved conductivity and breakdown performance of polypropylene film by parylene blending for power capacitor," *IEEE Trans. Dielect Elect. Insul.*, vol. 29, no. 3, pp. 997–1004, 2022.
8. M. E. Stournara and R. Ramprasad, "A first principles investigation of isotactic polypropylene," *J. Mater. Sci.*, vol. 45, pp. 443–447, 2010.
9. Y. P. Liang, D. M. Wang, J. X. Li and D. H. Liu, "Studies on switching transients and calculations of physical fields in power capacitors," *Int. Trans. Electr. Energy Syst.*, vol. 29, no. 1, 2019, Art. no. e2663.
10. R. Yao, H. Li, W. Lai, A. S. Bahman and F. Iannuzzo, "Lifetime analysis of metallized polypropylene capacitors in modular multilevel converter based on finite element method," *IEEE J. Emerg. Sel. Top. Power Electron.*, vol. 9, no. 4, pp. 4248–4259, 2021.
11. P. Karanja and Nath R, "Electrical conduction in biaxially-oriented polypropylene," *J. Electrostat.*, vol. 31, no. 1, pp. 51–63, 1993.
12. 13. Z. Osawa, "Role of metals and metal-deactivators in polymer degradation," *Polym. Degrad. Stabil.*, vol. 20, no. 3–4, pp. 203–236, 1988.
13. H. V. Nguyen and T. H. Pham, "Structural and electronic properties of defect-free and defect-containing polypropylene: a computational study by van der waals density-functional method," *Phys. Status Solidi B*, vol. 255, no. 3, 2018, Art. no. 1700036.
14. 15. E. Blatt, H. J. Griesser, J. H. Hodgkin et al, "Comparative studies of the chelation of transition metal ions with metal deactivators," *Polym. Degrad. Stabil.*, vol. 25, no. 1, pp. 19–29, 1989.
15. 16. N. S. Allen, M. Edge, "Perspectives on additives for polymers. 1. Aspects of stabilization," *J. Vinyl Addit. Technol.*, vol. 27, no. 1, pp. 5–27, 2021.
16. E. Martuscelli, M. Pracella adn L. Crispino, "Crystallization behaviour of fractions of isotactic polypropylene with different degrees of stereoregularity," *Polymer*, vol. 24, no. 6, pp. 693–699, 1983.
17. 18. P. Chammingkwan, F. Yamaguchi, M. Terano and T. Taniike, "Influence of isotacticity and its distribution on degradation behavior of polypropylene," *Polym. Degrad. Stabil.*, vol. 143, pp. 253–258, 2017.
18. 19. T. Konishi, K. Nishida, T. Kanaya and K. Kaji, "Effect of isotacticity on formation of mesomorphic phase of isotactic polypropylene," *Macromolecules.*, vol. 38, no. 21, pp. 8749–8754, 2005.
19. A. Kahouli, O. Gallotlavallée, P. Rain, O. Lesaint, C. Guillermin, and J. M. Lupin, "Relationship between structure and dielectric properties of Bi-oriented isotactic polypropylene films for capacitors," 2013 IEEE International Conference on Solid Dielectrics (ICSD), Bologna, Italy, 2013, pp. 1068–1071.

20. 21. L. Cheng, Z. Y. Li, J. R. Wang, Z. Xu, W. F. Liu and S. T. Li, "Degradation behavior and mechanism of metalized film capacitor under ultrahigh field," *IEEE Trans. Dielect Elect. Insul.*, vol. 30, no. 2, pp. 509–517, 2023.
21. V. Ianniello et al, "Evaluating the molecular weight distribution of ultrahigh molecular weight polypropylene through rheology," *Phys. fluids*, vol. 35, no. 6, 2023, Art. no. 063108.
22. 23. H. Lei, T. Xu, C. T. Gao, "Effect of molecular weight and its distribution on the crystalline morphology of polypropylene materials," *Polym. Polym. Compos.*, vol. 18, no. 1, pp. 45–51, 2010.
23. 24. O. R. de Ballesteros, F. Auriemma and C. D. Rosa, "Phase diagram of syndiotactic polypropylene: influence of stereoregularity and temperature on the polymorphic behavior," *Macromolecules*, vol. 40, no. 3, pp. 611–622, 2007.
24. J. M. Yang, M. Z. Gao, H. Zhao, S. L. Liu, M. Hu and S. H. Xie, "Space charge characteristics of polypropylene modified by rare earth nucleating agent for β crystallization," *Materials*, vol. 12, no. 1, 2019, Art. no. 12010042.
25. 26. Z. Y. Ran, B. X. Du, M. Xiao, H. L. Liu and J. W. Xing, "Effect of crystallization regulation on the breakdown strength of metallized polypropylene film capacitors," *IEEE Trans. Dielect Elect. Insul.*, vol. 28, no. 1, pp. 175–182, 2021.
26. 27. Z. Y. Ran, B. X. Du and M. Xiao, "High-temperature breakdown performance improvement of polypropylene films based on micromorphology control," *IEEE Trans. Dielect Elect. Insul.*, vol. 28, no. 5, pp. 1547–1554, 2021.
27. Y. S. Zhao, M. Xiao and B. X. Du, "Crystallization time-dielectric property relationship in the modified polypropylene for metallized film capacitors," 2023 IEEE 4th International Conference on Electrical Materials and Power Equipment, Shanghai, China, 2023, pp. 1–4.
28. Z. Y. Ran, B. X. Du, H. L. Liu, M. Xiao and J. W. Xing, "Improved anti-aging performances based on doping of organic additives of PP films for capacitors," 2022 IEEE 4th International Conference on Dielectrics, Palermo, Italy, 2022, pp. 785–788.
29. Z. Y. Ran, B. X. Du, M. Xiao and H. L. Liu, "Improved breakdown performances of PP films based on molecular chain and aggregate structure design," 2022 IEEE 4th International Conference on Dielectrics, Palermo, Italy, 2022, pp. 676–679.
30. 31. Z. Y. Ran, B. X. Du, M. Xiao and J. Li, "Crystallization morphology-dependent breakdown strength of polypropylene films for converter valve capacitor," *IEEE Trans. Dielect Elect. Insul.*, vol. 28, no. 3, pp. 964–971, 2021.
31. 32. A. Immirzi and P. Iannelli, "Whole-pattern approach to structure refinement problems of fibrous materials: application to isotactic polypropylene," *Macromolecules*, vol. 21, no. 3, pp. 768–773, 1988.
32. 33. D. R. Ferro, S. V. Meille, S. Bruckner et al, "Energy calculations for isotactic polypropylene: A contribution to larify the β crystalline structure," *Macromolecules*, vol. 31, no. 20, pp. 6926–6934, 1998.
33. 34. J. P. Jones, J. P. Llewellyn and T. J. Lewis, "The contribution of field-induced morphological change to the electrical aging and breakdown of polyethylene," *IEEE Trans. Dielect Elect. Insul.*, vol. 12, no. 5, pp. 951–966, 2005.

Chapter 4
Dielectric Breakdown

4.1 Introduction

Metallized film capacitors (MFCs), as the core equipment of flexible DC converter valves, the insulating properties of their dielectric polypropylene (PP) films play an important role in improving the safety and stability of DC support capacitors and even flexible DC converter valves. With the development of high-voltage direct current (HVDC) transmission technology, MFCs towards the direction of high voltage, large capacity development, which puts forward more stringent requirements on the working conditions of the PP film [1].

The insulation failure mechanism and influencing factors of MFCs are extremely complex, and the capacitors are subjected to the combined effects of multiple harmonics superimposed on DC, transient over-voltage and high temperatures, which is a challenge to the insulation performance of PP film. The DC support capacitor in the flexible DC converter valve is subjected to DC superimposed harmonic voltages. During the long-term operation of power capacitors, internal temperature rises and stress pulling is also a factor that cannot be ignored. The rise in internal temperature of the capacitor leads to dielectric aging of the dielectric material, which plays a highly important role in its breakdown and dielectric properties. The heat generation of the electrode resistance and the presence of polymer dielectric losses will lead to an increase in the internal temperature of the capacitor, accelerating the aging of the insulating medium and even causing the breakdown phenomenon [2]. The combined effect of electric and thermal fields is the primary source of dielectric aging. The harmonic voltage content and frequency domain characteristics of damping capacitors are different. Short-circuit failures and even explosive fires caused by internal insulation breakdown of capacitors occur frequently, resulting in significant economic and social losses [3]. The dielectric properties of the existing domestic capacitor film cannot meet the needs of capacitors for converter valves, and the research on the issue of capacitor film insulation failure has failed to make breakthrough progress. Therefore, to improve the insulation performance of the film

B. Du and M. Xiao, *Polypropylene Film for HVDC Capacitors*,
https://doi.org/10.1007/978-981-96-3029-5_4

dielectric is the key to ensure the safe and reliable operation of ultra-high voltage DC transmission equipment, which is of great strategic significance to ensure the scale of new energy grid connection and sustainable development of electric power.

4.2 Effect of Electric Field

The most important advantage of HVDC capacitors is the high working field strength (> 200 kV/mm). At present, the voltage level of UHVDC engineering is increasing, which not only increases the number of capacitor components and the difficulty of protection reliability, but also raises the operating voltage of capacitors. The extreme operating conditions that capacitors are subjected to put forward more stringent requirements for their operational reliability [3].

PP film has become the main insulating dielectric material for HVDC capacitors due to its outstanding electrical properties, such as high dielectric strength, low dielectric loss, excellent self-healing performance, and good thermal stability, which directly affects the long-term operational stability of the MFCs [4]. However, the insulation degradation of PP film under the action of strong electric field should not be ignored [5]. The phenomena of charge accumulation, partial discharges and field-guided chemical degradation under strong electric fields may gradually damage the dielectric materials, thus limiting the performance and service life of capacitors [6]. Since PP materials have long been widely used in various types of electrical equipment, the performance evolution and insulation degradation mechanism of PP films under strong electric fields have been fully studied by scholars.

4.2.1 Degradation Process

(1) **Charge Injection**

In relation to the mechanism behind the insulation degradation of PP (polypropylene) films when subjected to a strong electric field, it has been discovered that there is a significant intensification in both interfacial charge injection and carrier transport within the medium. This enhanced activity results in the overall degradation of the insulation properties of the PP films. As the strength of the external electric field is increased, the slope of the potential line that it forms also becomes steeper, which further reduces the interfacial potential barrier.

The volatility of electrons plays a crucial role in this process. Specifically, when the width of the barrier is smaller than the De Broglie wavelength of the electrons, electrons can still cross the interfacial barrier despite not having enough energy to do so traditionally. This phenomenon results in charge migration across the barrier. As the external electric field becomes more intense, the height of the interfacial barrier decreases even further, leading to a corresponding reduction in the barrier width.

This scenario facilitates an increased number of electrons being injected into the interior of the dielectric material from the metal electrode through a process known as tunneling, particularly at high electric fields. This specific form of tunneling is referred to as Fowler–Nordheim tunneling [7]. Essentially, the decrease in both the height and width of the interfacial barrier under enhanced external electric fields significantly contributes to the increased rate of electron injection and subsequent charge migration, thereby exacerbating the degradation of the insulation properties of the PP films.

(2) **Charge Migration**

In addition to the space occupied by molecular chains, polymers have a large number of unoccupied three-dimensional pore space. In this part of the space, the molecular chains are loosely structured and free to move, so it is called the "free volume". Under the action of a strong electric field, the charge injected at the interface or excited by the body can be accelerated in the free volume [8]. When the free volume is large enough, the charges can accumulate enough energy to cross the intermolecular barriers, resulting in a steep increase in the leakage current of the insulating material [9]. High-energy electrons may even directly impact on the molecular chains of the PP film, causing macromolecular cleavage and insulation degradation, which further increases the free volume.

(3) **Partial Discharge**

Due to the existence of electrical weaknesses such as impurity defects and interface defects between crystalline and amorphous regions, space charge accumulates inside the PP films at the defects, triggering an electric field distortion at that location, which in turn leads to partial discharges. Local discharges produce high-energy photons and phonons in addition to ionizing substances such as air within the defect to form ions. High-energy photons impinge on the molecular chains near the defects, causing the polymer long chains to degrade to produce small molecule free radicals, which further catalyze the polymer degradation to form a chain reaction. The generated phonons exacerbate the lattice vibrations close to the discharge site, which promotes the thermal movement of the molecular chains and reduces the stability of the chemical bonds. Under the combined effect of sound and light, the material with massive defects gradually undergoes thermo-oxidative ageing and deterioration, with the growth in the size of defects, thus gradually forming insulation deterioration channels visible and ultimately leading to through-and-through insulation damage [10].

During the production process of PP films, micro-defects such as bubbles, cracks, impurities, are inevitably introduced inside the matric. Due to the difference in dielectric parameters between the defects and the dielectric materials, strong electric field concentration and electric field distortion occur inside the defects under strong electric field. When the strength of the local electric field exceeds the breakdown strength at the defect, the breakdown occurs at the defect, resulting in a partial discharge phenomenon, which ultimately leads to a breakdown channel through the positive and negative electrodes [11].

When the electric field strength reaches 300 kV/mm or more, the conductivity of PP increases by 2–3 orders of magnitude. When the pulse voltage was applied for 40 min, the carrier mobility increased significantly and the temperature rise of the capacitor was severe, leading to premature insulation failure. A large amount of bipolar space charge was detected to accumulate inside the BOPP film at room temperature and a DC field of 350 kV/mm. The accumulation of space charge leads to local distortion of the electric field in the insulating medium, and the resulting electromechanical stress deforms the polypropylene molecular chains, which accelerates the ageing of the dielectric material. After applying an applied electric field polarization of 150 kV/mm at 70 °C, the maximum local electric field is as high as 120 kV/mm, which is about 80% of the applied field. Therefore, insulation failure triggered by space charge accumulation inside the dielectric and serious distortion of electric field distribution under strong electric field is a key factor limiting the development of metallized film capacitors.

4.2.2 Harmonic Component

In addition to enduring strong electric fields, strong magnetic fields, high temperatures, and significant stress environments as previously mentioned, DC support capacitors also face the additional challenge of withstanding complex voltage stresses during operation. In the long-term operation, power capacitors face the complex working conditions superimposed by high-voltage DC, harmonics and transient overvoltage, which in turn affects the service life and work stability of high-voltage DC capacitors. The combination of high-voltage DC, harmonics, and transient overvoltages presents a formidable challenge for high-voltage DC capacitors. These conditions can lead to accelerated aging, increased energy losses, and reduced insulation performance. Consequently, the service life and operational stability of the capacitors are adversely affected.

In the simulation analysis of actual engineering examples, it is concluded that the harmonic components of DC support capacitor voltage mainly contain the 1st, 2nd and 3rd harmonic components. It has been shown that the maximum value of DC capacitor harmonic voltage can be up to about 20% of the amplitude of DC voltage. As an example, for the 500 MW Xiamen FDC converter valve project, the rated DC voltage of the DC support capacitor is 1600 V, the peak harmonic voltage is about 285 V, and the peak harmonic voltage accounts for about 17% of the amplitude of the DC voltage. The rated DC voltage of the domestic DC support capacitor is 4.4 kV, the harmonic voltage is 960 V, and the peak harmonic voltage accounts for about 22% of the amplitude of the DC voltage. In the practical implementation of projects, capacitors employed in DC applications demonstrate a service life that is merely one-third of capacitors used in AC applications. This stark contrast underscores the profound impact of voltage waveforms on capacitor durability and operational longevity.

Voltage fluctuations in the process, the electric field strength applied to the samples is also changing. The interfacial charge injection and charge migration are varied with the applied electric field as shown in Fig. 4.1 (a2)–(c2). According to the Schottky charge injection principle, when the electric field is in the rising stage, the increasing electric field leads to a decrease in the charge injection potential barrier at the interface of the electrode-film, and the electrode injects more charge into the interior of the specimen. And the charge is easy to get out of the trap after obtaining larger energy from the electric field, resulting in an increase in the number of free carriers. A large number of carriers collide with the PP molecular chains under the electric field, resulting in the destruction of the molecular chains.

When the electric field is in the decreasing stage, the electrode injects less charge into the interior of the sample. At the same time the weakening of the electric field decreases the charge energy, the charge is captured by the trap and then the difficulty of leaving the trap increases, and the charge entering the trap releases energy to destroy the PP molecular chain. In a DC superimposed harmonic electric field, the repeated process of charge entering and escaping from the trap leads to breakage of the PP molecular chain, accelerating the formation of the low-density region inside the matric and triggering the breakdown. As the number of harmonic voltages increases, as shown in Fig. 4.1 (b1–c1), the charge entrapment and decampment behaviors become more frequent and the damage to the molecular chains of PP is more severe, resulting in a decrease in the breakdown strength of PP films at higher number of harmonic voltage.

The breakdown field strength of polypropylene films under DC superimposed harmonic voltages is shown in Fig. 4.2. Under DC superimposed harmonic composite electric field, the breakdown field strength of polypropylene films decreases from 521 kV to 488.7 kV, 468.0 kV, and 453.8 kV as the amplitude of the 1st harmonic component is increased from 0 to 1, 2, and 4 kV [12]. Elevated harmonic components intensify the electric field variations, increase the number of migrating charges, and exacerbate polypropylene molecular chain breakage and dielectric degradation. It has been shown that when the number of harmonics increases, the in-trapping and de-trapping behaviors of the charges are more frequent and the damage to the molecular chains is more severe. The breakdown field strengths of polypropylene films are 500.0, 483.3, and 468.0 kV/mm for DC superimposed 1st, 2nd, and 3rd harmonic voltages, respectively. In summary, an increase in the harmonic amplitude or number will result in the formation of low-density zones in the film dielectrics, which will accelerate the failure of the insulation.

4.2.3 Impulse Voltage

The neutral bus impulse capacitor in HVDC transmission system is mainly subjected to transient overvoltage caused by lightning strike, operation and fault. The damping capacitor is connected in parallel to both ends of insulated gate bipolar transistor (IGBT) in the HVDC converter valve. It is applied to the overvoltage protection and

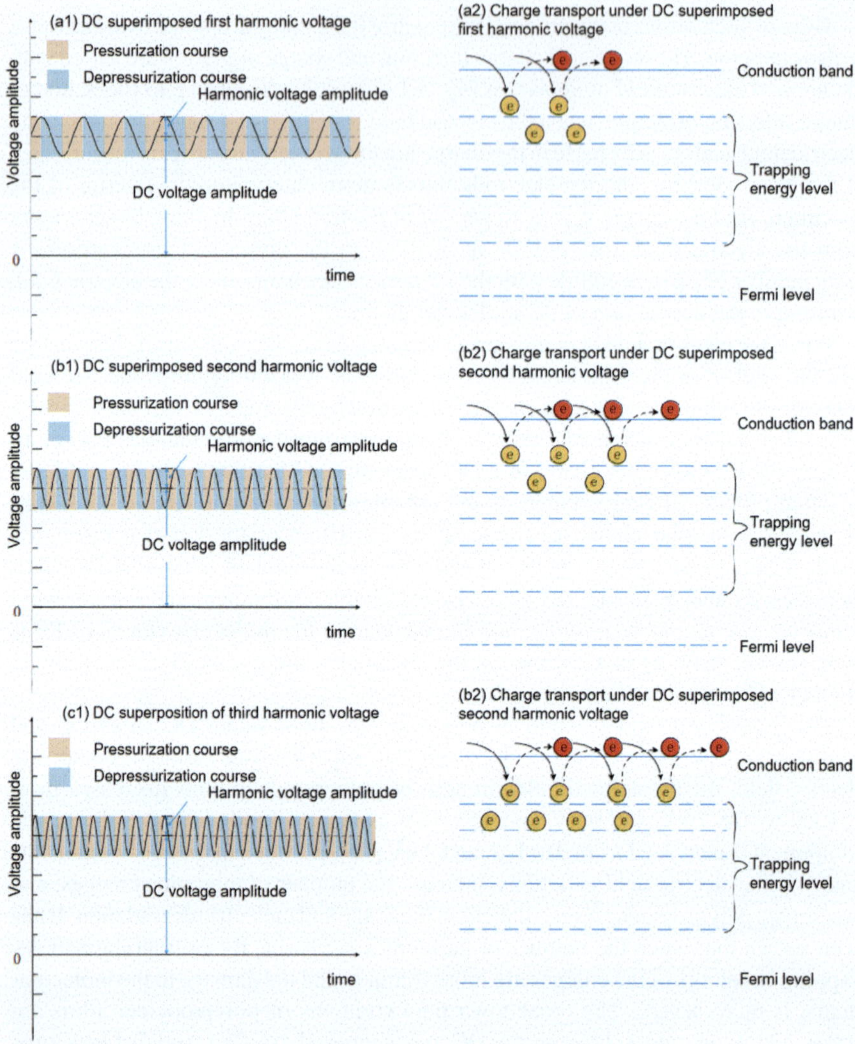

Fig. 4.1 Charge transport model under DC superimposed harmonic voltages

AC commutation absorption of the thyristor in the rectifier circuit, and also bears the transient overvoltage in actual operation.

Under the action of DC superimposed pulse, the maximum voltage of DC pulse is much higher than that of DC harmonic voltage, so the carriers can obtain extremely high potential energy and great kinetic energy in a very short time, and the carriers transition to the orbit much higher than the conduction band, and cause enough impact on the polypropylene molecular chain. At the same time, in the process of pulse drop and end, the energy level span of carrier fallback is larger, the energy released during the fallback process is more, and the energy transfer and loss are

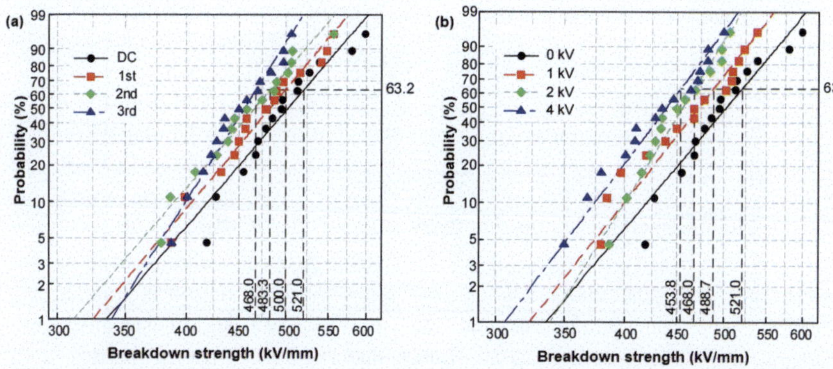

Fig. 4.2 Breakdown strength of PP film under DC superimposed the different harmonic voltages

more significant. The result is shown in Fig. 4.3b. Therefore, only a small amount of carriers can lead to rapid deterioration of insulation under DC superimposed pulse. The breakdown probability of the pure PP is over 50% when the superimposed electric field reaches 400 kV/mm as shown in Fig. 4.3a. It can be inferred that the damages caused by the pulse voltage to PP films are more severe than those caused by the DC or harmonic voltage. The overvoltage phenomenon needs more systematic concern in future researches.

Under non-DC electric field, the breakdown field strength of BOPP film decreases, as shown in Fig. 4.4. Under VA-VC (constant DC to AC) voltage, with the increase of temperature, the breakdown field strength of BOPP film decreases to below 300 kV/mm. With the increase of AC voltage frequency, the breakdown field strength is further reduced. Under VD-VF (constant DC variable pulse) voltage, the voltage decreases by 60–63% after the temperature is increased to 85 °C. With the increase of the pulse frequency, the breakdown field strength also decreases further, but the

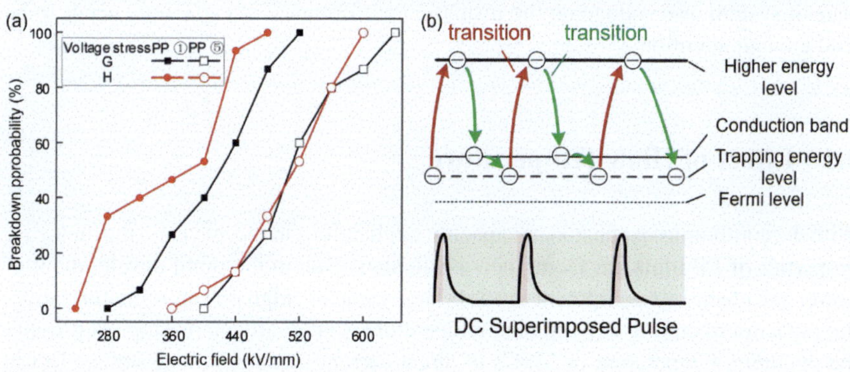

Fig. 4.3 **a** Relationship between the breakdown probability and the electric filed under the DC superimposed pulse voltage. **b** Carrier transport process under a DC superimposed pulse voltage

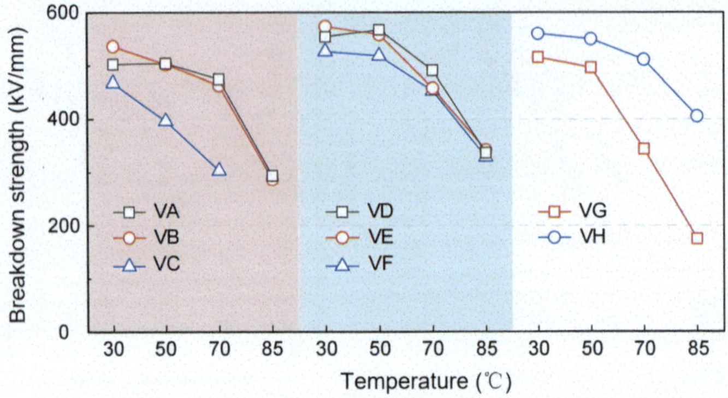

Fig. 4.4 Relationship between the breakdown strength and the temperature

influence of the pulse frequency is relatively small compared with the influence of the harmonic frequency. Under VG and VH (constant harmonic variable DC and constant pulse variable DC) voltages, the breakdown field strength of BOPP film also decreases with the increase of temperature.

The data show that after 600 times of 15 kV standard lightning wave (1.2/50 μs) accumulation, the molecular chain of polypropylene film with a thickness of 12 μm breaks and polar groups are formed, resulting in 1.651 times increase in the dielectric loss of the film [13]. The team of Tianjin University chose a pulse waveform with a rising edge of about 50 s to study the breakdown characteristics of polypropylene film under DC-pulse superimposed voltage [14]. Compared with DC alone, the insulation breakdown performance of polypropylene film decreases obviously under DC superimposed pulse voltage. When the pulsed electric field intensity is 120 kV/mm and the DC electric field intensity is 280 kV/mm, the film breakdown probability exceeds 50%. It can be inferred that impulse overvoltage causes more damage to insulation than DC or harmonic voltage. Therefore, in the future research, the influence of system overvoltage on the insulation performance of capacitors needs to be paid enough attention.

4.3 Effect of High Temperature

With the continuous expansion of capacitor applications in recent years, the dielectric properties of PP films are facing new challenges. Due to the rapid development of power and electronic equipment towards large capacity, high power, miniaturization, the capacitor operating environment temperature is gradually increasing. At present, the operating temperature of MFCs in direct current transmission systems, electric vehicles and aerospace applications can reach above 100 °C [15], the operating temperature of DC-DC converters for electric vehicles is more than 120 °C [16, 17]. However, PP films can only operate stably for a long time below 85 °C. In a high-temperature environment, the decrease of the breakdown strength and the increase of the conductivity loss of the PP film can easily accelerate the decay of the capacitance and service life, resulting in frequent failures of MFCs. The temperature of the application of high power density pulsed power supply ranges from 125–180 °C [18]. However, PP films allow a long-term operating temperature of less than 85 °C and a short-term maximum operating temperature of 105 °C, which is difficult to meet the demand for high-temperature applications of capacitors at this stage [17].

In addition to the demands of high-temperature working conditions, the DC support capacitor running in a high temperature environment of its own heat generation cannot be ignored. DC support capacitor electrode for nm level thickness of the evaporation electrode, thin electrode thickness will lead to electrode resistance, current flow through the evaporation electrode resistance loss heat caused by DC support capacitor internal temperature rise; in addition, DC support capacitor self-healing occurs when the heat cannot be dispersed in a timely manner, but also cause the DC support capacitor internal local temperature rise [17].

At present, the evolution law of dielectric properties of insulating materials under strong electric heat field and the research mechanism of insulation failure have been thoroughly investigated, and the basic dielectric parameter properties of PP energy storage medium under strong electric heat field have been widely discussed. In this section, the mechanism of insulation degradation of PP films at high temperatures is summarized from the perspectives of crystal dissociation, charge injection, impurity decomposition, and thermo-oxidative aging.

(1) **Impurity Dissociation**

Polypropylene can introduce impurities, known as ash, during processing, production and packaging. In the presence of a high temperature environment, ash not only leads to local electric field distortions within the film, inducing partial discharges and causing local breakdown failures. High temperatures provide the activation energy required for inerting reactions to occur within the polypropylene film, thereby lowering the reaction barrier for impurity decomposition. The ionization behavior of the impurities is promoted, generating a large number of ionic carriers, leading to an increase in the number of charges within the film. The carrier migration behavior in the film is promoted and therefore the probability of charge dispersion and migration within the polypropylene film is increased. Under high temperature conditions, the probability of the trapped electrons jumping to the conduction band increases, leading to an increase in conductivity loss.

(2) **Charge Injection**

Thermal field effects are the main factor leading to severe degradation of dielectric properties and capacitor failure. As the ambient temperature increases, the energy of carriers in the metal electrode and the dielectric increases. Gold. The charge in the metal electrode is easier to cross the interface barrier and inject into the dielectric. At the same time, the valence band and charge distribution in the dielectric are easier to cross the interface barrier and inject into the dielectric. The model of charge injection at high temperature is shown in Chap. 1.

(3) **Crystal Dissociation**

As the ambient temperature gradually increases, the thermal motion of the molecular chain intensifies. The molecular chain located at the inner boundary of the wafer gradually separates from the wafer under the action of high temperature, and the amorphous region gradually expands, as shown in Fig. 4.5. As the temperature continues to rise (more than 160 °C), the wafers and crystals gradually disappear, and the interior of the material is completely amorphous. Studies have shown that high crystallinity can block the carrier transport inside polypropylene, so that the insulation performance of polypropylene film is better. Therefore, with the increase of temperature and the dissociation of crystal segments, the blocking effect of crystal on carrier transport is weakened, and the insulation performance of polypropylene film decreases rapidly.

The influence mechanism of temperature on the microstructure and charge transport of PP film is shown in Fig. 4.6. Under high temperature polypropylene molecular chain slip and local chain segment relaxation triggered by an increase in the free volume, the average free range of the carriers increased, the energy gained by the carriers migrating in the free volume increased, resulting in carrier impact polypropylene molecular chain and the probability of fracture increased. Due to impurity defects, the interface between the crystalline region and the amorphous region of the defects and other electrical weaknesses exist, polypropylene film inside the accumulation of space charge caused by serious distortions in the electric field, partial discharge in the space charge sites to take the lead in the occurrence of local discharges, the gradual increase in local discharges caused by a reduction in the film dielectric strength [19].

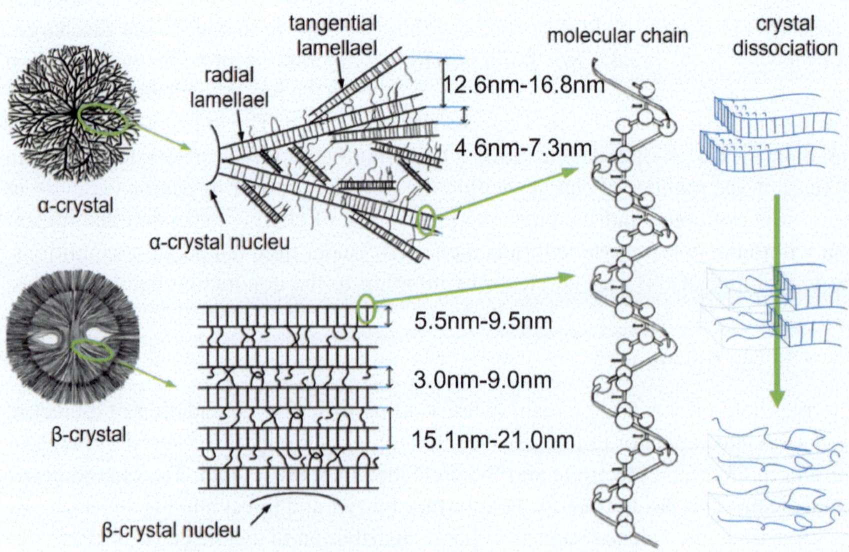

Fig. 4.5 Schematic diagram of the crystal dissociation

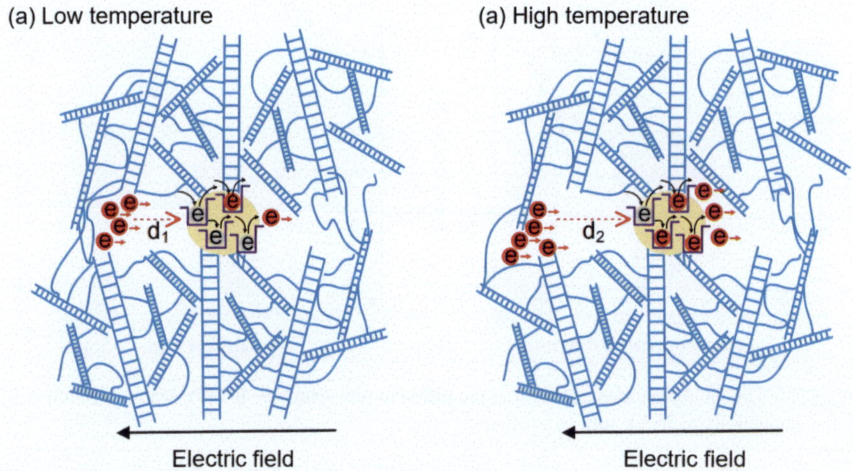

Fig. 4.6 Charge transport process at different temperatures

(4) **Thermo-Oxidative Aging**

In addition to providing energy for charge injection and excitation, high temperatures also provide reaction activation energy for chemical reactions. At high temperatures, the free energy of microscopic particles increases and the stability of chemical bonds decreases. At high temperatures and in the presence of oxygen, the tertiary carbon atoms in polypropylene tend to dehydrogenate and produce free radicals. It was shown that the carbonyl index of polypropylene increased dramatically from 1.2 to 8.0 after heat aging treatment at 135 °C for 500 h. Free radicals react with oxygen to form peroxyl radicals, which extract another unstable hydrogen atom from the macromolecular chain to form hydroperoxides. The hydroperoxide undergoes cleavage to generate two free radicals, which promotes the spontaneous progression of the reaction and is a key process affecting the rate of thermo-oxidative degradation of polypropylene, which usually makes the oxidative degradation process exhibit a typical induced phase.

It can be found that the crystallinity of polypropylene specimens increased from 48 to 51 and 55% after ageing at 80 and 120 °C for 1000 h, which was attributed to recrystallization during ageing. During the preparation of polypropylene samples, some molecules in the amorphous region are still in thermodynamic sub-stable states, and the high temperature provides these molecules with the energy for secondary rotation and alignment, and adjusts the molecules from local to global thermal stability during the annealing process, which further improves the degree of crystallinity. With the deepening of the degree of aging, the content of small molecules in the material gradually increases, the intermolecular friction in the process of polarization relaxation intensifies, and the dielectric loss increases. At the same time, the internal defects of the material expand, and the breakdown field strength decreases.

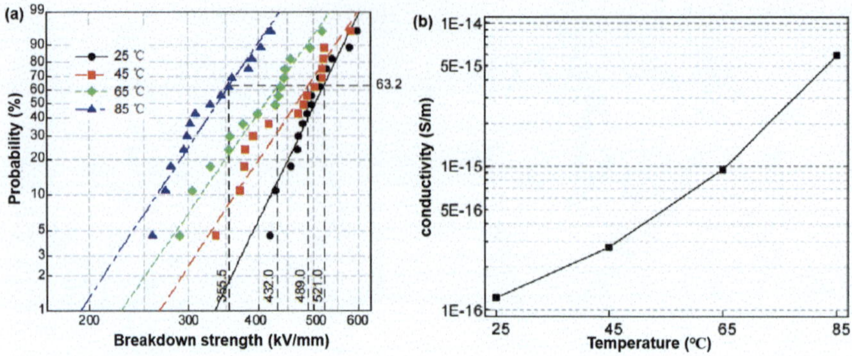

Fig. 4.7 **a** The DC breakdown strength at the different temperatures. **b** Temperature dependence of the conductivity at 25 kV mm

The conductance current of polypropylene films at high temperatures was investigated, and it was found that the conductivity activation energy of polypropylene films increased from 0.2–0.52 eV to 1.05 eV when the temperature was higher than 80 °C, and the conductance current increased [20]. The variation of the conductivity of the polypropylene film with temperature is shown in Fig. 4.7b. The drastic increase of the conductivity loss leads to the rapid decrease of the capacitor charging and discharging efficiency, and under the extreme conditions of 120 °C and 400 kV/mm, the capacitor charging and discharging efficiency will be reduced by more than 30%, and the maximum energy storage density will be reduced from 4 to 1.9 J/cm^3 [21]. At the same time, the increase of conductivity loss will also lead to a further increase in the internal temperature of the capacitor, and the vicious circle of loss and temperature rise seriously affects the stable operation of the capacitor. And the variation of the breakdown strength of the polypropylene film with electric field at different temperatures is shown in Fig. 4.7a. When the temperature rises from 25 to 85 °C, the polypropylene film breakdown strength decreases by more than 20%, which directly affects the operational safety and service life of the capacitor [22].

The vicious cycle between the temperature rise inside the DC support capacitor and the polypropylene film conductivity loss under high temperature environment will lead to the degradation of the polypropylene film dielectric properties. Polypropylene films will undergo electrothermal aging under the joint action of long-term electric field-temperature field, and the crystallinity and crystalline morphology of polypropylene films will change, leading to an increase in the dielectric loss of polypropylene films [23]. It has been shown that thermal ageing of polypropylene films for a certain period of time results in changes in the crystalline morphology of the polypropylene films and leads to a decrease in the breakdown field strength [24]. After two polypropylene films with different crystallinity were thermally aged at 65 °C and 19.2 kV/mm, it was found that the breakdown field strength of the polypropylene film with higher crystallinity before and after aging was higher, and that the breakdown field strength of the two polypropylene films with different crystallinity after thermally aging was lower than that of the polypropylene film without aging [25].

4.4 Effect of Mechanical Stress

The mechanical stresses generated by the winding process and temperature gradients and the electrical stresses generated by pulsed electromechanical overloads contribute to the formation of film defects, causing accelerated insulation degradation, resulting in reduced insulation performance, which is a bottleneck for the long-life application of polypropylene film pulse capacitors. The BOPP film is subjected to tensile stresses along the intra-layer direction of the film and extrusion pressures along the inter-layer direction in the process of winding and heat-setting [26]. Mechanical stretching destroys the spherical crystalline structure of polypropylene dielectrics, forming polymer low-density regions as well as crystalline-amorphous region interfacial defects, which exacerbate the charge transport behaviour and interfacial charge accumulation, leading to a degradation of the insulating properties of BOPP films. In addition, the pulse capacitor high-power discharge moment, instantaneous high-current flow through the metal electrode produces an induced magnetic field, between the electrode and the film to produce a strong electromotive force, affecting the insulating properties of BOPP film.

In addition to the pulse capacitor film in the manufacturing process of winding tensile stress and charging and discharging process of transient electromotive force, BOPP film is also affected by the large temperature gradient generated by the mechanical stress. Pulse current transient thermal pulse caused by the BOPP film dielectric surface and the internal formation of large temperature gradients, resulting in BOPP film inside and outside the thermal contraction and thermal expansion is extremely uneven, along the film thickness direction of the large gradient mechanical stress, further aggravated by the molecular chain of the pull and destruction, accelerating the insulation degradation process.

When the mechanical stress was increased from 0 to 35 N, the average partial discharge of polypropylene films increased by more than 40%. The effect of mechanical stress on the discharge of polypropylene films is shown in Fig. 4.3. When the tensile deformation is 30%, the BOPP conductivity increases by 540%, and the space charge injection leads to an increase in the electric field distortion rate by more than 20% and a decrease in the breakdown field strength by 38.5% [27]. The breakdown time of the medium under compressive stress of 10 MPa increased by 25%, while the breakdown time was reduced by 32% when the compressive stress was increased to 30 MPa [28].

Applying appropriate compressive stress can weaken the molecular chain movement ability and inhibit the development of insulation deterioration, while excessive compressive stress will cause the electrode metal particles embedded in the film to cause electric field distortion, which will reduce the insulating properties of thin film materials. The project team showed that the breakdown time of the dielectric increased by 25% at 10 MPa and decreased by 32% when the compressive stress was increased to 30 MPa [29].

4.5 Effect of Electromagnetic Field

Electromagnetic launch offers indispensable and substantial benefits, such as enhanced kinetic energy during launch, improved system efficiency, reduced launch intervals, accelerated response times, adjustable launch loads, and comprehensive control over the entire launch process. It epitomizes the forefront of next-generation advanced launch technology and is extensively utilized in various domains including catapult-assisted take-offs for shipboard aircraft and the implementation of magnetic levitation for ultra-high-speed trains, among others [30]. In order to push the load to a specific high speed in a very short time, the electromagnetic launch power supply system must be able to complete the rapid release of electrical energy, and polypropylene film capacitors are the best choice for the electromagnetic launch power supply system with millisecond current pulse width and GW-level discharge power. The magnetic bearing of the electromagnetic launch system operates within the powerful magnetic field produced along the launch trajectory. Therefore, there is a critical need to investigate the laws governing performance evolution and the mechanisms of insulation degradation in dielectric energy storage materials under these intense magnetic field conditions.

There are three primary factors contributing to the reduction in insulation strength under magnetic fields. Firstly, the presence of a robust magnetic field facilitates the excitation of charges and enhances the migration of carriers within the conduction band, leading to increased interaction with molecular chains. The charges excited under a magnetic field may come from within the chemical bonds, and the excitation and migration of these charges will weaken the stability of the polymer molecular chains, and the deterioration process of the material is accelerated by the magnetic field. Partial discharges are also enhanced under magnetic fields as defects are created or extended, further contributing to the deterioration of insulation. Secondly, deflection occurs during space charge transport under magnetic field, and the charge accumulated at the defects inside the capacitor film is redistributed, and the unevenly distributed charge further aggravates the electric field distortion of the dielectric defects, which further promotes partial discharge and insulation degradation as shown in the Fig. 4.8 [31]. Ultimately, the magnetic field imposes constraints on the dipole's orientation by heightening its rotational resistance, thereby limiting its movement. This restriction, coupled with the overlap of its local electric field with the primary electric field, induces distortion in the local electric field. This distortion contributes to the initiation of partial discharges.

The conductivities of the four films under different magnetic fields are shown in Fig. 4.9a. The conductivity of PI is the highest, and PVDF is slightly lower than that of PI, while the conductivities of PP and PE are within the same order of magnitude,

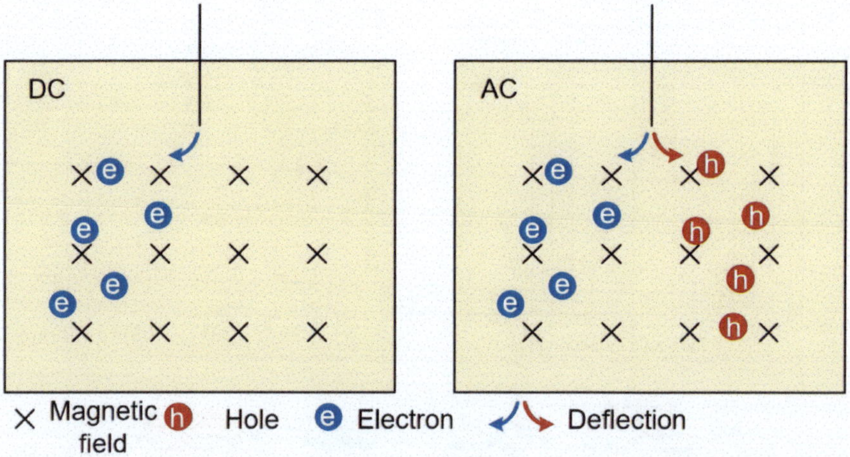

Fig. 4.8 Charge injection and deflection process under the electromagnetic coupling field

which are much lower than that of PI and PVDF. The conductivities of the four dielectrics are gradually increased with the enhancement of the magnetic field. The structure of the four dielectrics is further analyzed by quantum chemical calculation method, and the calculation model is shown in Fig. 4.9b. It can be found that the forbidden bandwidth of PI molecules is much smaller than that of the other three materials. The narrow forbidden bands of PI lead the electrons to jump from the valence band to the conduction band more easily under the strong electric field, and the electrons are continuously transported in the conduction band. The narrow forbidden band of PI film makes it easier for electrons to jump from the valence band to the conduction band under a strong electric field and to be transported continuously in the conduction band. As a result, the conductivity of PI films is higher than that of other dielectric materials.

It has been shown that the polarizability of polymers under a strong magnetic field of 30 T decreases up to 33.6% [29], while the electron displacement polarization strength will increase by a factor of several thousand. The inversion of dipole polarization and electron polarization behavior of energy storage materials under strong magnetic field makes the permittivity evolution law more and more complicated as shown in Fig. 4.10. Under the action of electromagnetic coupling strong field, the dipole polarization relaxation path inside the dielectric is shifted, and the friction between adjacent polarization units is intensified, leading to a significant increase in the dielectric loss of BOPP film. The dielectric loss of insulating materials generally rises by about 20% in the frequency range of 50–100 kHz under a 3T magnetic field [33]. The larmor progression of atomic nuclear and electronic magnetic moments by strong magnetic fields destabilizes the chemical bonding structure of BOPP under extreme conditions, and the re-establishment of steady state electron orbitals leads to significant changes in the Schottky barriers and P-F traps [34]. The electron spin magnetic moment is forced to rotate under a strong magnetic field,

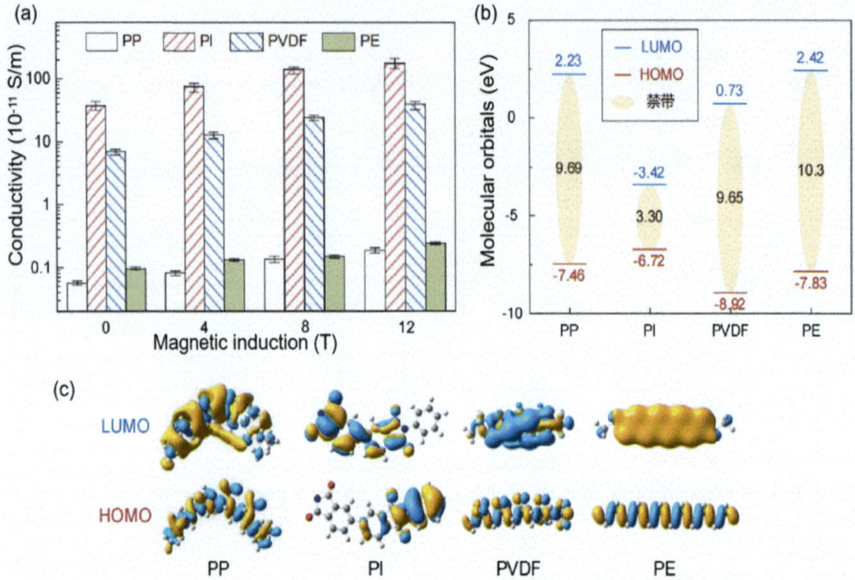

Fig. 4.9 **a** Conductivity under the different magnetic fields. **b** Energy levels and **c** Molecular orbital distribution of LUMO and HOMO [32]. Copyright Clearance Center's RightsLink© License Number: 5811930920083

leading to electronic excitation and increasing the leakage conductance loss, which increases the leakage conductivity of BOPP films by 230% under a 12 T magnetic field [32]. A research team from Tianjin University found that insulation degradation phenomena such as molecular chain breakage occurred locally in BOPP under the action of 400 mT orthogonal electromagnetic field, and the insulation breakdown time caused by electrical dendritic degradation decreased from 120 to 76 min [33], and the breakdown field strength of the BOPP film decreased by 16.8% under 12 [35].

As mentioned above, under the action of strong electromagnetic fields, the BOPP film loss increases and the insulation breakdown decreases, so that the life of the pulse capacitor is greatly shortened, and the energy storage density decreases rapidly after a limited number of discharges.

4.6 Summary

In this chapter, the operating environment of polypropylene films for capacitors in high-voltage DC systems is summarized, and the performance evolution of polypropylene films and insulation failure mechanism under complex electric field, strong thermal field, strong magnetic field and large mechanical stress environment are summarized. The specific research contents are as follows:

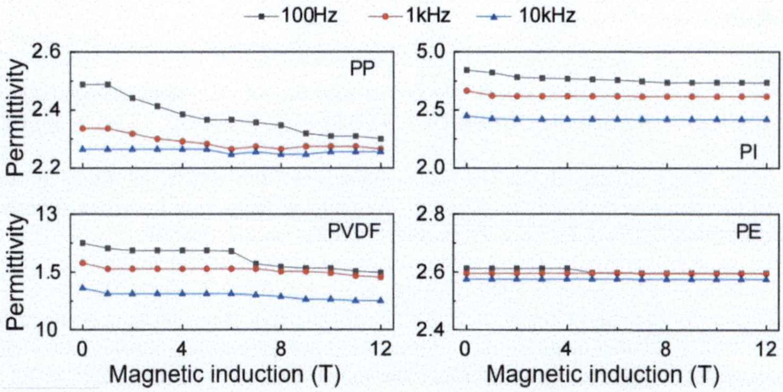

Fig. 4.10 Relationship between the permittivity and the magnetic field

The research significance and subject background of MFCs under complex electric fields are introduced, and the high voltage problems faced by dielectric capacitors in high-voltage DC transmission systems are described. The action mechanism of insulation degradation caused by high field strength is described from the perspectives of charge injection, charge migration and partial discharge. The failure characteristics of PP films under complex operating conditions such as DC superimposed harmonics or surge voltage are introduced.

PP films for capacitors cannot meet the temperature requirements in high-voltage DC transmission systems. This chapter reviews the insulation degradation mechanism of MFCs under high temperature environment, and describes the effects of high temperature on the dielectric properties of PP films from the perspectives of impurity decomposition, charge injection, crystal dissociation, and thermo-oxidative aging, respectively.

The changing rules of PP film dielectric properties and partial discharge properties under strong electromagnetic field are summarized, and the mechanism of partial discharge enhancement under strong magnetic field is revealed based on the charge deflection behavior and the evolution of dielectric properties. It is found that the permittivity, conductivity loss, insulation breakdown strength of BOPP film under strong magnetic field environment undergoes subversive changes, and the insulation performance decreases significantly with magnetic flux density.

Mechanical stresses generated by the winding process and temperature gradient and electrical stresses generated by pulsed electromechanical overload are presented. The stresses contribute to the formation of film defects, cause accelerated insulation degradation, and lead to the reduction of insulation performance, which are the bottlenecks restricting the long-life application of BOPP film pulse capacitors. This chapter summarizes the current research on the effect of mechanical stress on the dielectric properties of PP films.

Finally, the evolution of the properties of PP film energy storage media under strong electric, thermal, electromagnetic and stress fields is reviewed.

References

1. M. Xiao, Y. S. Zhao and B. X. Du, "Dielectric performance of furan-functionalized polypropylene film for HVDC capacitors," *IEEE Trans. Dielect Elect. Insul.*, vol. 31, no. 1, pp. 160–167, 2024.
2. M. Xiao, M. D. Zhang, B. X. Du, Z. Y. Ran, H. L. Liu and Y. W. Qin, "Breakdown strength improvement of polypropylene by introducing long-chain branched structures for film capacitors," *J. Phys. D: Appl. Phys.*, vol. 55, 2022, Art. no. 245501.
3. H. M. Umran, F. Wang and Y. He, "Ageing: causes and effects on the reliability of polypropylene film used for HVDC capacitor", *IEEE Access*, vol. 8, pp. 40413–40430, 2020.
4. M. Xiao, K. L. Fan, M. D Zhang, B. X. Du, Z. Y. Ran, H. L. Liu, and Z. L. Xing, "Effect of cyclic olefin copolymer on dielectric performance of polypropylene composites for metallized film capacitor," *IEEE Trans. Dielect Elect. Insul.*, vol. 29, no. 6, pp. 2266–2273, 2022.
5. B. X. Du, J. H. Zhang, M. Xiao, H. Liu and Z. Y. Ran, "Dielectric breakdown performance of deashed polypropylene for HVDC film capacitors," *IEEE Trans. Dielect Elect. Insul.*, vol. 29, no. 6, pp. 2209–2217, 2022.
6. B. X. Du, J. H. Zhang, M. Xiao, J. W. Xing, Z. Y. Ran and H. L. Liu, "Potential of metal passivators in improving the insulation performance of polypropylene films for capacitors," 2022 IEEE 4th International Conference on Dielectrics, Palermo, Italy, 2022, pp. 729–732.
7. S. T. Li, D. M. Min, W. W. Wang and G. Chen, "Modelling of dielectric breakdown through charge dynamics for polymer nanocomposites," *IEEE Trans. Dielect Elect. Insul.*, vol. 23, no. 6, pp. 3476–3485, 2016.
8. M. Xiao, Y. S. Zhao, Y. N. Song and B. X. Du, "Polypropylene films for HVDC capacitors with enhanced dielectric performance by irradiation crosslinking modification," *IEEE Trans. Dielect Elect. Insul.*, vol. 32, no. 2, pp. 666–673, 2024.
9. M. Xiao Z. Y. Zhang and B. X. Du, "Dielectric performance improvement of polypropylene film by hierarchical structure design for metallized film capacitors," *J. Phys. D: Appl. Phys.*, vol. 57, no. 38, pp. 5501–5509, 2024.
10. T. Asai et al, "Partial discharge deterioration of polypropylene films with different film-surface structure," 2015 IEEE Conference on Electrical Insulation and Dielectric Phenomena, Ann Arbor, MI, USA, 2015, pp. 779–782.
11. Y. Yoshida and M. Nishimatsu, "Power capacitors," *IEEE Trans. Dielect Elect. Insul.*, vol. EI-21, no. 6, pp. 963–972, 1986.
12. R. R. Xu, B. X. Du, M. Xiao and J. Li, "Dielectric properties dependent on crystalline morphology of PP film for HVDC capacitors," *Polymer,* vol. 213, 2020, Art. no. 123204.
13. W. Sima, J. Hua, P. Sun and Q. L. Chen, "Accumulative failure characteristic of polypropylene film in surge capacitor under repeated impulse voltage," *High Volt. Engineering*, vol. 44, no. 22, pp. 358–365, 2018.
14. Z. Y. Ran, B. X. Du, M. Xiao and J. Li, "Crystallization morphology-dependent breakdown strength of polypropylene films for converter valve capacitor," *IEEE Trans. Dielect Elect. Insul.*, vol. 28, no. 3, pp. 964–971, 2021.
15. M. Xiao, M. D. Zhang, H. L. Liu, B. X. Du and Y. W. Qin, "Dielectric property and breakdown dtrength performance of long-chain branched polypropylene for metallized film capacitors," *Materials*, vol. 15, 2022, Art. no. 3071.
16. J. S. Ho and S. G. Greenbaum, "Polymer capacitor dielectrics for high temperature applications", *ACS Appl. Mater. Interfaces*, vol. 10, no. 35, pp. 29189–29218, 2018.
17. Z. P. Li et al, "High dielectric constant polycarbonate/nylon multilayer films capacitors with self-healing capability," *ACS Appl. Energy Mater.*, vol. 1, no. 4, pp. 867–875, 2019.
18. Y. Zhou et al, "Interface-modulated nanocomposites based on polypropylene for high-temperature energy storage," *Energy Storage Mater.*, vol. 28, pp. 255–263, 2020.
19. 20. B. Dang, J. L. He, J. Hu and Y. Zhou, "Large improvement in trap level and space charge distribution of polypropylene by enhancing the crystalline-amorphous interfaces effect in blends," *Polym. Int.*, vol. 65, no. 4, pp. 371–379, 2016.

20. A. Kahouli, O. Gallot-Lavallée and P. Rain, "Dielectric features of two grades of bioriented isotactic polypropylene," *J. Appl. Polym. Sci.*, vol. 132, no. 28, 2015, Art. no. 42224.
21. M. Horak and P. Mach, "Changes of loss factor of polypropylene film capacitors during long time thermal ageing," 2018 International Conference on Diagnostics in Electrical Engineering, Pilsen, Czech Republic, 2018, pp. 1–4.
22. M. Rabuffi and Picci G, "Status quo and future prospects for metallized polypropylene energy storage capacitors," *IEEE Trans. Plasma Sci.*, vol. 30, no. 5, pp. 1939–1942, 2002.
23. T. Umemura and D. Couderc, "Thermal aging of biaxially oriented polypropylene films part II morphology and dielectric properties," Conference on Electrical Insulation & Dielectric Phenomena - Annual Report 1981, Whitehaven, PA, USA, 1981, pp. 314–321.
24. 25. M. Teixeira, I. D. Hoyo, F. Wandrowelsti, V. Swinka-Filho and M. Munaroet, "Evaluation of thermal degradation in isotactic polypropylene films used in power capacitors," *J. Therm. Anal. Calorim.*, vol. 130, pp. 997–1002, 2017.
25. J. Kurimsky, M. Kosterec and B. Vargova, "Breakdown voltage of polypropylene film during DC and thermal ageing," 2017 18th International Scientific Conference on Electric Power Engineering, Kouty nad Desnou, Czech Republic, 2017, pp. 1–4.
26. 27. G. Picci and M. Rabuffi, "Pulse handling capability of energy storage metallized film capacitors," *IEEE Trans. Plasma Sci.*, vol. 28, no. 5, pp. 1603–1606, 2000.
27. 28. B. X. Du, Y. Zhang, X. X. Kong and H. C. Liang, "Morphological feature analysis of electrical tree growth in epoxy resin under tensile and compressive stress," *IEEE Trans. Dielect Elect. Insul.*, vol. 29, no. 1, pp. 343–346, 2022.
28. R. A. ALI-Zade, "Influence of magnetic field on dielectric permittivity of nanocomposites on the base of polymeric Matrix: collagen, polystyrole, and magnetite nanoparticles," *IEEE T. Magn.*, vol. 51, no. 7, pp. 1–8, 2015.
29. 30. R. A. Ali-zade, "Influence of magnetic field on dielectric permittivity of nanocomposites on the base of polymeric matrix: Collagen, polystyrole, and magnetite nanoparticles," *IEEE T. Magn.*, vol. 51, no. 7, pp. 1–8, 2015.
30. H. L. Liu, B. X. Du and Meng Xiao, "Improved energy density and charge discharge efficiency of polypropylene capacitor film based on surface grafting", *IEEE Trans. Dielect Elect. Insul.*, vol. 28, no. 5, pp. 1539–1546, 2021.
31. M. Dawber and Scott J F, "Modelling of leakage current in ferroelectric thin film capacitors," *Ferroelectrics*, vol. 260, no. 1, pp. 487–492, 2001.
32. 33. H. Liu, B. Du, M. Xiao, Z. Ran and Y. Ma, "Dielectric properties of polymer films in strong electromagnetic field for energy storage capacitor," *IEEE Trans. Dielect Elect. Insul.*, vol. 29, no. 5, pp. 1745–1753, 2022.
33. 35. M. Y. Wang, B. X. Du, X. X. Kong, Z. L. Li, M. Xiao and Y. W. Ma, "Effects of gradient magnetic field on charge behavior and electrical tree growth in epoxy resin," *IEEE Trans. Dielect Elect. Insul.*, vol. 28, no. 5, pp. 1686–1693, 2021.
34. 36. M. Y. Wang, B. X. Du, X. X. Kong, Z. L. Li, T. Han and Y. W. Ma, "Effects of high magnetic field on partial discharge and dielectric breakdown behavior of silicone rubber," *IEEE Trans. Dielect Elect. Insul.* vol. 29, no. 3, pp. 915–923, 2022.
35. 38. Y. Yu, B. X. Du, J. X. Jin, T Han and J. G. Su, "Effect of magnetic field on electrical treeing behavior of silicone rubber at low temperature," *IEEE Trans. Appl. Supercond.*, vol. 26, no. 7, pp. 1–4, 2016.

Part II
Microstructure Modification

Chapter 5
Deash Treatment

5.1 Introduction

The approach to reducing the ash content in PP is by minimizing the amount of ash. The production of PP involved a slurry polymerization process, utilizing the first and second generation of Ziegler Nata catalyst systems for polymerization. However, due to the low activity of catalyst, a significant amount of catalyst is required during the production process, resulting in a high level of catalyst residue and ash content in the final product. Therefore, it is essential to enhance the deashing process to eliminate catalyst residues and impurities [1, 2].

The removal of ash from polypropylene products was mainly achieved through solvent washing, which was a complex and time-consuming process that required a large amount of solvent consumption. A solvent washing and ash removal method is proposed for polypropylene products prepared by slurry process [3]. The polypropylene is washed with isopropanol or n-butanol solvent to dissolve the ash into the solvent, and then centrifuged to achieve ash removal. The removal efficiency of Ti, Cl, and Mg ash is 67–79%, but this method does not mention the removal efficiency of other ash contents, such as Al. In order to promote the removal effect of polypropylene ash, adding functional substances to washing solvents has gradually become a more widespread choice. A solution washing and ash removal method for random polypropylene was proposed in [4]. The complexing ability of hydroxysulfonic acid reacts with the metal components in ash to achieve ash removal, but this process is extremely complex and the washing time is relatively long, which is not conducive to industrial application.

With the introduction and development of carriers and electron donors, the activity of Ziegler Nata catalysts has significantly increased, and the ash content in polypropylene products has also been greatly reduced. Due to the complexity and low efficiency of the solution washing and ash removal process, it has been phased out in the industrial production of polypropylene [5]. However, numerous studies have shown that

B. Du and M. Xiao, *Polypropylene Film for HVDC Capacitors*,
https://doi.org/10.1007/978-981-96-3029-5_5

trace amounts of ash (ppm) in polypropylene can lead to a significant decrease in long-term stability, especially with residual catalysts. Therefore, in order to develop high-purity and high-performance polypropylene raw materials, washing and deashing treatment of polypropylene is an essential step, and designing a simple and efficient solution washing and deashing method has gradually become a hot research topic for many scholars. A polyolefin chemical washing system was proposed in [6], which includes a liquid-phase washing device, impurity separation device, and degassing device. The polypropylene was mainly washed and de ashed using alcohols or aqueous epoxy compounds. The generated hydrogen chloride gas made the catalyst residue fluffy, resulting in a better washing and de ashing effect, with an ash removal efficiency of up to 81%. However, this process not only has extremely high complexity and energy consumption, but may also introduce high boiling point impurities into polypropylene during the washing process. A method for removing polypropylene ash by washing with an "alkane + alcohol + complexing agent" solution was proposed in Ref. [7]. After washing and deashing treatment, the ash content in polypropylene was significantly reduced, and the ash removal efficiency was 33–50%.

At present, in the washing and ash removal methods of polypropylene powder, the advantages and disadvantages of the methods are mainly evaluated by testing the ash removal efficiency. Organic solvents are often used in the washing process, and the swelling phenomenon of polypropylene in the solvent is used to promote the removal of ash [8, 9]. It is worth noting that during the swelling process of polypropylene, organic solvents can damage the crystalline/amorphous interface and introduce structural defects, leading to a decrease in dielectric properties, which is extremely detrimental to polypropylene materials for electrical purposes [10]. For polypropylene raw materials used in capacitors, the main purpose of washing and deashing treatment is to suppress the adverse effects of ash content on dielectric properties and long-term stability. Therefore, on the basis of considering the efficiency of polypropylene ash removal, more attention should be paid to changes in dielectric properties.

5.2 Ash Reduction Method

5.2.1 Basic Principle

The microscopic process of washing deashing polypropylene with metal ion passivation agent is shown in Fig. 5.1. Polypropylene has good chemical resistance to water, alkali, and acid. However, polypropylene powder may experience a certain degree of swelling in solvents such as ethanol. In the study of polymer dissolution and swelling processes, the free volume theory is commonly used to describe and understand the permeation of small molecules into the interior of polypropylene. This theory suggests that the volume of polymers consists of two parts, namely occupied

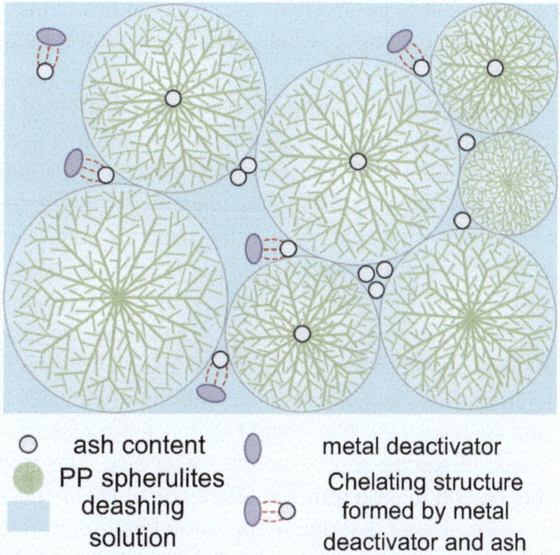

Fig. 5.1 Schematic diagram of deashing method

volume and free volume [11]. Occupation volume refers to the volume occupied by polymer molecular chains themselves, while free volume refers to the gaps between molecular chains, which can provide space for molecular chain activity. During the washing and ash removal process of polypropylene powder, the washing solution gradually infiltrates into the interior of the polypropylene, resulting in significant swelling of the polypropylene and an increase in its free volume. Therefore, the metal ion passivation agent in the washing solution migrates in the free volume of polypropylene and chelates with the ash content, forming a stable chelating structure, thereby promoting the migration of ash content into the washing solution. Under a certain washing time, most of the ash in polypropylene has formed a stable chelating structure with metal ion passivators. After filtration and deashing treatment, the removal of ash in polypropylene can be achieved.

The removal effect of polypropylene ash largely depends on the distribution position of ash. Most of the ash in polypropylene will be squeezed to the boundaries of the spherulites as they grow. After complete crystallization, the ash will aggregate in the amorphous region. In the amorphous region of polypropylene, the arrangement of macromolecular chains is disorderly, and the distance between chains is large. The washing solution is easy to penetrate and interact with ash, thereby promoting the removal of ash. It should be pointed out that a small portion of ash in polypropylene can play a role in heterogeneous nucleation during the matrix crystallization process, and ultimately remain inside the crystallization zone as the spherical crystals grow [12]. In the crystallization zone of polypropylene, the molecular chains are arranged in an orderly manner with high regularity, which to some extent hinders the infiltration of the washing solution into the matrix. The ash content in the crystallization zone is difficult to interact with the washing solution, and the removal effect of ash content is poor. Therefore, residual ash is inevitably present in the deashing polypropylene sample, which also reveals the limitations of the solution deashing method.

In addition, the washing and ash removal effect of polypropylene also depends on the stability of the chelating structure formed by the metal ion passivation agent and ash. In order to study the influencing factors of solution washing and ash removal efficiency at the micro level, the chelating structure formed by metal ion passivators and ash was analyzed. Generally speaking, the stability of chelates is related to the structural properties of ligands (metal ion passivators) and coordination ions (ash content). For metal ion passivators, their chelating ability increases with the increase of electron pair donors. When there are more electron pair donors, metal ion passivators can form more coordination bonds with ash, increase the number of chelating rings, and enhance the stability of chelates [13]. Compared to MarkCDA-1 and Chelsea 180, NaugradXL-1 has more electron pair donors and can form up to three chelating ring structures with ash. The stability of chelates is relatively stronger, and the ash removal efficiency of NX-0.05–1.5–30 is also relatively better. It is worth noting that due to the steric hindrance effect of ligands, electron pairs on the five membered heterocycles in MarkCDA-1 hinder their chelation reaction with ash, weakening the stability of the chelation structure. Therefore, the washing and deashing effect of MC-0.1–2.5–40 is relatively poor [14]. In addition, the amount of charge carried by ash also affects the stability of the chelating structure. When ash carries more charge, its coordination with metal ion passivators is enhanced, and the stability of the formed chelates is improved. The ash content in polypropylene usually exists in the form of ions, including Ti^{4+}, Fe^{3+}, Al^{3+}, and Mg^{2+}. Among them, Ti^{4+} and Fe^{3+} belong to transition metal elements, have empty d-orbitals, carry a large amount of charge, and form chelating structures with good stability [15]. Therefore, the metal ion passivation agent washing and deashing treatment has a good removal effect on Ti, Fe, and Al ash content in polypropylene.

5.2.2 *Influencing Factors*

The main purpose of the solution washing and ash removal method for polypropylene raw materials used in high-voltage DC capacitors is to reduce the ash content of polypropylene and improve its dielectric performance. Among them, breakdown strength is the most important dielectric performance parameter, which directly affects the reliability of the entire capacitor operation. In order to study the influence of solution washing parameters on the breakdown performance of deashing polypropylene and optimize the solution washing deashing method, the influence of anhydrous ethanol solvent washing deashing treatment on the breakdown and electrical conductivity of deashing polypropylene was first studied as a control group for the "metal ion passivation agent + anhydrous ethanol" solution washing deashing treatment. Then, the correlation between the washing parameters of the "metal ion passivation agent + anhydrous ethanol" solution and the breakdown and conductivity properties of deashing polypropylene was established around the types and concentrations of washing solutions, washing time, and washing temperature, and the optimal solution washing parameters were obtained. In addition, the influence of physical parameters of polypropylene (powder particle size and initial ash content) on the breakdown and conductivity performance after washing and deashing treatment was studied, revealing the application characteristics and limitations of the solution washing and deashing method.

(1) **Polypropylene Particle Size**

The effectiveness of solution washing and ash removal is related to the degree of infiltration of the washing solution and the distribution of ash in polypropylene. In general, the particle size of polypropylene directly affects the infiltration degree of washing solution, and the distribution of ash in polypropylene may be related to the ash content. The optimal washing and deashing parameters were used to wash and deash polypropylene powder with different physical parameters (particle size and initial ash distribution), and breakdown and conductivity tests were conducted. The influence of particle size and initial ash content of polypropylene on the washing and deashing effect was studied, revealing the application characteristics and limitations of the solution washing and deashing method.

Using the optimal washing and deashing parameters, the polypropylene powder with particle sizes of 10, 20, 30, 40, and 50 was subjected to washing and deashing treatment, and electrical conductivity tests were conducted. The results are shown in Fig. 5.2. It can be observed that as the particle size of polypropylene decreases, the conductivity of the deashing sample shows a pattern of first decreasing and then increasing. When the particle size of polypropylene is large (10 mesh), the infiltration degree of the washing solution is poor, and the washing and deashing effect is poor. The conductivity of deashing polypropylene is relatively high, at 1.12×10^{-14} S/m. When the particle size decreases to 30 mesh, the conductivity of deashing polypropylene decreases to 0.99×10^{-14} S/m, which is due to the enhanced infiltration of washing solution promoting the swelling phenomenon of polypropylene, and the washing and ash removal effect is significantly improved. However, when the particle size is reduced to 50 mesh, the conductivity of deashing polypropylene increases to 1.20×10^{-14} S/m, which is attributed to the structural damage caused by the washing solution on the interface between the crystalline and amorphous regions of polypropylene. The particle size of polypropylene has a significant impact on the washing and ash removal efficiency, which may be related to the degree of infiltration of the washing solution. When the particle size of polypropylene is large, it will weaken the infiltration degree of the washing solution, and the washing and ash removal effect will be poor; When the particle size of polypropylene is small, the infiltration degree of the washing solution is too strong, and the structural defects at the interface between the crystalline and amorphous regions of polypropylene increase, leading to a decrease in the dielectric properties of polypropylene. The above test results indicate that the washing and deashing effect is best when the

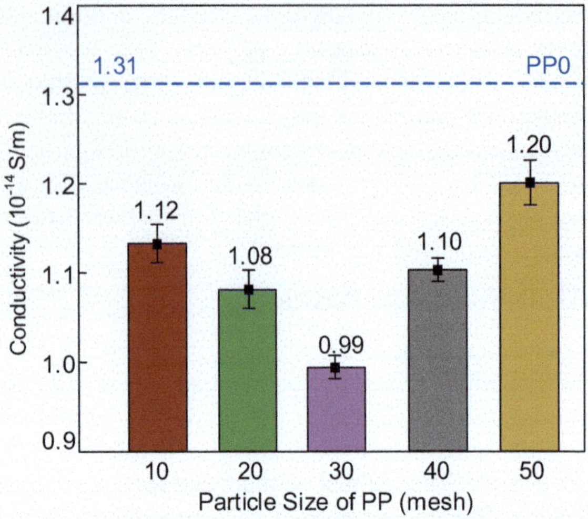

Fig. 5.2 Relationship between the conductivity and the particle size

particle size of polypropylene is 30 mesh, and deashing polypropylene exhibits the best electrical conductivity performance.

The ash distribution of polypropylene has a significant impact on the ash removal efficiency during the solution washing and ash removal process. Due to the difficulty of the washing solution penetrating into the crystallization zone of polypropylene, the washing and ash removal efficiency in the crystallization zone is poor; The amorphous region of polypropylene has a relatively good washing and ash removal effect due to the loose arrangement of molecular chains. It is generally believed that the distribution of ash in polypropylene largely depends on its ash content. In order to study the effect of initial ash content on the washing and deashing effect of polypropylene, the optimal washing parameters were used to wash and deash polypropylene powder with ash content of 200, 300, and 400 ppm, and the ash content of the deashing samples was tested. The results showed that the residual ash content of polypropylene with ash content of 200, 300, and 400 ppm after washing and deashing treatment was extremely close, reaching 148.5, 165.0, and 142.5 ppm, respectively. Due to the different distribution of ash in polypropylene, the washing and deashing effect has a certain degree of unevenness. Part of the ash in polypropylene can play a heterogeneous nucleation role and be distributed in the crystallization zone, but the deashing effect in the crystallization zone is relatively poor; Most of the ash will aggregate to the boundaries of the spherical crystals as they grow, and ultimately gather in the amorphous region, which has a relatively good ash removal effect. The remaining ash content in the deashing polypropylene sample is relatively close, and it can be considered that the difference in initial ash content of polypropylene is mainly reflected in the amorphous zone, while the ash content in the crystalline zone is basically the same. Therefore, in the crystallization process of polypropylene, the amount of ash that can play a heterogeneous nucleation role is limited, and most of the ash tends to aggregate in the amorphous region [16].

The breakdown strength and conductivity of deashing polypropylene were tested, and the results are shown in Fig. 5.3. It can be observed that as the ash content increases, the breakdown strength of polypropylene shows a monotonic decrease trend, while the conductivity shows a monotonic increase trend. When the ash content of polypropylene increases from 142.5 to 405.0 ppm, its breakdown strength decreases from 655.3 to 485.9 kV/mm, and the conductivity increases from 0.99×10^{-14} S/m increases to 1.31×10^{-14} S/m. Therefore, by studying the changes in the breakdown and conductivity of polypropylene before and after solution washing and deashing treatment, the effectiveness of washing and deashing can be characterized to a certain extent. In this study, the goal of washing and deashing polypropylene is to enhance its dielectric performance by reducing ash content, thereby improving the operational reliability and economy of high-voltage DC capacitors. Employing dielectric performance as a criterion for assessing the efficacy of washing and deashing processes holds practical significance. This approach provides a practical method to evaluate the effectiveness of these procedures, leveraging dielectric properties as an indicator of their impact.

(2) **Detergent Types and Concentrations**

In the solution washing and ash removal method, the washing solution serves as the main medium and plays a crucial role in the ash removal effect of polypropylene. In order to study the effect of washing solution on the deashing effect of polypropylene, different washing solutions were prepared by changing the type and content of metal ion passivators. The polypropylene powder was washed and deashing treated, and the breakdown performance test was carried out. The results are shown in Fig. 5.2, where the blue and black dashed lines represent the breakdown strength results of

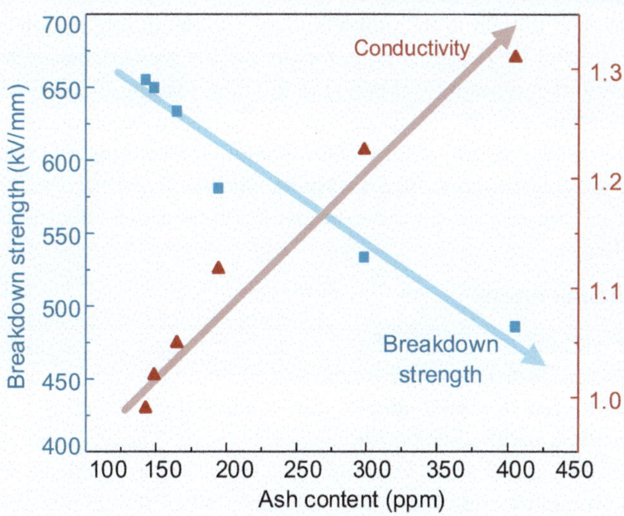

Fig. 5.3 Relationship between the ash content, the breakdown strength, and the conductivity

PP0 and ET-1.5, respectively. It can be found that the type of washing solution has a significant impact on the breakdown performance of deashing polypropylene. When the concentration of washing solution is 0.1 g/100 ml, the breakdown strength of deashing polypropylene samples is 484.6–596.2 kV/mm, with MC-0.1 breakdown strength of 596.2 kV/mm, which is 22.7% higher than PP0, while CH-0.1 breakdown strength is only 484.6 kV/mm.

In order to further analyze the influence of solution washing and deashing treatment on the breakdown performance of polypropylene, Fig. 5.2 shows the Weibull distribution parameters of the deashing polypropylene sample. The results showed that regardless of adjusting the concentration of the "metal ion passivation agent + anhydrous ethanol" washing solution, the Weibull shape parameters of the deashing polypropylene sample significantly decreased, but slightly increased compared to the anhydrous ethanol deashing sample, ranging from 8 to 17. In Fig. 5.4a–c, it can be observed that the concentration of the washing solution also has a significant effect on the breakdown strength of the deashing polypropylene sample. When the concentration of the washing solution is 0.01–1.00 g/100 ml, the breakdown strengths of MarkCDA-1, Chelsea 180, and NaugradXL-1 deashing polypropylene samples are 538.9–596.2, 472.3–544.2, and 489.1–565.4 kV/mm, respectively. It is worth noting that as the concentration of the washing solution increases, the Weibull scale parameters and shape parameters of the deashing polypropylene sample show a trend of first increasing and then decreasing. When the concentration of the washing solution is low, the content of metal ion passivation agent is insufficient to interact with all the ash in polypropylene, resulting in relatively poor deashing effect and lower breakdown strength of deashing polypropylene. As the concentration of the washing solution increases, more ash can interact with metal ion passivators, enhancing the deashing effect and enhancing the breakdown strength of deashing polypropylene. However, when the concentration of the washing solution is too high, the metal ion passivator may agglomerate and remain in the polypropylene powder during the filtration and deashing process. After the deashing polypropylene is hot pressed, the agglomerated metal ion passivator becomes a local defect in the film, leading to a decrease in the breakdown performance of the deashing sample. Therefore, when using a solution of "metal ion passivation agent + anhydrous ethanol" for washing and deashing of polypropylene, the concentration of the washing solution is not necessarily higher, and comprehensive consideration needs to be given to the washing effect and filtration and deashing effect.

(3) **Washing Temperature**

Usually, the washing temperature has a significant impact on the infiltration of the washing solution. In order to improve the uniformity of the solution washing and deashing effect, different deashing samples were prepared by controlling the washing temperature, and their breakdown performance was tested. The results are shown in Fig. 5.5. The results show that the washing temperature has a significant impact on the breakdown strength of deashing polypropylene. When the temperature is 20–60 °C, the breakdown strengths of MC-0.1–2.5, CH-0.05–1.5, and NX-0.05–1.5 are 591.3–629.1, 583.5–655.3, and 572.9–638.7 kV/mm, respectively. In order to further

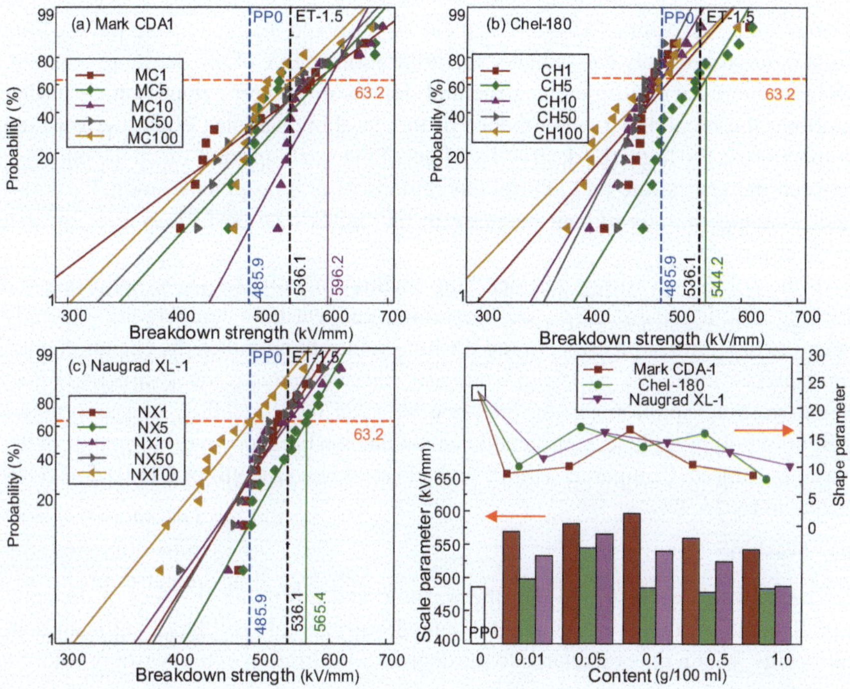

Fig. 5.4 Relationship between the **a–c** breakdown performance, **d** Weibull distribution parameters and the deashing solution

investigate the evolution of the breakdown performance of deashing polypropylene under different washing temperatures, Fig. 5.5 shows the Weibull distribution parameters of the deashing samples. The results showed that as the washing temperature increased, the breakdown strengths of MC-0.1–2.5, CH-0.05–1.5, and NX-0.05–1.5 all showed a slight increase followed by a significant decrease. Compared with the deashing samples at 20 °C, the breakdown performance of MC-0.1–2.5–40, CH-0.05–1.5–30, and NX-0.05–1.5–30 is significantly enhanced, with breakdown strengths of 629.1655.3 and 638.7 kV/mm, respectively, and Weibull shape parameters of 22.05, 18.87, and 18.69. However, when the washing temperature increased to 60 °C, the breakdown strength of deashing polypropylene significantly decreased, with breakdown strengths of MC-0.1–2.5–60, CH-0.05–1.5–60, and NX-0.05–1.5–60 being 591.3, 583.5, and 572.9 kV/mm, respectively. In general, the infiltration of washing solution into polypropylene increases with the increase of washing temperature, leading to more obvious swelling phenomenon of polypropylene. The ash content in it is more prone to chelation reaction with metal ion passivators, which significantly improves the washing and deashing effect [17]. However, as the infiltration degree of the washing solution increases, its destructive effect on the interface between the crystalline and amorphous regions of polypropylene intensifies, leading to an increase in structural defects in polypropylene and a decrease in breakdown performance. According to Fig. 5.5d, it can be observed that increasing the washing temperature reasonably can enhance the infiltration degree of the deashing solution, while not introducing too many structural defects in polypropylene, which further improves the breakdown performance of the deashing sample. When the washing temperature is too high, the destructive effect of the washing solution on the interface between the crystalline and amorphous regions of polypropylene gradually dominates, leading to a significant decrease in the breakdown performance of the ash removal sample.

In the solution washing and deashing method of polypropylene, adjusting the washing solution parameters, washing time, and washing temperature not only enhances the infiltration effect of the washing solution, promotes the chelation reaction between metal ion passivators and ash, but also reduces the destructive effect of the washing solution on the interface between the crystalline and amorphous regions of polypropylene, significantly enhancing the breakdown performance of the deashing sample. Compared with PP0, the breakdown strength of CH-0.05–1.5–40 increased by 34.9%.

(4) **Washing Time**

The above research found that the washing and ash removal effect of the solution of "metal ion passivation agent + anhydrous ethanol" still has significant non-uniformity, which may be related to the degree of infiltration of the washing solution

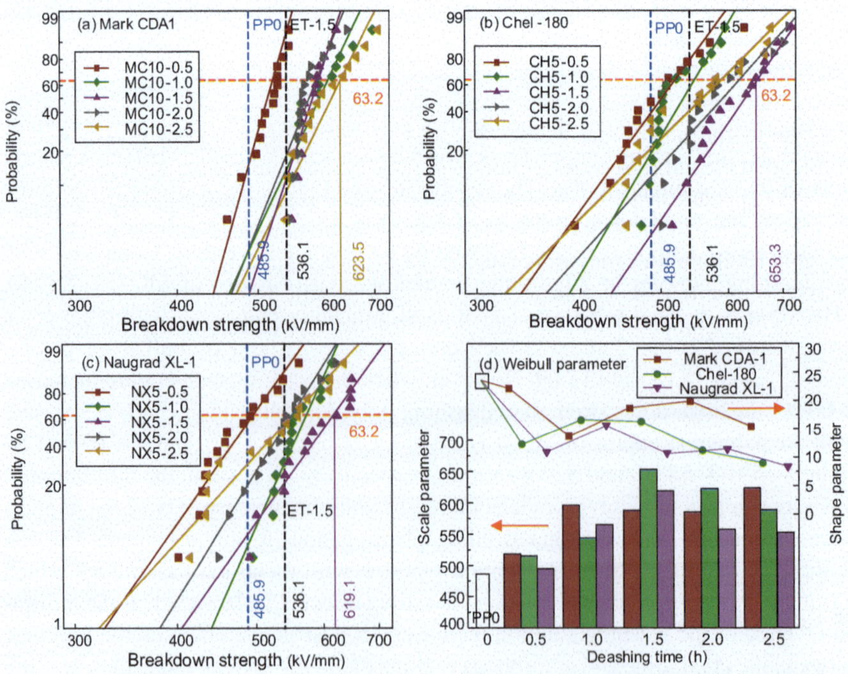

Fig. 5.5 Relationship between the **a**–**c** breakdown performance, **d** Weibull distribution parameters and the deashing time

into polypropylene. Normally, with the extension of washing time, the infiltration of washing solution into polypropylene gradually increases and eventually stabilizes. Therefore, in the process of solution washing and deashing, in order to promote the infiltration effect of washing solution into polypropylene, different deashing samples were prepared by adjusting the washing time, and breakdown performance tests were carried out. The results are shown in Fig. 5.6.

The results showed that under the optimal washing solution parameters, the breakdown strength of deashing polypropylene was further improved by adjusting the washing time. The breakdown strengths of MC-0.1–2.5, CH-0.05–1.5, and NX-0.05–1.5 were increased to 623.5, 653.3, and 619.1 kV/mm, respectively. Compared with MC-0.1, CH-0.05, and NX-0.05 before optimizing the washing time, the breakdown strengths were increased by 4.5%, 20.0%, and 9.5%, respectively. It is worth noting that when the washing time is short, the increase in breakdown strength of polypropylene after washing and deashing treatment is relatively small. The breakdown strengths of MC-0.1–0.5, CH-0.05–0.5, and NX-0.05–0.5 are 519.3, 518.4, and 494.6 kV/mm, respectively. As the washing time prolongs, the breakdown strengths of MC-0.1, CH-0.05, and NX-0.05 show a trend of first increasing and then decreasing. At shorter washing times, the deashing effect of polypropylene increases with the prolongation of washing time. On the one hand, the infiltration

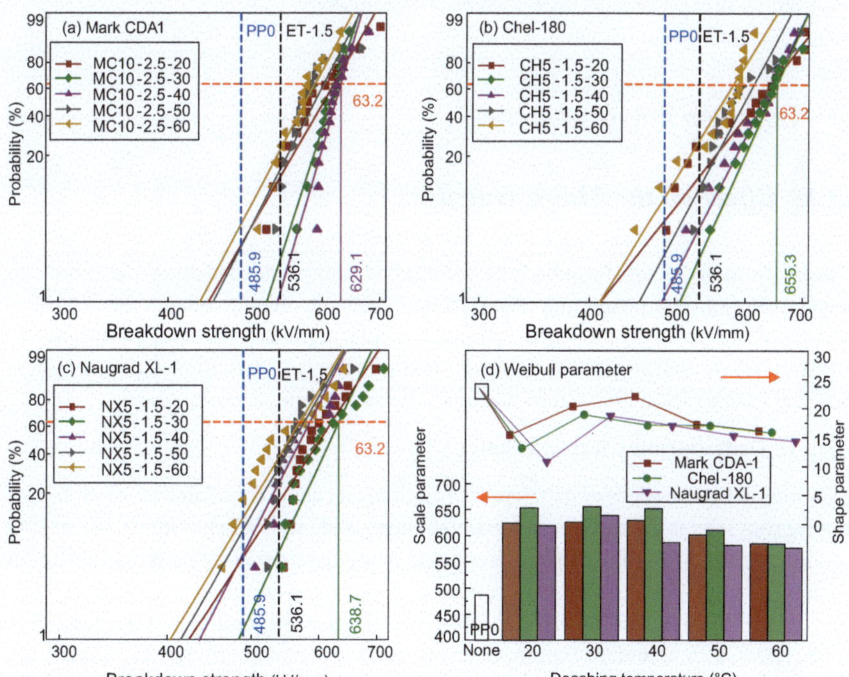

Fig. 5.6 Relationship between the **a**–**c** breakdown performance, **d** Weibull distribution parameters and the deashing temperature

effect of washing solution increases with the prolongation of washing time, leading to an increase in the swelling degree and free volume of polypropylene. Therefore, metal ion passivators are more likely to chelate with ash, promoting the migration and removal of ash [18]; On the other hand, with the extension of washing time, more ash can undergo chelation reaction with metal ion passivators, forming chelating structures, which increases the probability of ash removal. However, under prolonged washing time, the infiltration of washing solution can cause damage to the crystal/amorphous interface of polypropylene, introducing a large number of structural defects and reducing the breakdown performance of polypropylene. Therefore, a reasonable washing time is an important condition for improving the washing and ash removal efficiency of polypropylene.

The washing time has a significant impact on the washing and deashing effect of polypropylene. The optimal washing time should be selected during the solution washing and deashing process, mainly based on the following two factors: on the one hand, as the washing time prolongs, the infiltration of washing solution into polypropylene and the chelation reaction between metal ion passivators and ash are enhanced, which is conducive to improving the washing and deashing effect; On the other hand, when the washing time is too long, the infiltration of the washing solution will cause damage to the interface between the crystalline and amorphous regions of polypropylene, leading to an increase in structural defects and having an adverse effect on the breakdown performance of polypropylene.

5.3 Low Ash Polypropylene

5.3.1 Effect on the Microstructure

In order to analyze the influence of solution washing and deashing treatment on the physicochemical properties of polypropylene, this chapter selects the deashing polypropylene sample under the optimal washing parameters to carry out a series of physicochemical properties analysis, including ash distribution characteristics, chemical composition characteristics, and crystallization characteristics.

(1) **Ash Distribution Characteristics**

By analyzing the types of ash in polypropylene, it can be found that most of the ash in polypropylene is caused by the residual main catalyst ($TiCl_4$), with Ti ash in PP0 accounting for 85.8% of the total ash content. The ash removal efficiency of different deashing polypropylene is mainly reflected in the removal effect of Ti ash. The Ti ash content in ET-1.5, MC-0.1–2.5–40, CH-0.05–1.5–40, and NX-0.05–1.5–30 is 280.5, 190.5, 126.5, and 96.0 ppm, respectively, the removal efficiencies were 19.3, 45.2, 63.6, and 81.4%, respectively. Furthermore, it can be observed from Fig. 5.7b that there are also Fe, Mg, and Al ash contents present in the polypropylene. The ash content of Fe, Mg, and Al in PP0 is 25.5, 22.5, and 9.5 ppm, respectively. On the one hand, the ash removal efficiency of Mg and Al by washing with anhydrous ethanol is not ideal. The Mg and Al ash content of ET-1.5 is 20.1 and 8.1 ppm, respectively, and the removal efficiency is 10.7 and 8.4%. It is worth noting that anhydrous ethanol has a relatively good removal efficiency for Fe ash, as Fe ash mainly comes from the mechanical equipment or external environment in contact with polypropylene production and processing, making it easier to remove catalyst residues inside polypropylene [19]; On the other hand, the metal ion passivation agent washing and deashing treatment has a relatively ideal removal effect on Fe and Al ash content. The Fe ash content in MC-0.1–2.5–40, CH-0.05–1.5–40, and NX-0.05–1.5–30 is completely removed, while the Al ash content is 7.0, 0.0, and 0.0 ppm, respectively. It is worth noting that the Mg ash content in MC-0.1–2.5–40, CH-0.05–1.5–40, and NX-0.05–1.5–30 is relatively high, at 15.5, 16.0, and 8.5 ppm, respectively. In the polymerization reaction of propylene, the main catalyst TiCl4 is usually loaded on the $MgCl_2$ carrier. After polymerization, $MgCl_2$ will remain inside the polypropylene, making it difficult to remove it. Therefore, the removal efficiency of Mg ash is relatively poor. In addition, metal ion passivators exhibit certain differences in chelating ability towards different metal ions, which may also be the reason for the significant differences in ash removal efficiency for Ti, Fe, Mg, and Al [20].

(2) **Chemical Composition**

In the washing and deashing process of polypropylene, filtration and deashing treatment is used to achieve the removal of ash from polypropylene powder. When the filtration effect is not ideal, there may be residual metal passivators in deashing polypropylene, which has a significant impact on the macroscopic properties of polypropylene. Therefore, it is necessary to conduct research on the residual situation of metal ion passivators in deashing polypropylene.

Fourier transform infrared spectroscopy is commonly used to study the chemical composition and characteristic functional groups of substances. Fourier transform infrared spectroscopy is performed on deashing polypropylene, and the results are shown in Fig. 5.8. The results showed that the infrared spectra of all ash removed samples exhibited typical polypropylene characteristic peaks, with 2957 and 2871 cm^{-1} corresponding to the asymmetric and symmetric stretching vibration peaks of methyl ($-CH_3$), 2916 and 2843 cm^{-1} corresponding to the asymmetric and symmetric stretching vibration peaks of methylene ($-CH_2-$), 1460 and 1378 cm^{-1} corresponding to the bending vibration peaks of $-CH_2-$ and $-CH_3$, respectively, and 1157 cm^{-1} and 972 cm^{-1} corresponding to the rocking vibration peaks of $-CH_3$. In addition, it is worth noting that MC-0.1–2.5–40 and CH-0.05–1.5–30 did not show any new infrared characteristic peaks compared to PP0, indicating that their composition is only polypropylene, and there is no residual metal ion passivation agent after washing and deashing treatment, which has a good filtering effect. However, new infrared characteristic peaks appeared in NX-0.05–1.5–30 compared to PP0, with the newly added characteristic peak of 3700 cm^{-1} corresponding to the stretching vibration of hydroxyl (–OH), the characteristic peak of 3400 cm^{-1} corresponding to the stretching vibration of secondary amine (–NH–), and the characteristic peak

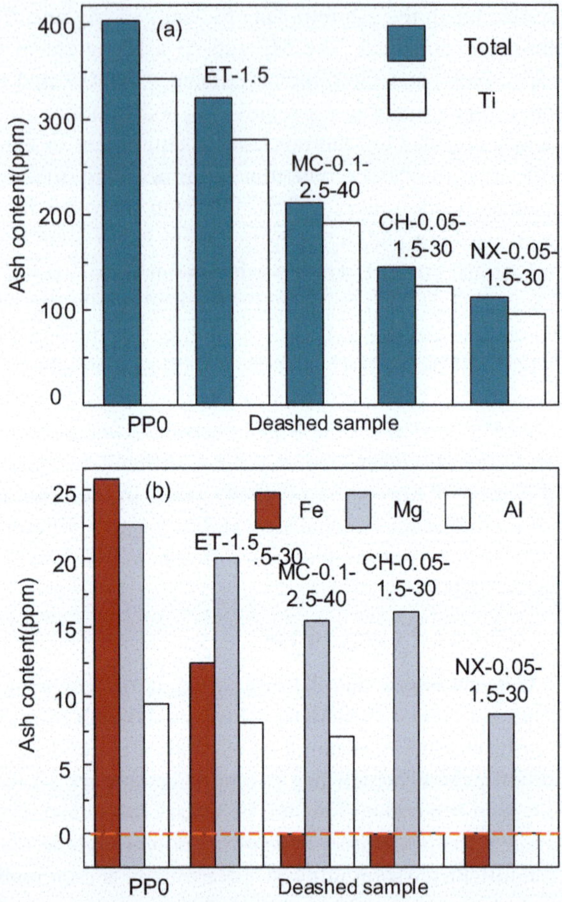

Fig. 5.7 **a** Total metal, Ti, **b** Fe, Mg and Al contents after optimizing the deashing parameters [19]. Copyright Clearance Center's RightsLink© License Number: 5812480798017

of 1740–1660 cm^{-1} corresponding to the stretching vibration of carbonyl. These characteristic groups all come from NaugradXL-1 used in the solution washing process, indicating poor filtration efficiency of NaugradXL-1 [21, 22]. Compared to MarkCDA-1 and Chelsea 180, NaugradXL-1 has a complex molecular structure, with longer molecular chains that are prone to local aggregation, resulting in residual NaugradXL-1 in deashing polypropylene. When the deashing sample is hot pressed, the aggregated NaugradXL-1 is highly likely to become a local defect in the thin film, thereby adversely affecting the dielectric properties.

(3) **Crystallization Characteristics**

Polypropylene is a typical semi crystalline polymer, and the presence of ash inevitably leads to changes in its condensed structure (crystallization characteristics), which has a significant impact on its dielectric properties. In order to study the evolution law of crystallization characteristics of delisted polypropylene, the cooling

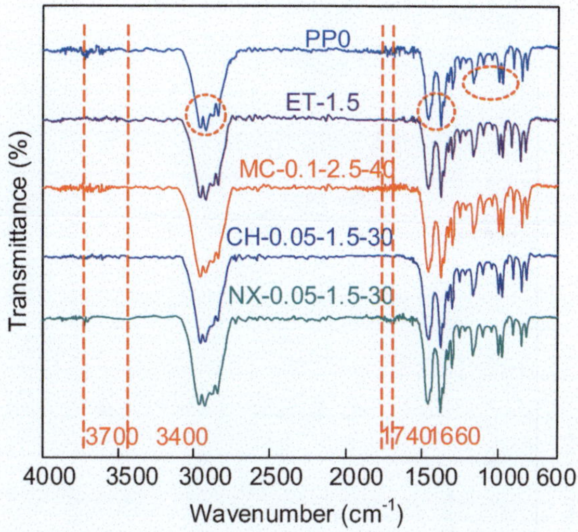

Fig. 5.8 Infrared spectrogram after optimizing the deashing parameters

crystallization process of the sample was observed using a polarizing microscope. The results are shown in Fig. 5.9, where the vertical axis represents the crystallization process of different delisted samples, and the horizontal axis represents the crystallization morphology at different stages of the crystallization process, including nucleation, growth, and completion stages.

According to Fig. 5.9, during the crystallization nucleation stage, PP0 exhibits a large number of evenly distributed nucleation sites. In contrast, the number of crystal nuclei in ET-1.5, MC-0.1–2.5–40, CH-0.05–1.5–30, and NX-0.05–1.5–30 significantly decreased. In the crystallization nucleation process of polypropylene, molecular chains will form locally ordered "crystal clusters" due to thermal motion at a certain moment. When the size of the "crystal clusters" increases to a certain value, they become primary crystal nuclei, which is the homogeneous nucleation process of polypropylene. In addition, heterogeneous nucleation occurs during the crystallization process of polypropylene, which is attributed to the smaller heterogeneous structures, such as ash content. During the crystallization process of polypropylene, ash can act as heterogeneous nucleation sites, thereby promoting the crystallization behavior of polypropylene [23, 24]. After washing and deashing polypropylene, its ash content significantly decreases, which to some extent weakens the heterogeneous nucleation phenomenon in the crystallization process of polypropylene, leading to a significant reduction in the number of crystal nuclei. It is worth noting that additives that can play a heterogeneous nucleation role in the melting and recrystallization process of polypropylene generally need to have a high melting point. Due to the low melting point range of NaugardXL-1 (170–180 °C), it does not play a significant role in heterogeneous nucleation during the crystallization process of polypropylene, and usually forms intergranular defects. Therefore, the residual NaugardXL-1 in NX-0.05–1.5–30 will form more intergranular defects during the matrix crystallization process, as shown in the area marked by the black solid line in the figure.

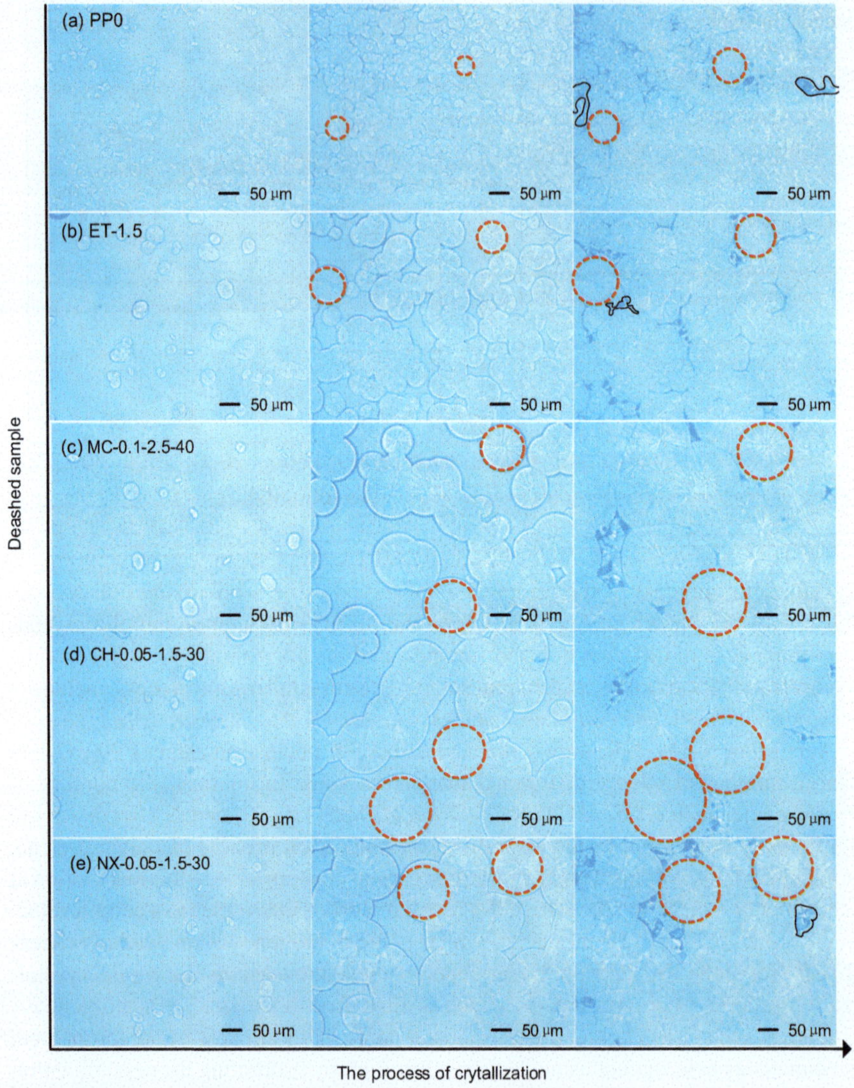

Fig. 5.9 Crystal morphology during the isothermal crystallization

After solution washing and deashing treatment, the spherical crystal size of polypropylene significantly increased and the number of spherical crystals significantly decreased. Among them, the spherical crystal size of PP0 was 60–80 μ m, and the spherical crystal size of ET-1.5 was 90–130 μ m. The spherical crystal size of the deashing polypropylene sample with metal ion passivation agent increased to 150–180 μ m. The solution washing and deashing treatment can significantly reduce the ash content of polypropylene, resulting in a reduction of heterogeneous nucleation sites and an increase in the distance between crystal nuclei in the deashing sample during the nucleation stage; During the crystal growth stage, the stacking and squeezing phenomena between adjacent spherulites decrease; In the complete crystallization stage, the spherical crystal size of deashing polypropylene increases. In the crystallization process of polypropylene, a portion of the ash can play a heterogeneous nucleation role, while the majority of the ash will aggregate at the boundaries of the spherulites with the growth of the spherulites, forming intergranular defects, which to some extent weakens the macroscopic dielectric properties of polypropylene [25, 26]. In order to quantitatively analyze the intergranular defects of the ash removal sample, the intergranular defect coefficient (M) was defined, and the calculation formula is as follows:

$$M = \frac{S_d}{S} \times 100\% \tag{5.1}$$

S and S_d respectively represent the total area and intergranular defect area observed by a polarizing microscope after complete crystallization of polypropylene. This article uses the defect recognition theory based on threshold segmentation to calculate the intergranular defect coefficient of polypropylene. In order to reduce calculation errors, 10 observation areas are selected for each sample for calculation, and the average value is taken as the final result. The intergranular defect coefficient of the sample is shown in Fig. 5.10. The results showed that the solution washing and deashing treatment could significantly reduce the intergranular defects of polypropylene, leading to a decrease in the intergranular defect coefficient. The M of PP0 was 4.27%, while the M of ET-1.5, MC-0.1–2.5–40, CH-0.05–1.5–30, and NX-0.05–1.5–30 decreased to 3.44, 1.79, 1.26, and 3.21%, respectively. During the crystallization process of polypropylene, most of the ash is excluded from the spherulites and aggregates into the gaps between the spherulites as the spherulites grow, forming intergranular defects. However, due to the significant decrease in ash content in deashing polypropylene, the intergranular defects during the crystallization process are reduced, and the intergranular defect coefficient decreases. It is worth noting that the residual NaugardXL-1 in NX-0.05–1.5–30 hinders the growth of spherulites during the crystallization process of the sample, and aggregates at the boundaries of spherulites to form intergranular defects. Therefore, NX-0.05–1.5–30 has a higher intergranular defect coefficient compared to other metal ion passivation agent deashing samples [27].

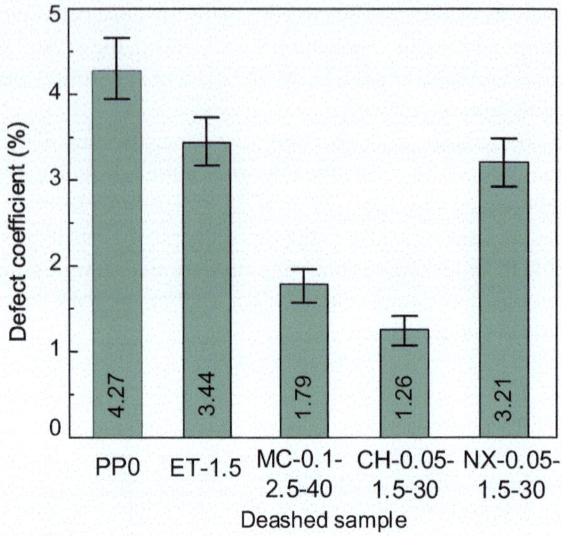

Fig. 5.10 Defect coefficient after the crystallization

5.3.2 *Effect on the Dielectric Properties*

In high-voltage DC capacitors, the dielectric properties of polypropylene film have a significant impact on the operational reliability and economy of the system, and are extremely important performance parameters. This chapter conducts dielectric performance tests on deashing polypropylene under optimal washing parameters, including breakdown strength, conductivity, dielectric constant, loss factor, and energy storage characteristics.

(1) **DC Breakdown Strength**

The breakdown performance of polypropylene film is an important parameter for measuring its withstand voltage level and is the foundation for stable operation of voltage DC capacitors. The breakdown strength of deashing polypropylene was tested under the optimal washing parameters, and the results are shown in Fig. 5.11. The results showed that within the testing temperature range, deashing polypropylene showed significantly enhanced breakdown performance compared to PP0, with the breakdown strengths of MC-0.1–2.5–40, CH-0.05–1.5–30, and NX-0.05–1.5–30 increased to 629.1–655.3, 585.3–604.4, and 428.9–472.6 kV/mm at 25, 55, and 85 °C, respectively. In addition, as the testing temperature increased, the breakdown strength of the polypropylene sample significantly decreased. The breakdown strength of PP0 at 25 °C was 485.9 kV/mm. When the temperature increased to 55 °C and 85 °C, the breakdown strength decreased by 11.9% and 24.5%, respectively. Generally speaking, under the action of an electric field, a large number of charge carriers are generated in polypropylene due to factors such as ash decomposition and electrode injection, which will migrate within the matrix and have an adverse impact on the breakdown performance of polypropylene. As the temperature increases, on the

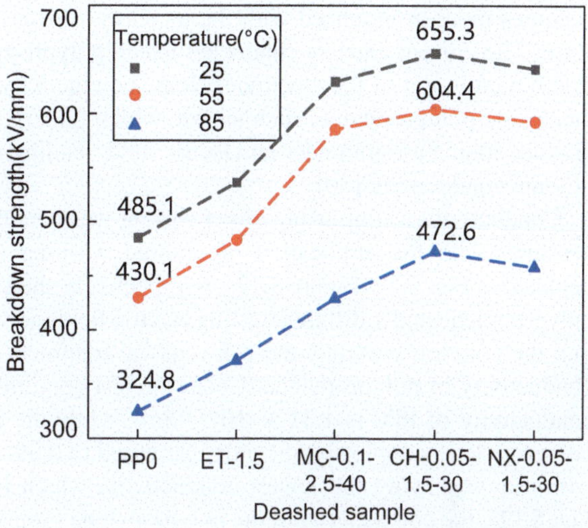

Fig. 5.11 Relationship between the breakdown strength and the temperature

one hand, the thermal motion of polypropylene macromolecular chains is enhanced, the intermolecular arrangement becomes loose, and the free volume of the matrix rapidly increases, leading to an enhanced directional migration behavior of charge carriers; On the other hand, the phenomenon of ash decomposition and dissociation intensifies, and the number of charge carriers in polypropylene increases. These two reasons lead to a significant weakening of the breakdown strength of polypropylene samples at high temperatures [28, 29].

It is worth noting that compared to pure polypropylene samples, the increase in breakdown strength of deashing samples gradually increases with the increase of testing temperature. When the testing temperature is 25, 55, and 85 °C, the breakdown strength of MC-0.1–2.5–40, CH-0.05–1.5–30, and NX-0.05–1.5–30 increased by 28.8–34.8, 31.4–40.5, and 32.1–45.5% compared to PP0, respectively. This indicates that using metal ion passivators to wash and deash polypropylene can enhance the temperature stability of breakdown strength.

(2) **DC Conductivity**

Conductivity is commonly used to characterize the conductivity loss of polypropylene, which is closely related to the heating and temperature rise in thin film capacitors. Generally speaking, the DC conductivity of polypropylene mainly includes electronic conductivity and ionic conductivity. The sources of electronic conductivity include intrinsic electron excitation and electrode electron emission, where intrinsic electron excitation refers to the phenomenon of electrons transitioning from the valence band to the conduction band and becoming free electrons. Due to the large bandgap width of polypropylene itself (~ 7 eV), the intrinsic excitation of electrons is extremely small. Electrode electron emission refers to the phenomenon of electrodes injecting electrons into the interior of a dielectric under high field conditions, which becomes more pronounced with the increase of external electric field and temperature. The current carriers conducted inside polypropylene usually come from the electron emission of the electrodes. Ionic conductivity mainly comes from impurities inside polypropylene, such as ash, which undergo thermal decomposition and produce ionic carriers under an electric field, leading to a significant enhancement of ionic conductivity [30].

Conductivity testing using direct current was conducted on polypropylene post-deashing with the application of optimal washing parameters, with the results depicted in Fig. 5.12. Figure 5.12a, b respectively show the conductivity of deashing polypropylene under different testing electric fields and temperatures. It can be found that the solution washing and ash removal treatment can significantly reduce the conductivity of polypropylene. Under the test conditions of 25 °C, 20 kV/mm, the conductivity of PP0 is 1.31×10^{-14} S/m, while the conductivity of the deashing polypropylene sample significantly decreased to $0.99–1.15 \times 10^{-14}$ S/m, with CH-0.05–1.5–40 having the lowest conductivity, which is 24.4% lower than PP0. In Fig. 5.12a, the conductivity of the polypropylene sample significantly increases with the increase of the test electric field. When the test electric field increases from 10 kV/mm to 20 kV/mm and 30 kV/mm, the conductivity of PP0 increases by 26.0% and 97.0%, respectively, while the conductivity of the deashing polypropylene sample increases by 13.8–24.1% and 25.0–83.8%, respectively. As the test electric field increases, on the one hand, the phenomenon of electrode electron emission is significantly enhanced, and the number of charge carriers injected into the polypropylene interior increases; On the other hand, the mobility of charge carriers is improved, which leads to a significant enhancement of the conductivity process of polypropylene under high fields. It is worth noting that the solution washing and deashing treatment can significantly reduce the ash content of polypropylene, leading to a decrease in the number of charge carriers generated by ash decomposition and a significant weakening of the carrier transport process, which to some extent enhances the electric field stability of polypropylene conductivity [31].

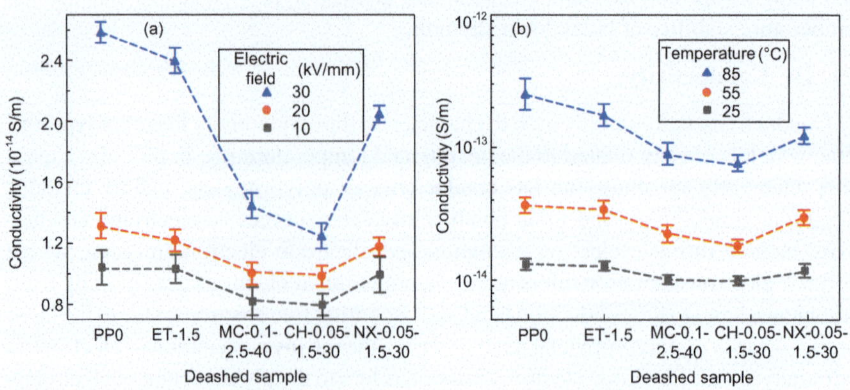

Fig. 5.12 Relationship between the DC conductivity and the **a** electric fields, **b** temperature

In Fig. 5.12b, the conductivity of the polypropylene sample significantly increased with the increase of testing temperature. When the testing temperature was raised from 25 °C to 55 °C and 85 °C, the conductivity of PP0 increased by 181.7% and 576.7%, respectively, while the conductivity of the ash removed sample increased by 83.8–153.4% and 294.6–313.7%, respectively. Generally speaking, as the temperature increases, the DC conductivity of polypropylene increases exponentially. As the temperature increases, on the one hand, the phenomenon of hot electron emission from the electrode becomes more significant; On the other hand, the dissociation phenomenon of polypropylene ash intensifies, and the number of charge carriers significantly increases. Therefore, at high temperatures, the carrier transport process in polypropylene is enhanced, and the conductivity significantly increases [32, 33]. It is worth noting that solution washing and deashing treatment can significantly enhance the temperature stability of polypropylene conductivity, which is attributed to the reduced ash content.

(3) **Energy Storage Density**

The energy storage density of polypropylene is of great significance for the operation economy and volume miniaturization of thin film capacitors. As a typical nonlinear material, the energy storage density of pure polypropylene is usually obtained through theoretical calculations. The theoretical energy storage density (U_e) is very close to the actual value, and the calculation formula is as follows:

$$U_e = \frac{1}{2} \times \varepsilon_0 \varepsilon_r E^2 \tag{5.2}$$

where ε_0 represents the vacuum dielectric constant, $\varepsilon_0 = 8.85 \times 10^{-12}$ F/m; ε R represents the relative dielectric constant of polypropylene; E represents the breakdown strength of polypropylene. Within the testing temperature range of 25–85 °C, the dielectric constant of polypropylene remains almost unchanged [34, 35]. The dielectric constant of the polypropylene sample at 25 °C was taken to calculate the theoretical energy storage density Ue, and the results are shown in Fig. 5.13. In Fig. 5.13, it is found that the solution washing and deashing treatment can significantly increase the theoretical energy storage density of polypropylene, which is mainly attributed to the significantly improved breakdown strength of the deashing polypropylene sample. When the temperature is 25, 55, and 85 °C, the U_e of PP0 is 2.31, 1.81, and 1.04 J/cm^3, respectively, while the Ue of the deashing sample increases to 2.81–4.15, 2.25–3.53, and 1.39–2.16 J/cm^3, and the energy storage density improvement rate is 21.6–79.7, 24.3–95.0, and 33.6–107.7%, respectively. It is worth noting that as the temperature increases, the increase in energy storage density of the deashing polypropylene sample increases, which is attributed to the enhanced breakdown performance and temperature stability of the deashing sample.

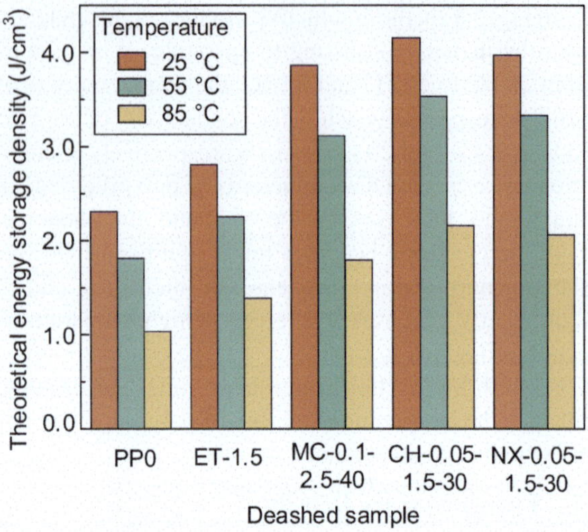

Fig. 5.13 Relationship between the theoretical energy storage density and the temperature [36]. Copyright Clearance Center's RightsLink© License Number: 5812481508684

5.4 Summary

Aiming at the insulation failure of polypropylene film for high-voltage DC capacitors, the ash in polypropylene was inhibited by solution washing and deashing and inhibitor micro-filling, and the synergistic improvement of breakdown, conductivity and energy storage performance was realized. By studying the evolution of physical and chemical properties and dielectric properties, the theoretical model of polypropylene ash inhibition method was established, and the microscopic evolution process of dielectric properties was revealed, which provided a theoretical and experimental reference for the development of high-quality polypropylene film for capacitors. The main conclusions of this paper are as follows:

Based on the ash removal method of polypropylene powder, a solution washing deashing scheme of 'metal ion passivator + anhydrous ethanol' was proposed. Under the appropriate washing parameters, the ash content of polypropylene can be reduced to 27.7% by the solution washing and deashing treatment, resulting in the breakdown strength of polypropylene film increased by 34.8%, the conductivity decreased by 24.4%, and the theoretical energy storage density increased by 79.7%. IDC experimental analysis shows that the solution washing deashing treatment reduces the number of shallow traps in polypropylene, which weakens the trapping and detrapping behavior of carriers. At the same time, due to the reduction of the number of carriers formed by ash decomposition, the charge transport process is weakened, which promotes the improvement of dielectric properties.

From the perspective of polypropylene breakdown and conductivity properties, the influencing factors of solution washing deashing effect were explored. The relationship between the washing parameters and the breakdown and conductivity properties of polypropylene was established around the type and concentration of the washing solution, the washing time and the washing temperature. The optimal solution washing and deashing scheme was obtained: 0.05 g/100 ml Chel-180 absolute ethanol solution was used to wash polypropylene at 40 °C for 1.5 h. The effects of particle size and ash content of polypropylene on the breakdown and conductivity properties after washing and deashing were studied. It was found that the deashing effect was the best when the particle size of polypropylene was 30 mesh, and the initial ash content had no significant effect on the deashing effect. The limitations of the solution washing deashing method are revealed. It is considered that the solution washing deashing method is highly dependent on the particle size of polypropylene, and the deashing effect has an upper limit.

References

1. H. M. Umran, F. Wang and Y. He, "Ageing: Causes and effects on the reliability of polypropylene film used for HVDC capacitor," *IEEE Access*, vol. 8, pp. 40413–40430, 2020.
2. J. Flicker, R. Kaplar, M. Marinella and J. Granata, "Lifetime testing of metallized thin film capacitors for inverter applications," 2013 IEEE 39th Photovoltaic Specialists Conference, Tampa, FL, USA, 2013, pp. 3340–3342.
3. W. Sarjeant, "Capacitors," *IEEE Trans. Dielect Elect. Insul.*, vol. 25, no.5, pp. 861–922, 1990.
4. B. X. Du, J. W. Xing, Z. Y. Ran, H. L. Liu, M. Xiao and J. Li, "Dielectric and energy storage properties of polypropylene by deashing method for DC polymer film capacitors," *IEEE Trans. Dielect Elect. Insul.*, vol. 28, no.6, pp. 1917–1924, 2021.
5. A. Schneuwly, P. Gröning, L. Schlapbach, P. Brüesch, M.W Carlen and R Gallay, "Temperature-dependent dielectric breakdown strength of oil impregnated polypropylene foils," *Mater. Sci. Eng. B*, vol. 54, no.3, pp. 182–188, 1998.
6. H. Li et al, "Electric field and temperature dependence of electrical conductivity in biaxially oriented polypropylene films," *IEEE Trans. Plasma Sci.*, vol. 42, no. 11, pp. 3585–3591, 2014.
7. M. Ritamäki, I. Rytöluoto and K. Lahti, "Performance metrics for a modern BOPP capacitor film," *IEEE Trans. Dielect Elect. Insul.*, vol. 26, no. 4, pp. 1229–1237, 2019.
8. M. Makdessi, A. Sari and P. Venet, "Metallized polymer film capacitors ageing law based on capacitance degradation," *Microelectron. Reliab.*, vol. 54, no. 9–10, pp. 1823–1827, 2014.
9. T. D. Huan, S. Boggs and G. Teyssedre, "Advanced polymeric dielectrics for high energy density applications," *Prog. Mater. Sci.*, vol. 83, pp. 236–269, 2016.
10. M. Nemamcha, J. P. Gosse and B. Gosse, "Discharges in all-PP film model capacitors," *IEEE Trans. Dielect Elect. Insul.*, vol. 28, no.3, pp. 315–323, 1993.
11. M. E. Stournara and R. Ramprasad, "A first principles investigation of isotactic polypropylene," *J. Mater. Sci.*, vol. 45, pp. 443–447, 2010.
12. H. V. Nguyen and T. H. Pham, Structural and electronic properties of defect-free and defect-containing polypropylene: a computational study by van der waals density-functional method, *Phys. Status Solidi B*, vol. 255, no. 3, 2018, Art. no. 1700036.
13. P. Karanja and R. Nath, "Electrical conduction in biaxially-oriented polypropylene," *J. Electrostat.*, vol. 31, no. 1, pp. 51–63, 1993.
14. Z. Osawa, "Role of metals and metal-deactivators in polymer degradation," *Polym. Degrad. Stabil.*, vol. 20, no. 3–4, pp. 203–236, 1988.
15. E. Blatt et al, "Comparative studies of the chelation of transition metal ions with metal deactivators," *Polym. Degrad. Stabil*, vol. 25, no. 1, pp. 19–29, 1989.
16. A. P. Mar′in, M. Bonora and Greci L, "Washing out oligomeric triazinic-hindered amine from polypropylene," *J. Appl. Polym. Sci.*, vol. 78, no.12, pp. 2158–2165, 2000.
17. L. Luciani and G. Nava, Process for purifying polypropylene: US, 4051313, 1977–09–27.

18. C. L. Willis and L. H. Slaugh, Method for removing catalyst residues from atactic polypropylene: US, 4316008, 1982–02–16.
19. X. B. Du, J. H. Zhang, M. Xiao, L. H. Liu and Z. Y. Ran, "Dielectric breakdown performance of deashed polypropylene for HVDC film capacitors," *IEEE Trans. Dielect Elect. Insul.*, vol. 29, no. 6, pp. 2209–2217, 2022.
20. Z. Osawa and K. Tsurumi, "Evaluation of some commercial copper deactivators by chemiluminescence," *Polym. Degrad. Stabil.*, vol. 26, no.2, pp. 151–160,1989.
21. N. S. Allen, M. J. Harrison and M. Ledward, "Thermal and photo-chemical degradation of nylon 6, 6 polymer: Part III—Influence of iron and metal deactivators," *Polym. Degrad. Stabil.*, vol. 23, no.2, pp. 165–174, 1989.
22. A. Mar′in, L. Greci and P. Dubs, Physical behavior of 3-aryl-benzofuran-2-ones (Irganox® HP-136) in polypropylene," *Polym. Degrad. Stabil.*, vol. 78, no. 2, pp. 263–267, 2002.
23. A. Xu, S. Roland and Colin X, "Physico-chemical characterization of the blooming of Irganox 1076® antioxidant onto the surface of a silane-crosslinked polyethylene," *Polym. Degrad. Stabil.*, vol. 171, 2020, Art. no. 109046.
24. S. V. Suraci, D. Fabiani, A. Xu, S. Roland and X. Colinet, "Ageing assessment of XLPE LV cables for nuclear applications through physico-chemical and electrical measurements," *IEEE Access*, vol. 8, pp. 27086–27096, 2020.
25. L. Petersson, J. Viertel, H. Hillborg, G. Zhang and T. C. M. Chung, "Electrical properties of polypropylene-bonded hindered phenol blends," *IEEE Trans. Dielectr. Electr. Insul.*, vol. 27, no. 2, pp. 590–596, 2020.
26. R. R. Xu, J. W. Xing, B. X. Du, M. Xiao and J. Li, "Improved breakdown strength of polypropylene film by polycyclic compounds addition for power capacitors," *Materials*, vol. 14, no. 5, 2021, Art. no. 1185.
27. J. G. Simmons and M. C. Tam, "Theory of isothermal currents and the direct determination of trap parameters in semiconductors and insulators containing arbitrary trap distributions," *Phys. Rev. B*, vol. 7, no. 8, 1973, Art. no. 3706.
28. J. A. Waynick, "The development and use of metal deactivators in the petroleum industry: a review," *Energy Fuels*, vol. 15, no. 6, pp. 1325–1340, 2001.
29. S. Morlat-therias et al, "Photochemical stabilization of linear low-density polyethylene/clay nanocomposites: Towards durable nanocomposites," *Polym. Degrad. Stabil.*, vol. 93, no. 10, pp. 1776–1780, 2008.
30. C. Parlak, Ö. Alver and M. Şenyel, "Vibrational spectroscopic study on some Hofmann type clathrates: M (1-Phenylpiperazine) 2Ni (CN) 4· 2G (M= Ni, Coand Cd; G= aniline)," *J. Mol. Struct.*, vol. 919, no. 1–3, pp. 41–46, 2009.
31. G. Zhang, C. Nam, L. Petersson, J. Jämbeck, H. Hillborg and T. C. Mike Chung, "Increasing polypropylene high temperature stability by blending polypropylene-bonded hindered phenol antioxidant," *Macromolecules*, vol. 51, no. 5, pp. 1927–1936, 2018.
32. Z. L. Li, Z. Y. Zhong and B. X. Du, "Dielectric relaxation and trap-modulated DC breakdown of polypropylene blend insulation," *Polymer*, vol. 185, 2019, Art. no. 121935.
33. A. Nasir, T. Yasin, A. Islam, "Thermo-oxidative degradation behavior of recycled polypropylene," *J. Appl. Polym. Sci.*, vol. 119, no. 6, pp. 3315–3320, 2011.
34. Y. Zhou, C. Yuan, C. Y. Li, P. F. Meng, J. Hu, Q. Li and J. L. Liang, "Temperature dependent electrical properties of thermoplastic polypropylene nanocomposites for HVDC cable insulation," *IEEE Trans. Dielect Elect. Insul.*, vol. 26, no. 5, pp. 1596–1604, 2019.
35. M. Xiao, M. D. Zhang, B. X. Du, Z. Y. Ran, H. L. Liu and Y. W. Qin, "Breakdown strength improvement of polypropylene by introducing long-chain branched structures for film capacitors," *J. Phys. D: Appl. Phys.*, vol. 55, no. 24, 2022, Art. no. 245501.
36. B. X. Du, J. H. Zhang, M. Xiao, H. L. Liu, Z. Y. Ran and J. W. Xing "Improvement of dielectric properties of polypropylene films for capacitors based on metal-ash suppression," *Mater. Today Commun.*, vol. 32, 2022, Art. no. 104189.

Chapter 6
Long-Chain Branching

6.1 Introduction

Polypropylene metallized film capacitors are the core devices of HVDC converter valves, electric vehicle DC/DC converters, pulse power supplies and other devices. In recent years, electrical equipment has been developing towards large capacity, high power and miniaturization, and the operating temperature of capacitors has been rising. At high temperature, the conductivity loss of polypropylene film increases sharply and the breakdown strength decreases significantly, which has become the bottleneck problem restricting the high temperature application of metallized film capacitor.

The molecular chain slip and chain segment relaxation of PP at high temperature are important reasons for the degradation of PP properties. Adjusting the molecular chain characteristics is an important method to improve the dielectric properties of PP at high temperature. The long chain branches in long chain branched polypropylene can enhance chain entanglement, increase frictional resistance, inhibit the movement and slip of molecular chains, and improve the structural stability. In addition, the long-branched chain can promote the heterogeneous nucleation, regulate the crystal shape and number, and limit the carrier migration in the polypropylene film under high temperature and strong electrical field.

In the process of preparing long chain branched polypropylene, the free radicals generated are generally stabilized by adding functional monomers to inhibit the degradation reaction and promote the formation of long chain branched polypropylene. It is an important method to improve the dielectric properties of polypropylene films by introducing polar groups to form deep traps and enhance the carrier trapping ability. Therefore, the introduction of long branched chains and functional monomers containing polar groups can enhance the stability of molecular structure and enhance the carrier trapping ability, and improve the dielectric properties of polypropylene films from two aspects.

B. Du and M. Xiao, *Polypropylene Film for HVDC Capacitors*,
https://doi.org/10.1007/978-981-96-3029-5_6

The long chain branching structure has two different effects on crystal properties of polypropylene. On the one hand, long and long branched chains can nucleate during crystallization, resulting in higher initial nucleation density and promoting heterogeneous nucleation. On the other hand, the enhancement of molecular chain entanglement limits the movement of molecular chains and inhibits the growth of spherulites. There is a competitive relationship between the two influences of long chain branches on crystallization, and it is difficult to achieve the ideal control effect. Therefore, the dielectric properties of long chain branched polypropylene can be further improved by regulating the crystallization of the polypropylene.

This chapter introduces the preparation methods of long chain branched polypropylene, including reactor synthesis method and post-reactor method. The effect of long chain branching modification on the molecular chain structure and aggregation structure of polypropylene was discussed. The effects of the introduction of long branched chain structure, the grafting of functional monomers and the crystallization control on the dielectric properties of polypropylene films were discussed. Finally, the problems faced by long chain branched modification in the application of capacitor polypropylene film are summarized, and its future development direction is prospected.

6.2 Long-Chain Branching Method

6.2.1 Basic Principle

Long-chain branched polypropylene (LCBPP) is a modified polypropylene (PP) with more than 2.5 times entangled molecular weight long-chain branched structure introduced into the main chain of linear PP. The molecular structures of linear PP and LCBPP are shown in Fig. 6.1. The main preparation methods of long chain branched polypropylene are post-reactor synthesis and reactor method.

The preparation of LCBPP directly by polymerization reaction in the reactor is the reactor synthesis method. The reactor synthesis method includes macromolecular monomer method and direct polymerization method.

According to the "terminal branching" mechanism, the long-branched chain structure in reactor synthesis is formed by inserting vinyl-terminated macromolecular

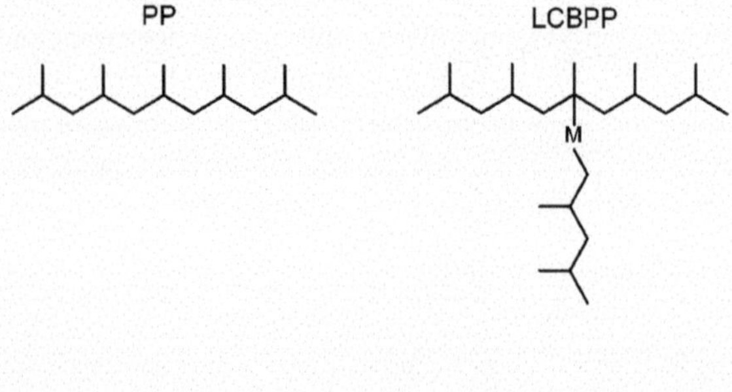

Fig. 6.1 Molecular structures of the linear PP and the LCBPP

monomers into the polymer growth chain [1]. According to the generation method of macromolecular monomer, macromolecular monomer method can be divided into in situ generation of macromolecular monomer method and additional macromolecular monomer method. In the method of in situ macromolecular monomer formation, the polymer chain with vinyl ends is generated in the original position of the polymerization system, and the polymer chain is directly copolymerized with propylene to produce LCBPP. The rac-Me2Si (2-Me-4-Ph-1-Ind)$_2$ZrCl$_2$ was used as catalyst to form PP chains with vinyl ends, and the macromolecular monomer was inserted into the polymer growth chain to synthesize LCBPP [2]. For the additional macromolecular monomer method, the polymer chain containing ethylene terminal structure with more than 2.5 times the entangled molecular weight is prepared separately, and then it is added to the reaction system for polymerization reaction with propylene. Macromolecular monomers containing vinyl were prepared by continuous reaction stirring system using CGC-Ti as catalyst. Then, under the catalysis of rac-Me$_2$Si (2-MeBenz[e]Ind)$_2$ZrCl$_2$(MBI), the macromolecular monomer was polymerized with propylene to obtain LCBPP [3].

Direct polymerization method refers to the synthesis of LCBPP by polymerization of propylene and non-conjugated α, ω-diolefin under the action of catalyst. The principle of direct polymerization includes: propylene and diolefin copolymerize to produce polyolefin containing unsaturated vinyl; H-type LCBPP was synthesized by polymerization of polyolefin and diolefin [4]. LCBPP was prepared by polymerization of 1,7-octodiene and 1,9-dodecene with propylene under the catalysis of rac-Me$_2$Si(2-Me-4-phenyl)$_2$ZrCl$_2$/modified methylaluminoxane (MMAO) [5].

The reaction principle of LCBPP prepared by post-reactor method includes: (1) H atom on tertiary carbon of PP is captured by high-energy electron or initiator to form tertiary carbon free radical. (2) Unstable tertiary carbon free radical β fracture. (3) The fractured molecular chain is coupled with the main chain to form LCBPP [6]. Since the β fracture of the secondary carbon free radical is faster, and the coupling reaction between the broken molecular chain and the main chain is relatively slow, functional monomers (such as styrene, maleic anhydride, etc.) can be added in the reaction process to stabilize the generated free radicals, in order to inhibit the degradation reaction [7]. The reaction principle is shown in Fig. 6.2. The post-reaction methods mainly include high energy radiation method, melting reaction method, solid phase grafting method and so on.

High energy radiation method is to stimulate polymer molecules through high energy radiation and cause molecular ionization, decomposition of positive and negative ions and charge neutralization, and then lead to various chemical reactions, especially the free radical reaction is the most common. Radiation sources used for high-energy radiation need to have enough energy to penetrate the radiated PP, ionize the molecular structure, and excite the atomic structure simultaneously without affecting the nucleus. At the same time, the radiation dose should be strictly controlled to prevent PP degradation and crosslinking caused by excessive radiation dose. Electron radiation and X, γ and other radiation are more commonly used sources of radiation. It was found that when LCBPP was prepared by electron irradiation, some long branched chains could be formed on the main chain of polypropylene

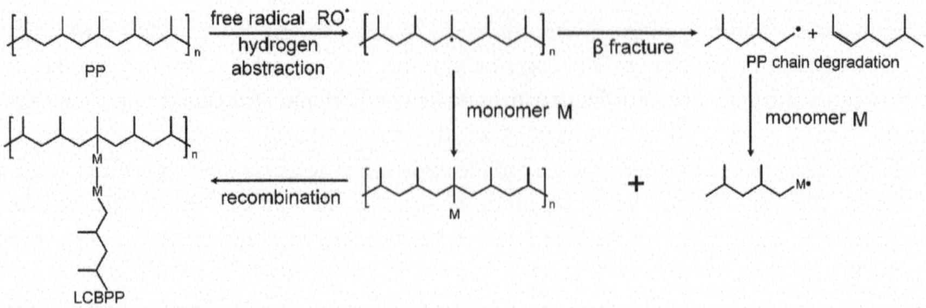

Fig. 6.2 Reaction principle of LCBPP preparation

under low dose irradiation when functional monomers were present, and the modified polypropylene had strain hardening effect during uniaxial stretching, thus LCBPP was prepared [8, 9]. This is due to the fact that functional monomers are linked to macromolecules after excitation to form branched structures.

Melt reaction processing refers to the use of reaction processing equipment (such as reaction extruder, mixer, etc.) to melt blending reaction grafting PP with other materials to obtain LCBPP method. In addition to the base resin polypropylene, the system generally includes initiators, graft monomers and some other auxiliaries. Peroxide initiators have been widely used in initiating polymer grafting reactions. Common peroxide initiators include benzoyl peroxide (BPO), diisopropylbenzene peroxide (DCP) and so on. In the preparation of long-chain branched polypropylene, 2, 5.dimethyl-2, 5.di-tert-butylhexane peroxide was used as initiator, and pentaerythritol triacrylate (PETA) was used as graft monomer [10]. The results show that the long chain branched structure has been formed in the polypropylene system, and the long chain branched structure is increasing with the increase of monomer, and the efficiency of monomer conversion to long chain branched structure is not more than 20%.

Solid phase peroxide treatment method refers to the peroxide treatment of solid or between solid and molten polymer at a temperature below the melting point of the polymer, so that it produces a long-branched chain structure. The peroxide used to initiate the branching reaction needs to have a suitable decomposition temperature, decomposition rate and suitable half-life. Himont [11] and Chisso [12] prepared LCBPP by initiating a branch reaction of solid phase PP with peroxide at low temperature, which has a high melt strength and no gel generation.

In general, the post-reactor method uses commercial PP as raw material to form long chain branched structure through free radical reaction or group reaction, so as to obtain LCBPP. The post-reactor method is suitable for industrial production because of its advantages of simple operation and low cost. Most of the LCBPP on the market are produced by the post-reactor method.

6.2.2 *Long-Chain Branched Polypropylene*

(1) Microstructure

Long-chain branches can enhance molecular entanglement and affect the rheological and thermodynamic properties of PP. The relationship between tensile force and tensile rate (V) of H-type LCBPP prepared by direct copolymerization [13, 14] and PP at 200 °C is shown in Fig. 6.3. The melt strength corresponds to the tensile force when the melt breaks, reflecting the resistance of the melt to tensile deformation. When the tensile rate is small, the tensile force of PP increases slowly with the increase of the tensile rate. When the tensile rate is high, the melt viscous flow occurs, and the curve appears a plateau. With the further increase of the tensile rate, the curve appears unstable oscillation until the melt breaks. Different from linear PP, the tensile force of LCBPP increases rapidly with the increase of the tensile rate, and unstable oscillation occurs at a relatively low tensile rate. The tensile resistance of LCBPP is significantly enhanced, and the melt strength is 16.3 cN higher than that of linear PP. Compared with PP, the friction resistance of LCBPP unentanglement under tensile force is greater, and its ability to resist tensile deformation is stronger, so themelt strength is higher [16, 17].

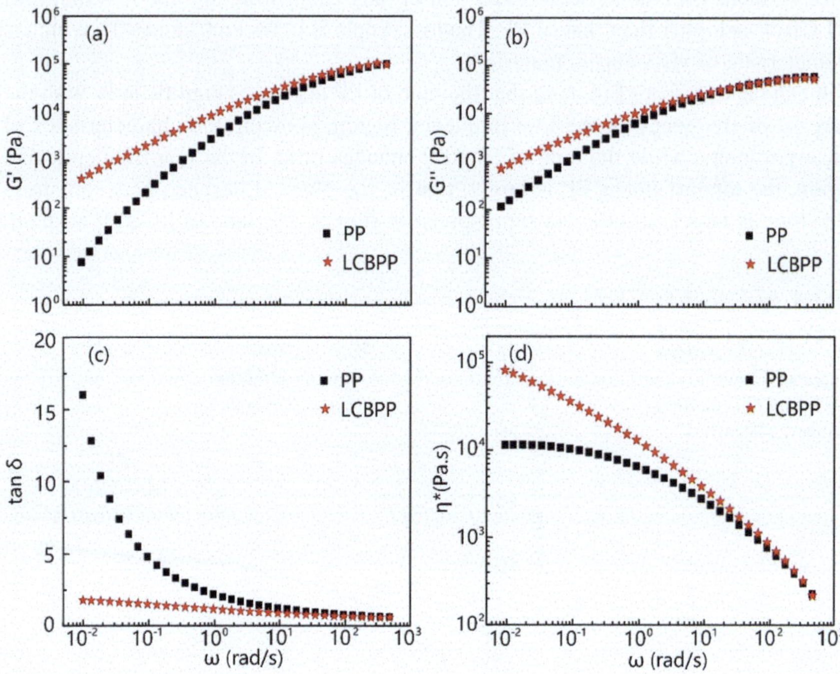

Fig. 6.3 Relationship between the tensile force and the tensile rate at 200 °C [15]. Copyright Clearance Center's RightsLink© License Number: 1495404-1

The relationship curves between the energy storage modulus (G′), loss modulus (G″), loss Angle tangent (tanδ), complex viscosity (η*) and frequency of PP and LCBPP in the range of 0.01–500 rad/s frequency (ω) at 200 °C are shown in Fig. 6.4. From the observations in Fig. 6.4a, it is evident that the energy storage modulus (G′) of PP exhibits a prominent peak at the lower end of the frequency spectrum, demonstrating a linear correlation with ω2. This behavior is characteristic of linear polymers and reflects their inherent structural linearity under these conditions. In contrast to PP, the energy storage modulus (G′) and slope of LCBPP are noticeably reduced in the low-frequency region of Fig. 6.4a, indicating a departure from typical end-rheological behavior. The relationship between G′ and ω2 for LCBPP shows deviations from the linear pattern observed in linear polymers, suggesting nonlinear structural characteristics in this material. Linear PP exhibits a short relaxation time, enabling it to untangle effectively at low frequencies. In contrast, LCBPP requires a longer relaxation time due to increased interchain entanglement. Consequently, LCBPP does not untangle easily at low frequencies, resulting in a higher elasticity of the melt compared to linear PP. This difference underscores the influence of molecular structure on the material's rheological properties, particularly in terms of relaxation behavior.

It can be seen from Fig. 6.4b that the energy storage modulus of PP in the low frequency region is significantly lower than the loss modulus, showing the characteristics of linear PP. Due to the introduction of long chain structure, the loss modulus of LCBPP is higher than that of PP. The loss Angle is between 0° and 90°, which is characteristic of viscoelastic materials.

It can be seen from Fig. 6.4c that the tanδ of PP decreases significantly with the increase of frequency in the low frequency region, showing the characteristics of linear polymers, while the tanδ of LCBPP changes little. In the low frequency end region, the tanδ of linear PP is larger, that is, the phase difference between stress and strain is larger, and the elastic response is slower. The tanδ of LCBPP is small,

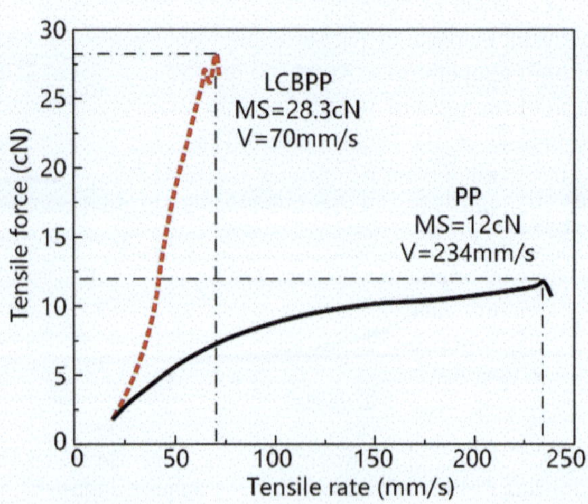

Fig. 6.4 **a** G′ ~ ω, **b** G″ ~ ω, **c** tan δ ~ ω, **d** η* ~ ω at 200 °C [15]. Copyright Clearance Center's RightsLink© License Number: 1495404-1

which indicates that the melt elasticity of LCBPP is large. The long-branched chain structure in LCBPP enhances interchain entanglement, resulting in longer relaxation time and smaller loss Angle tangent in the low frequency end region.

It can be seen from Fig. 6.4d that, due to the linear molecular chain structure of PP, a Newtonian plateau region appears in the curve at low frequencies. The complex viscosity of LCBPP in the low frequency region is higher than that of linear PP due to the introduction of long branched chain structure, and the Newtonian plateau region disappears, showing significant shear thinning behavior. The long-branched chain structure with more than 2.5 times entangled molecular weight in LCBPP leads to the enhancement of chain entanglement, the increase of molecular chain relaxation time, and the obvious thinning behavior of melt shear.

The tensile rheological behavior of linear PP and LCBPP showed obvious differences. Figure 6.5 shows the temporal curve of the transient tensile viscosity (ηE) of PP and LCBPP at the tensile rates of 0.01 s^{-1}, 0.1 s^{-1} and 1 s^{-1}, respectively. It can be seen from Fig. 6.5a that the transient tensile viscosity of linear PP first increases and then decreases with time. At the beginning of stretching, the deformation of PP is small and the entanglement between the molecular chains hinders the orientation movement of the molecular chains, and the transient tensile viscosity increases. With the increase of tensile time, the deformation of PP increases, the molecular chain unentanglement occurs, and the transient tensile viscosity decreases. In addition, there is little difference in the growth rate of the transient tensile viscosity of PP with time at different tensile rates. At the higher tensile rate, the faster unentanglement rate of the molecular chain leads to the decrease of the transient tensile viscosity of PP at an earlier time.

As can be seen from Fig. 6.5b, the tensile viscosity of LCBPP increases with the increase of drawing time. When the tensile deformation of LCBPP increases, the long-branched chain structure shows tensile orientation and entanglement with the main chain, which increases the frictional resistance between molecular chains and leads to a significant increase in the transient tensile viscosity, showing obvious

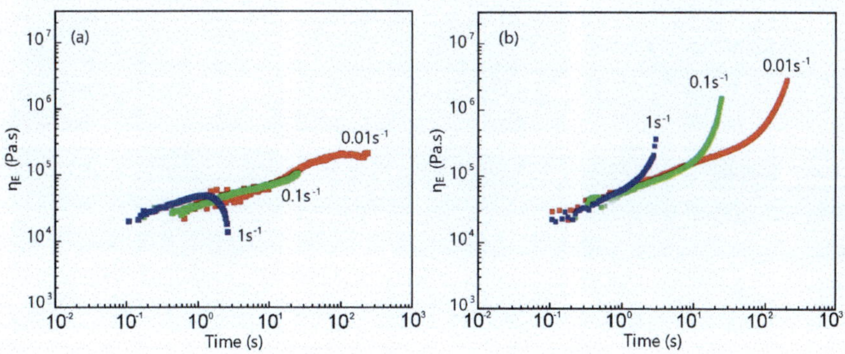

Fig. 6.5 Extensional rheological curve of the **a** PP and the **b** LCBPP [15]. Copyright Clearance Center's RightsLink© License Number: 1495404-1

tensile strain hardening behavior. When the tensile rate increases, LCBPP shows strain hardening in a relatively short time.

The crystallization properties of PP were affected by long chain branching modification. LCBPP has a branch chain length almost equal to the main chain, and its motility is strong, so it can be discharged into the lattice to promote heterophase nucleation. The crystal morphology of LCBPP prepared by PP and reactor method during the crystallization process is shown in Fig. 6.6. With the decrease of temperature, the crystallization process of the film underwent three stages: nucleation, crystal growth and crystal equilibrium. The crystal nucleus of PP is formed by molecular chain folding. It can be seen from the figure that the crystal nucleus density of PP is low, the spherulite size of PP is large and the boundary between spherulites is clear. Compared with linear PP, the initial nucleation density of LCBPP is higher and crystallization begins at higher temperature due to the heterogeneous nucleation of long branched chains. The crystallization temperature of LCBPP is 6.8 °C higher than that of PP, and the crystallinity is 2.3% higher. In addition, the large number of crystal nuclei and the high melt viscosity of LCBPP cause spherulites to reach a stable state at a small size, which hinders the growth of spherulites [18]. The size of spherulite is small, and a large number of overlapping microcrystals are formed in the film, and the boundary between spherulites is fuzzy.

(2) **Dielectric Properties**

Dielectric constant is a macroscopic parameter to characterize polarization, which can reflect the ability of dielectric to store charge under the action of electric field, and is an important parameter for insulating dielectric of capacitor. The dielectric constants and losses of PP and LCBPP are shown in Fig. 6.7a. The dielectric constant of the films remained stable with the increase of frequency, and the dielectric constant of LCBPP was slightly lower than that of linear PP. Compared with linear PP, the

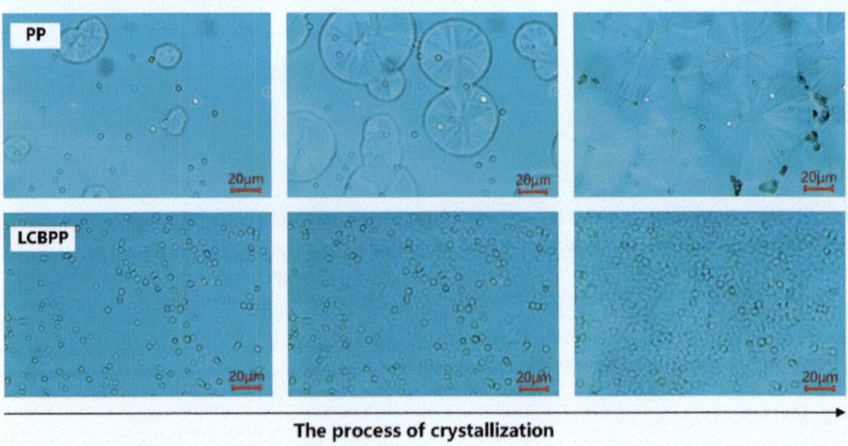

Fig. 6.6 Crystallization morphology of the PP and the LCBPP [15]. Copyright Clearance Center's RightsLink© License Number: 1495404-1

higher crystallinity and spherulite density in LCBPP lead to the reduction of internal defects, the reduction of interfacial polarization, and therefore the reduction of dielectric constant. The dielectric loss Angle tangent of LCBPP is slightly lower than that of PP, which is because the molecular chain motion of LCBPP with higher crystallinity is weak and the internal defects of the film are less, which reduces the relaxation polarization and interface polarization in the film. By comparing the dielectric constant and loss of linear PP and LCBPP, it is found that the long branch chain structure has little effect on the polarization of PP.

In the DC electric field, the dielectric loss in the dielectric is mainly conductance loss, and the conductive loss of the thin film can affect the charge and discharge efficiency of the capacitor. As shown in Fig. 6.7b, the conductivity of polypropylene film increases significantly with the increase of temperature. When the ambient temperature rises from 25 to 125 °C, the DC conductivity of PP film rises from 7.99×10^{-17} S/m to 7.40×10^{-14} S/m, an increase of nearly three orders of magnitude. Compared with linear PP, the conductivity of LCBPP is relatively low at high temperature. At the ambient temperature of 125 °C, the conductivity of LCBPP film is 1.76×10^{-14} S/m, which is about 1/4 of PP. The strong entanglement between molecular

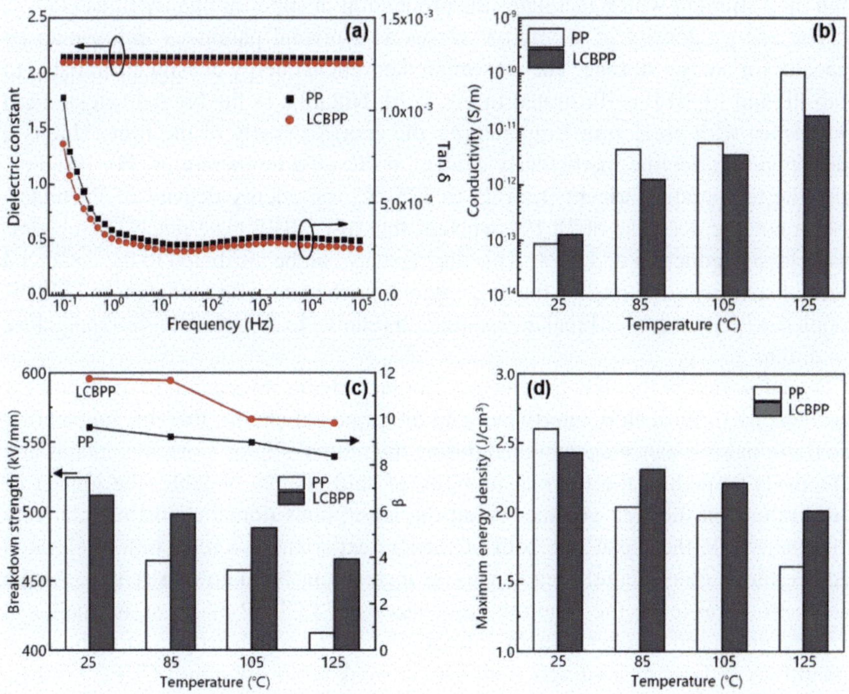

Fig. 6.7 **a** Relationship between the relative permittivity, the dissipation factor and the frequency, **b** relationship between the conductivity and the temperature, **c** relationship between the breakdown strength and the temperature, **d** relationship between the maximum energy density and the temperature [15]. Copyright Clearance Center's RightsLink© License Number: 1495404-1

chains in LCBPP restricts the movement and mutual slip of molecular chains at high temperature, thus improving the structural stability and reducing the conductivity at high temperature.

The energy density and operational lifespan of capacitors are significantly influenced by the breakdown strength of the films. As depicted in Fig. 6.7c, the breakdown strength of the film demonstrates a decreasing trend as the temperature rises. For instance, when the temperature elevates from 25 to 125 °C, the breakdown strength of PP experiences a notable decline of 112.1 kV/mm. This decline can be attributed to the heightened thermal motion of the molecular chain within the amorphous region of PP, leading to a reduction in breakdown strength. Upon comparing the breakdown characteristics of PP and LCBPP, it becomes evident that the breakdown strength of LCBPP exhibits relatively minimal decrease with the rise in temperature. At an ambient temperature of 125 °C, the breakdown strength of LCBPP film reaches 465.3 kV/mm, marking a substantial 12.9% increase over PP. The long-branched chain structure of LCBPP serves to enhance thermal stability by impeding the thermal movement of the molecular chain. Additionally, the mean free path of charge carriers in LCBPP remains relatively restricted at elevated temperatures. Moreover, the shape parameters of LCBPP tend to be slightly higher than those of linear PP, owing to the high melt strength which facilitates the production of superior quality films [19].

The energy density of a polymer serves as a pivotal parameter delineating its capacity for energy storage. The maximum theoretical energy density attributable to both PP and LCBPP is illustrated in Fig. 6.7d. Notably, as the breakdown strength diminishes with escalating temperatures, the energy density of the film exhibits a corresponding decline, particularly evident at elevated temperatures. For instance, with the temperature ascent from 25 to 125 °C, the energy density of PP undergoes a notable decrease of 38.1%, whereas that of LCBPP experiences a comparatively lesser reduction of 17.3%. This discrepancy can be attributed to the enhanced thermal stability conferred by the long-chain branched structure inherent in LCBPP, coupled with its superior breakdown strength relative to PP under high-temperature conditions.

As delineated in Fig. 6.8, the elongated branch chains inherent in LCBPP possess the capacity to intricately intertwine with the principal chains, thereby augmenting the frictional resistance encountered during the mutual slip of molecular chains and effectively impeding the thermal mobility of said chains. Notably, the impact of temperature on the free volume within the amorphous domain remains relatively minimal, while the mean free path of carriers experiences a reduction at elevated temperatures, consequently leading to an increase in breakdown strength. At the macrostructural level, the spherulite dimensions of LCBPP manifest as smaller in size and greater in number, thereby intensifying the spatial density existing between spherulites and correspondingly diminishing weak regions within LCBPP. Furthermore, the superimposition of microcrystals disrupts the continuity of carrier transport pathways, thereby mitigating the accumulation of carriers particularly under high-temperature conditions. The fortification of molecular entanglement alongside the augmentation of spherulite density emerges as pivotal factors contributing to the amelioration of dielectric properties exhibited by PP at elevated temperatures.

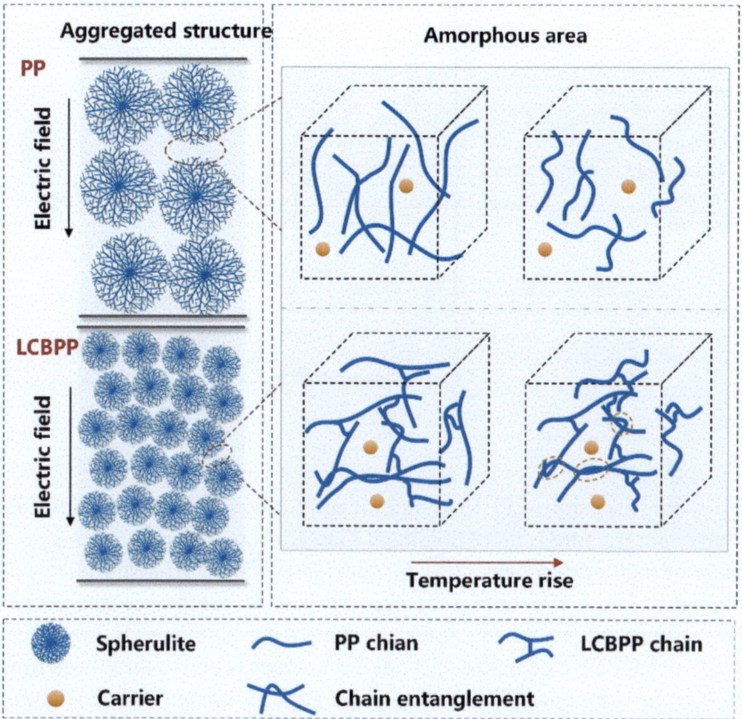

Fig. 6.8 Long-chain branched structure on the high temperature dielectric properties [15]. Copyright Clearance Center's RightsLink© License Number: 1495404-1

6.3 Long-Chain Branched Polypropylene Optimizing

6.3.1 Functional Monomers Doping

In the reaction process of LCBPP prepared by the post-reactor method, there exists the competition of β chain breaking and branching of PP main chain. Generally, by adding some functional monomers (such as styrene, acrylamide, etc.) to the free radical reaction, it combines with PP tertiary carbon free radicals, stabilizes the free radicals that have been formed through conjugation, inhibits the degradation of PP, and thus promotes the branching reaction of PP [7]. In addition, functional monomers containing polar groups or benzene ring structure can form deep traps and reduce the number and mobility of carriers inside the material [20].

The complex viscosity and storage modulus of the LCBPP melt prepared with Furfuryl Sulfide (FS) as the functional monomer (0.05, 0.1 and 0.15 wt%) and diamino peroxide (DCP) as the initiator (0.02 wt%) are shown in Fig. 6.9. The rheology of PP05 was significantly enhanced, indicating the formation of LCBPP.

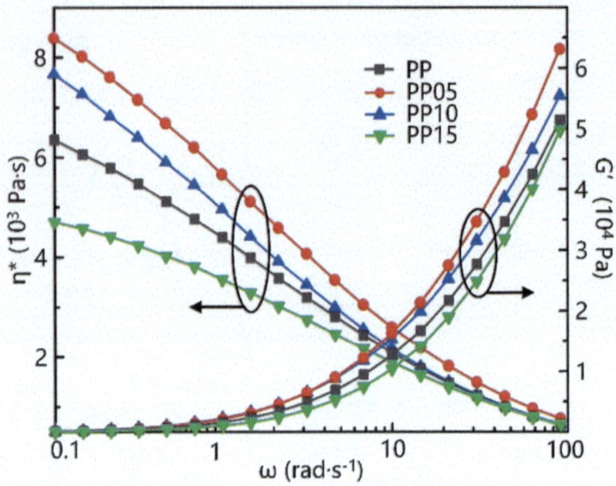

Fig. 6.9 Relationship between the complex viscosity, the storage modulus and the angular frequency at 190 °C [21]. Copyright Clearance Center's RightsLink© License Number: 5811970073314

The zero shear viscosity of PP15 is significantly lower than that of pure PP, indicating that chain fracture is more dominant than grafting. Degradation may occur in the process of melt grafting, resulting in a decrease in melt viscosity.

As shown in Fig. 6.10a, taking the pure PP film as a reference, the permittivity of LCBPP films with FS as functional monomer are significantly improved. The permittivity of LCBPP with 0.05 wt% FS is 2.33, which is higher than the permittivity of 2.195 measured by pure PP film. The polar graft monomer FS contributes to the improvement of the permittivity.

As shown in Fig. 6.10b, compared with pure PP, the conductivity of LCBPP films with FS as functional monomer decreased significantly. At 125 °C, the conductivity of the modified film is 86% lower than that of PP. This shows that the entanglement between the long chain branch structure hinders the sliding and movement of the PP main chain in the lattice, effectively improving the thermal stability. In addition, the furfural functional groups carried by the grafted functional monomer FS introduce a chemical deep trap that is almost unaffected by high temperatures, trapping charge carriers and thereby reducing electrical conductivity.

As shown in Fig. 6.10c, the breakdown strength of LCBPP films containing FS functional monomers is higher than that of pure PP films. The breakdown strength of the modified film at 125 °C is 17.45% higher than that of PP. This indicates that the long link dendrimers effectively improve the microstructure stability of PP at high temperatures, and the functional monomers introduce deep traps capable of trapping charge carriers, thereby increasing the breakdown strength at high temperatures. In addition, appropriate functional monomer grafting rate can increase the high temperature breakdown strength of the film, while too high functional monomer content may lead to the increase of electrical weakness. Local electric field distortion will occur around the electrical weakness inside the film, which will aggravate the degradation phenomenon.

Due to the influence of the breakdown strength of the film, the energy density of the film can be improved by introducing functional monomers containing polar

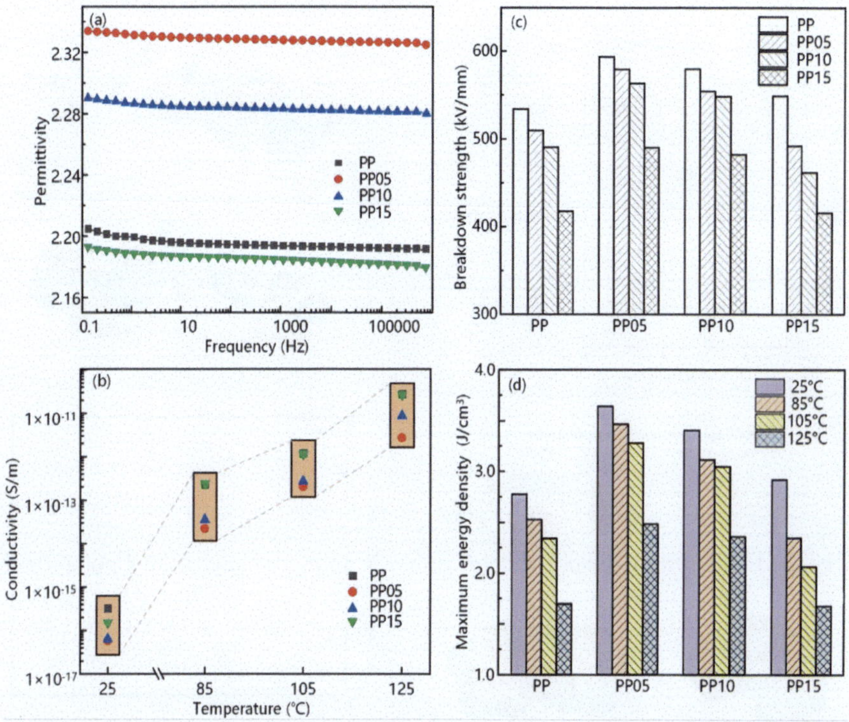

Fig. 6.10 **a** Relationship between the relative permittivity and the frequency, **b** relationship between the conductivity and the temperature, **c** relationship between the breakdown strength and the temperature, **d** relationship between the maximum energy density and the temperature [21]. Copyright Clearance Center's RightsLink© License Number: 5811970073314

groups. As shown in Fig. 6.10d, at 125 °C, the energy density of LCBPP film with FS functional monomer increased by 46.7% compared with pure PP film.

FS energy level distribution, DOS and MOs are shown in Fig. 6.11a, b. The lower LUMO level and the higher HOMO level enhance the likelihood that the carrier will be trapped in the trap instead of migrating into the dielectric. The three-dimensional electrostatic potential distribution of PP and FS is shown in Fig. 6.11c. PP can hardly absorb electric charge. The core of FS framework has a strong negative potential that can form a hole trap, and the edge has a strong positive potential that can form an electron trap. The trap energy level distribution of LCBPP with functional monomer FS is shown in Fig. 6.11d. The successful introduction of FS with long branched chains in PP05 and PP10 significantly improves the intensity and density of trap levels.

Therefore, it can be concluded that grafted furfural functional groups can form deeper chemical traps to trap electrons and holes, thus limiting the fast transport of high-energy carriers. The improvement of the breakdown strength of the modified film is mainly due to the improvement of the microstructure of the long chain branch

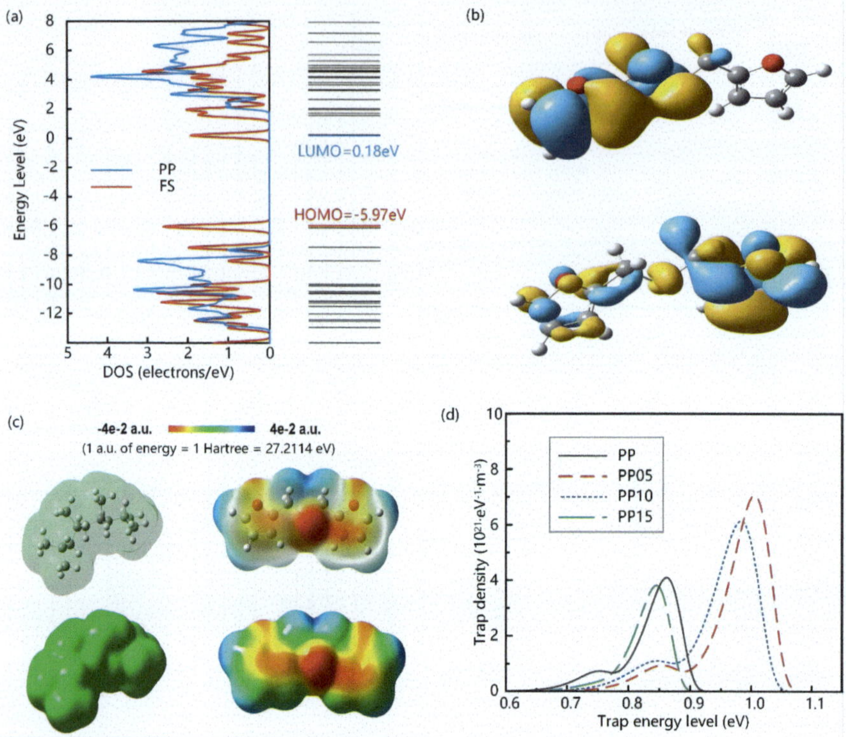

Fig. 6.11 **a** Energy level of the PP and the FS, **b** molecular orbitals of the FS, **c** 3D electrostatic potential distribution of the PP and the FS, **d** trap distribution of LCBPP [21]. Copyright Clearance Center's RightsLink© License Number: 5811970073314

structure and the deep trap introduced by the functional monomer. On the one hand, the long chain branched structure can enhance the thermal stability of the molecular chain and improve crystallization through heterogeneous nucleation. On the other hand, the introduction of furan groups can provide a deep well that inhibits charge migration. The synergistic effect of the two provides an effective means to solve the problem of low breakdown strength of PP films at high temperature.

6.3.2 Aggregate Structure Regulating

The crystallization characteristics of polymers are also an important factor affecting their dielectric properties, and it is difficult to achieve ideal crystallization control only by introducing LCBPP. Therefore, regulating the crystallization characteristics of LCBPP by micro-filling with nucleating agents is an important method to improve the dielectric properties of films.

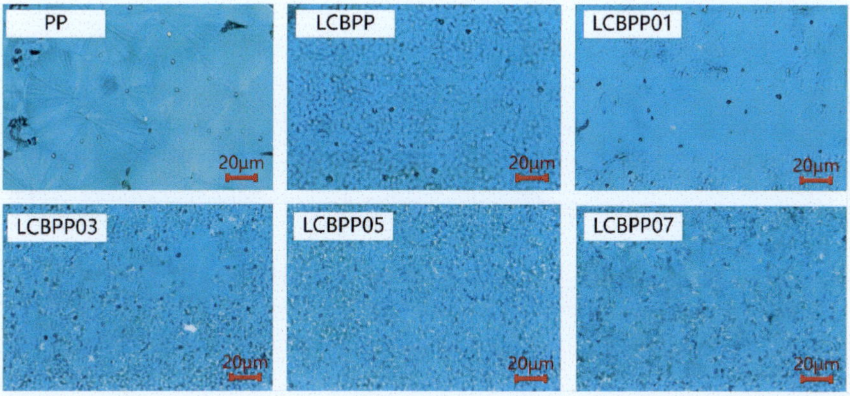

Fig. 6.12 Crystal morphology of modified LCBPP [22]

As shown in Fig. 6.12, NA-21 nucleating agent induced the crystallization of LCBPP strand segments, so the initial nucleation density in the modified film increased, and the strand segments of LCBPP grew along the surface perpendicular to the edge of the nucleating agent. The addition of nucleating agents reduces the critical size of nuclei suitable for crystal growth. With the addition of nucleating agent increasing, the number of spherulite in LCBPP films increases, the size of spherulite decreases and the boundary of spherulite becomes blurred. In addition, due to the high nucleation density of the modified LCBPP films, the spherulite growth process is not obvious. When the amount of nucleating agent is higher, the spherulite distribution is uneven, which is caused by the agglomeration of nucleating agent in LCBPP.

When the frequency is 1 kHz, the dielectric constant and dielectric loss Angle tangent of the crystal-controlled LCBPP modified film are shown in Fig. 6.13a. After adding a certain amount of nucleating agent, the dielectric constant of the modified LCBPP film decreased. Nucleating agent promoted the crystallization of LCBPP, the arrangement of spherulites in the modified LCBPP was denser, the internal defects of the film were reduced, and the interfacial polarization was reduced. In a certain range, the addition of nucleating agent causes the crystallinity of LCBPP to increase and the space between spherulites to become denser, which reduces the weak region in the film and reduces the medium loss. When nucleating agent content is higher, aggregation occurs in LCBPP, and the dielectric loss Angle tangent of LCBPP4 is slightly larger than that of LCBPP3. In general, the addition of nucleating agent has little effect on the dielectric constant and loss of LCBPP.

The conductivity of LCBPP films based on crystallization regulation is shown in Fig. 6.13b. The addition of nucleating agent can regulate the crystallization of LCBPP and reduce the conductivity loss of the film. At 125 °C, the conductivity of long chain branched polypropylene films modified by nucleators is more than one order of magnitude lower than that of pure LCBPP and linear PP. Nucleating agents promote the crystallization of LCBPP and increase its crystallinity, and the

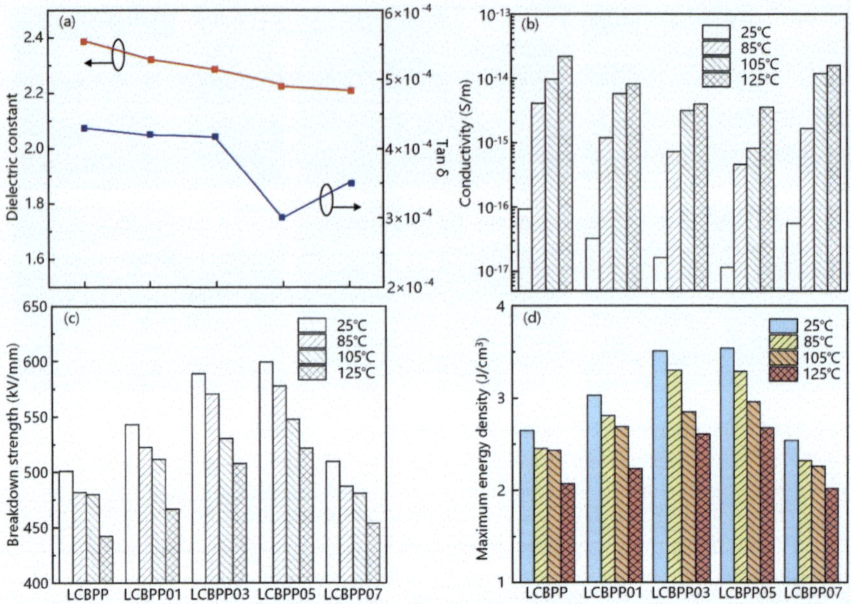

Fig. 6.13 **a** Relative permittivity and Tan δ at room temperature, **b** relationship between the conductivity and the temperature, **c** relationship between the breakdown strength and the temperature, **d** relationship between the maximum energy density and temperature [22]

crystal structure of the modified LCBPP films is dense, which reduces the mobility of carriers and leads to a decrease in the conductivity loss of the films. In addition, the interface between spherulite and amorphous in the film increases, which affects the trap energy level and density, and limits the carrier migration. The conductivity of LCBPP film is relatively large when the nucleating agent content is higher, which may be due to the uneven dispersion of nucleating agent content in the film.

The breakdown strength of LCBPP films based on crystallization regulation is shown in Fig. 6.13c. The breakdown strength of LCBPP films can be increased by adjusting the crystallization of LCBPP films. The increase in crystallinity, the increase in the number of spherulites and the tighter crystal arrangement of the modified films limit the transport of charge carriers in the films. When the ambient temperature is 125 °C, the breakdown strength of the modified film based on crystallization control is increased by 79.2 kV/mm and 109.2 kV/mm, respectively, compared with pure LCBPP and linear PP. Because the nucleating agent enhances the breakdown strength of the film by regulating the crystallization of LCBPP, the energy density of the modified film is 29.5% higher than that of pure LCBPP and 62.1% higher than that of linear PP at 125 °C. The breakdown strength of the modified films with higher nucleating agent content decreased, because the nucleating agent was unevenly dispersed in the LCBPP films with the increase of nucleating agent content.

The energy density of LCBPP films based on crystallization regulation is shown in Fig. 6.13d. The dielectric constant of nucleating agent modified LCBPP film decreased slightly, but because nucleating agent increased the breakdown strength of the film by regulating the crystallization of LCBPP, the energy density of nucleating agent modified film showed an upward trend with the increase of nucleating agent content when the amount of nucleating agent added was less than 0.05 wt%. The energy density of LCBPP05 is 33.6% higher than that of pure LCBPP. When the content of nucleating agent is high, the energy density of the film decreases due to the decrease of the breakdown strength and dielectric constant. With the increase of temperature, the energy density of the film decreases due to the decrease of the breakdown strength. When the temperature rises from 25 to 125 °C, the energy density of LCBPP film decreases by 21.9%, while that of linear PP decreases by nearly 38.1%, which is due to the introduction of long branch chain structure, which improves the thermal stability of the film. At 125 °C, the energy density of LCBPP3 film is 29.5% higher than that of pure LCBPP and 62.1% higher than that of linear PP.

The mechanism of influence of crystallization regulation on dielectric properties of long-chain branched polypropylene is shown in Fig. 6.14. The addition of nucleating agent promoted the crystallization of LCBPP, and the size and number of spherulites in the modified film were smaller and more, and the space between the spherulites was closer, which reduced the weak region in the film. With the increase of crystallinity, the internal structure of the modified film becomes denser and the proportion of amorphous region decreases, which affects the carrier transport. In addition, the formation of a large number of small spherulites in the modified films increases the spatial density between spherulites, and the interface between spherulites and amorphous can affect the trap energy level and density and limit the carrier migration in the films [23].

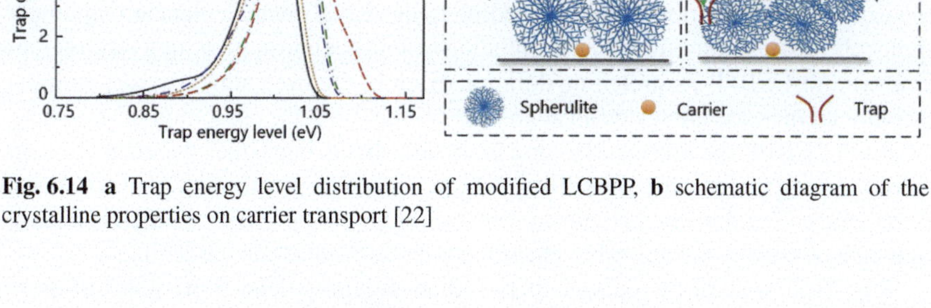

Fig. 6.14 **a** Trap energy level distribution of modified LCBPP, **b** schematic diagram of the crystalline properties on carrier transport [22]

6.4 Long-Chain Branched Polypropylene Stretching

6.4.1 Stretching Ratio Adjusting

The method of modifying polypropylene through long chain branching presents a viable solution to address the challenges of heightened conductivity loss and diminished breakdown strength encountered under conditions of elevated temperature and intense electric fields. However, in practical manufacturing processes, the polypropylene film intended for capacitor applications necessitates undergoing biaxial stretching to achieve suitability for industrial use. Presently, the impact of synchronous biaxial stretching processes on the dielectric properties of long-chain branched polypropylene films remains ambiguous, and the optimal configuration of stretching parameters, as well as the implementation of multi-performance collaborative control methods tailored to long-chain branched polypropylene, remains elusive. Consequently, there exists an urgent need to comprehensively comprehend the repercussions of biaxial stretching on the microstructural and dielectric properties of long-chain branched polypropylene, and to ascertain appropriate stretching process parameters conducive to the characteristics of these novel materials. This chapter endeavors to investigate the influence of stretching ratios on the crystallization behavior of long-chain branched polypropylene, subsequently delving into the underlying mechanisms governing the influence of stretching ratios on the dielectric properties exhibited by films composed of long-chain branched polypropylene.

Stretch viscosity (ηE) represents the ratio of stress exerted by a viscoelastic material on a unit cross-sectional area perpendicular to the direction of flow to the rate of tensile strain. It serves to characterize the material's viscosity during the process of tensile deformation. The temporal evolution curves depicting the tensile viscosity of both PP and LCBPP are delineated in Fig. 6.15. Upon examination of the graph, it becomes apparent that at lower stretching rates, the tensile viscosity of PP initially experiences an increment followed by a marginal decrement. Conversely, at higher stretching rates, the tensile viscosity of PP exhibits a gradual augmentation succeeded by a sharp decline over time. This phenomenon can be attributed to the initial stages of stretching wherein the molecular chain deformation of PP remains minimal, thereby augmenting chain entanglement and consequently elevating stretching viscosity. However, with prolonged stretching durations, the extent of deformation escalates, leading to increased entanglement among linear molecular chains, consequently precipitating a rapid decline in stretching viscosity. In contrast, for LCBPP, as stretching duration extends, the stretching viscosity exhibits a continual increase alongside an augmented slope. With the escalation of stretching extent in LCBPP, the presence of long branched chains intensifies the entanglement effects among molecular chains. This, in turn, engenders greater frictional resistance during the molecular chain alignment process, culminating in a pronounced increase in tensile viscosity and a notable phenomenon of tensile strain hardening.

The XRD spectra of long-chain branched polypropylene films with different stretching ratios are shown in Fig. 6.16. Both PP and LCBPP materials are made

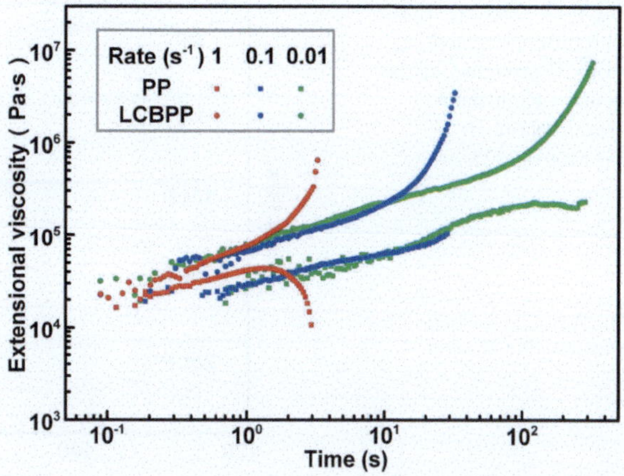

Fig. 6.15 Relationship between the extensional viscosity and time [23]. Copyright Clearance Center's RightsLink© License Number: 5811400103321

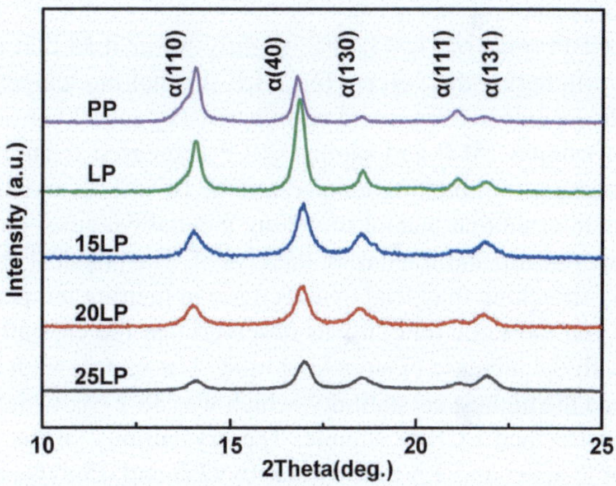

Fig. 6.16 XRD spectra with different stretching ratios [23]. Copyright Clearance Center's RightsLink© License Number: 5811400103321

of α crystalline, with five crystal faces corresponding to monoclinic crystals. For the three types of biaxially stretched samples, stretching did not change the distribution of diffraction peaks, indicating that there was no crystal transformation in long-chain branched polypropylene at all stretching rates. From the graph, it can be observed that as the stretching ratio increases, the intensity of diffraction peaks tends to weaken, which may be related to the oriented arrangement of grains caused by stretching.

The melting curves of long-chain branched polypropylene samples with different stretching ratios were tested using DSC differential scanning calorimetry, as shown in Fig. 6.17. The melting temperature of LCBPP material is 5.2 °C higher than that of PP, because the entanglement between long branched chains inhibits the movement of molecular chains, thereby improving thermal stability. At the same time, the presence of long branched chain structures and the increase of molecular entanglement will hinder the migration of molecular chain segments from the amorphous region to the crystal surface, inhibiting the crystallization process. Therefore, the crystallinity of

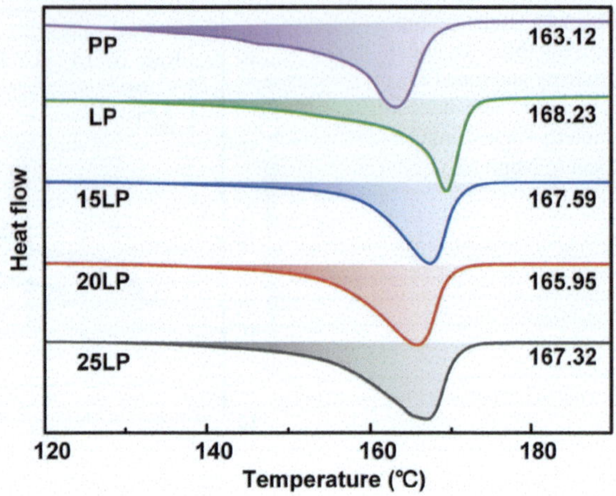

Fig. 6.17 Melting curves with different stretching ratios. Copyright Clearance Center's RightsLink© License Number: 5811400103321

LCBPP material is 43.58%, slightly lower than that of PP, which is 44.69%. It is worth noting that, except for 25LP, the melting temperature of the stretched sample shows a decreasing trend with increasing stretching rate. The melting temperatures of sample 15LP and sample 20LP decreased from 168.32 °C of the LCBPP hot pressing film melting temperature to 167.59 °C and 165.95 °C, respectively. It can be inferred that biaxial stretching inevitably causes damage to the crystal structure, such as slip and fracture of the crystal. The physical damage and defects introduced by stretching may lead to a decrease in melting temperature.

At the same time, it was observed that the crystallinity of the sample increased with the increase of stretching ratio. The sample with a stretching ratio of 2.5 times had the highest crystallinity, which was 48.27%, which was 10.76% higher than the unstretched LCBPP sample. The crystallinity of the samples at a stretching ratio of 2 times and 1.5 times were 46.92% and 45.87%, respectively, which increased by 7.66% and 5.25% compared to the unstretched LCBPP samples. This can be attributed to the occurrence of tensile stress induced crystallization phenomenon. Semi crystalline polymers can be seen as an interpenetrating network composed of crystalline frameworks and amorphous regions. During biaxial stretching, the crystalline framework bears the main stress, allowing for uniform distribution of strain in the sample. Under the action of a tensile force field, molecular chains extend and orient along the direction of force, and during this process, the conformational entropy of the molecular chains decreases, making it easier to arrange into the lattice. Stress induced crystallization can form crystals with more complete structures and higher melting points, and it also causes the melting temperature of the 25LP sample to rise from the original downward trend to 167.32 °C. It can be inferred that the slip, dislocation, damage, and tensile stress induced crystallization phenomena caused by the stretching process on the crystal structure coexist within the film. The higher the stretching ratio, the greater the damage to the crystal caused by stretching. However, an increased stretching ratio also helps to enhance the effect of tensile stress induced crystallization. Therefore, as the stretching ratio increases, the melting temperature shows a trend of first decreasing and then increasing.

At the same time, it was observed that the crystallinity of the sample increased with the increase of stretching ratio. The sample with a stretching ratio of 2.5 times had the highest crystallinity, which was 48.27%, which was 10.76% higher than the unstretched LCBPP sample. The crystallinity of the samples at a stretching ratio of 2 times and 1.5 times were 46.92% and 45.87%, respectively, which increased by 7.66% and 5.25% compared to the unstretched LCBPP samples. This can be attributed to the occurrence of tensile stress induced crystallization phenomenon. Semi crystalline polymers can be seen as an interpenetrating network composed of crystalline frameworks and amorphous regions. During biaxial stretching, the crystalline framework bears the main stress, allowing for uniform distribution of strain in the sample. Under the action of a tensile force field, molecular chains extend and orient along the direction of force, and during this process, the conformational entropy of the molecular chains decreases, making it easier to arrange into the lattice. Stress induced crystallization can form crystals with more complete structures and higher melting points, and it also causes the melting temperature of the 25LP sample to rise from the original downward trend to 167.32 °C. It can be inferred that the slip, dislocation, damage, and tensile stress induced crystallization phenomena caused by the stretching process on the crystal structure coexist within the film. The higher the stretching ratio, the greater the damage to the crystal caused by stretching. However, an increased stretching ratio also helps to enhance the effect of tensile stress induced crystallization. Therefore, as the stretching ratio increases, the melting temperature shows a trend of first decreasing and then increasing.

The examination of the final crystalline morphology of the sample was conducted using a polarizing microscope, with the results presented in Fig. 6.18. In the case of the unstretched materials, it was observed that the spherical size of linear PP was notably larger, exhibiting clear boundaries and a distinct polarized cross structure. On the other hand, the LCBPP material displayed a multitude of small-sized spherulites, with densely distributed fragmented crystals interspersed among them. The presence of heterogeneous nucleation effects stemming from the side chain branching points of LCBPP, coupled with its high melt viscosity, facilitated the attainment of stable spherical crystals at reduced sizes, thereby giving rise to an abundance of overlapping microcrystals characterized by indistinct boundary margins. Similar to the unstretched sample, the 15LP sample featuring a stretching ratio of 1.5 exhibited difficulties in discerning the boundaries of the spherulites, with the gaps between the spherulites forming small circular cavities. In contrast, for the 20LP sample, the boundaries of the spherical crystals became more conspicuous, and the fragmented crystals were no longer observable within the field of view. Notably, the regularity of crystal appearance exhibited marked improvement at this stretching ratio, indicative of the heightened significance of the recrystallization effect induced by stretching in comparison to the structural cracking and damage caused by the stretching process. Furthermore, the crystal of the 25LP sample was distinctly elongated into a spindle shape, resulting in an overall compact fibrous structure. It is noteworthy that the biaxial stretching in 25LP led to the occurrence of typical wavy slip bands due to

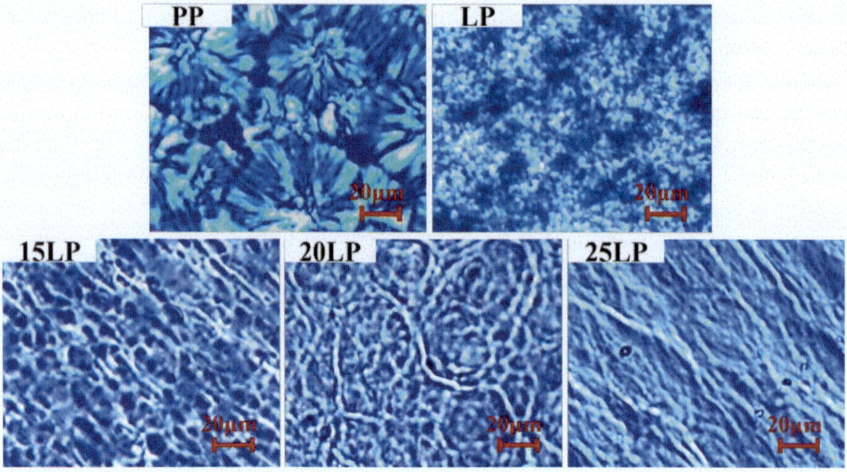

Fig. 6.18 Crystalline morphology with different stretching ratios [23]. Copyright Clearance Center's RightsLink© License Number: 5811400103321

the cross sliding of crystals, resulting in the formation of perfect and dense crystals. This observation suggests that tensile stress can induce an ordered rearrangement of the crystal structure during the simultaneous biaxial stretching process.

The arrangement of lattices and the presence of crystal defects play crucial roles in influencing the ease of free charge carrier and hole transport within a material. To explore the relationship between stretching rate, crystal morphology, and charge transport, this section employed the previously introduced isothermal discharge current method to examine the trap energy level distribution characteristics across five sample groups. Figure 6.19 illustrates the trap energy level distribution characteristics of the samples at varying testing temperatures.

At 25 °C, PP exhibits a deep trap characteristic peak at the 0.85 eV level, with a density of 5.18×10^{21} $eV^{-1}{\cdot}m^{-3}$. However, the deep trap level and density of LCBPP are slightly higher than PP, with values of 0.86 eV and 5.90×10^{21} $eV^{-1}{\cdot}m^{-3}$, respectively. This can be attributed to LCBPP's smaller spherical crystal size and more interface traps generated due to the intersection of lamellar dislocations at the boundaries, making carrier migration more difficult. As shown in the figure, as the testing environment temperature increases, the deep trap energy level of the sample gradually deepens and the density slightly decreases. At 125 °C, the deep trap energy levels of PP and LCBPP hot pressed films are 0.96 eV and 0.98 eV, respectively, with deep trap densities of 3.84×10^{21} $eV^{-1}{\cdot}m^{-3}$ and 4.30×10^{21} $eV^{-1}{\cdot}m^{-3}$. Biaxial stretching deepens the trap energy level and increases the trap density of the thin film. When the stretching rate is 1.5 times at 25 °C, the deep trap energy level and density of the sample are the highest, 0.89 eV and 8.44×10^{21} $eV^{-1}{\cdot}m^{-3}$, respectively. Afterwards, as the stretching rate increases, the deep trap energy level remains almost unchanged, while the density decreases. The deep trap densities of 20LP and 25LP are 7.60×10^{21} $eV^{-1}{\cdot}m^{-3}$ and 7.10×10^{21} $eV^{-1}{\cdot}m^{-3}$, respectively.

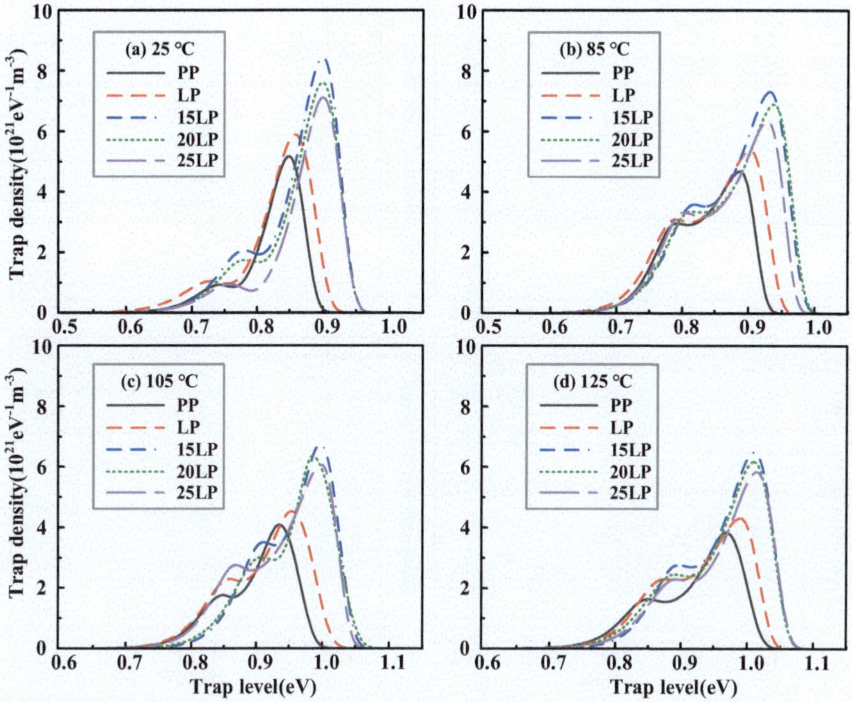

Fig. 6.19 Trap energy level distribution with different stretching ratios

According to local energetics, some defects and defects caused by crystallization during the stretching process, such as stacking faults and dislocations, can introduce traps in the bandgap. Carriers are captured by deep traps during transmission, making it difficult for trapped carriers to escape and staying in the deep traps for a long time, resulting in a decrease in carrier mobility. At the same time, deep traps can act as scattering centers to disperse carriers, which is beneficial for improving the insulation performance of thin films. For 20LP and 25LP, the recrystallization effect induced by tensile stress to some extent eliminates crystal defects, leading to a decrease in deep trap density compared to 15LP. Due to the higher stretching ratio, the better the effect of inducing recrystallization. Therefore, when the deep trap energy level remains unchanged, the deep trap density of 25LP is less than 20LP. The variation pattern of trap distribution in stretched samples at high temperatures is consistent with that at 25 °C. A stretching rate of 1.5 times mainly brings defects and damage to the crystal structure of the film, while the effect of tensile stress induced crystallization is not good, resulting in the highest deep trap energy level and density of 15LP, and the strongest ability to capture charge carriers. At 85 °C, 105 °C, and 125 °C, it reaches 0.94 eV, 0.99 eV, 1.02 eV, and 7.31×10^{21} $eV^{-1} \cdot m^{-3}$, 6.73×10^{21} $eV^{-1} \cdot m^{-3}$, and 6.50×10^{21} $eV^{-1} \cdot m^{-3}$, respectively. Under the stress induced crystallization effect, the deep trap energy levels of samples 20LP and 25LP at high temperatures are

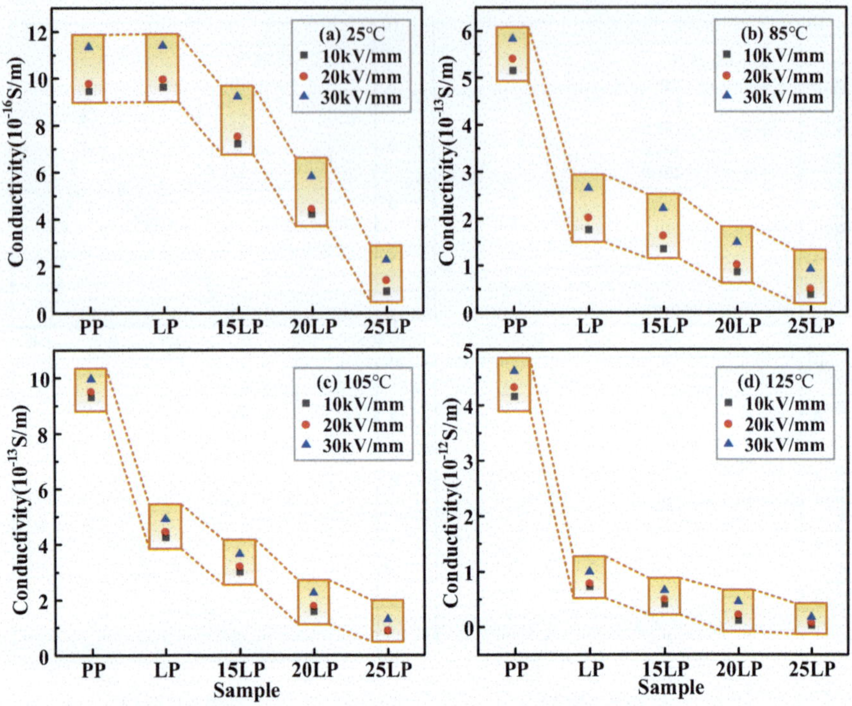

Fig. 6.20 Relationship between the conductivity and the temperature

almost the same as those of 15LP, while the deep trap density maintains a decreasing trend with increasing stretching rate.

In order to further understand the charge transfer characteristics, the three-electrode method was used to test the bulk conductivity of biaxial oriented long-chain branched polypropylene films with different stretching ratios under different environmental temperatures and electric strengths. The results are shown in Fig. 6.20.

As shown in the figure, as the temperature increases, the DC conductivity of all samples also increases. The conductivity of PP hot pressed film increased by four orders of magnitude from 25 to 125 °C. The DC conductivity of PP at 10 kV/mm, 20 kV/mm, and 30 kV/mm field strengths at 25 °C was 9.44×10^{-16} S/m, 9.78×10^{-16} S/m, and 1.14×10^{-15} S/m, respectively. The DC conductivity of PP at 10 kV/mm, 20 kV/mm, and 30 kV/mm field strengths at 125 °C was 4.15×10^{-12} S/m, 4.33×10^{-12} S/m, and 4.61×10^{-12} S/m, respectively. This can be attributed to the increased thermal motion, slip and relaxation of molecular chains caused by high temperature, resulting in an increase in free volume and an increase in the average free path of charge carriers. The migration and transport of charge carriers inside the film lack obstacles, leading to an increase in conductivity current and a decrease in high-temperature dielectric properties. As the electric field intensity increases from 10 kV/m to 30 kV/mm, the conductivity of PP also increases. This is because the increase in electric field can also enhance the hot electron emission phenomenon of the electrode, leading to an increase in the number and mobility of charge carriers injected into the film, resulting in an increase in leakage current and conductivity. For the LCBPP hot pressed film, the conductivity at 25 °C is slightly higher than that of PP, which may be due to the increased ion conductivity inside the film caused by the residual catalyst and dissociation products of impurities during the preparation process of LCBPP. At high temperatures, the conductivity loss of LCBPP is significantly reduced compared to PP. At a field strength of 20 kV/mm, the DC conductivity of LCBPP at 85 °C, 105 °C, and 125 °C is 2.03×10^{-13} S/m, 4.47×10^{-13} S/m, and 7.90×10^{-13} S/m, respectively. This is because the introduction of a long-branched chain structure enhances the entanglement between molecular chains, suppresses the slip between molecular chains at high temperatures, and gives LCBPP films higher thermal stability compared to linear molecular PP.

As shown in the figure, the DC conductivity of the stretched sample decreases with the increase of the stretching ratio at all four testing temperatures, especially at high temperatures. At 25 °C, the DC conductivity of sample 25LP is the lowest, at field strengths of 10 kV/mm, 20 kV/mm, and 30 kV/mm, which are 9.20×10^{-17} S/m, 1.41×10^{-16} S/m, and 2.27×10^{-16} S/m, respectively. This is because the lattice arrangement determines the efficiency of carrier drift and diffusion. During the stretching process, the long-branched chain structure is prone to orientation and stretching due to its long relaxation time. A higher stretching ratio tends to enhance the biaxial orientation and orderly alignment of molecular chains within the material. This, in turn, leads to the formation of a dense microstructure post-stretching, which effectively impedes the transport of charge carriers, consequently causing a reduction in conductivity.

With the rise in testing temperature, there is a corresponding increase in the conductivity of the sample. At 125 °C, the DC conductivity of samples with stretching ratios of 1.5, 2.0, and 2.5 times are 5.01×10^{-13} S/m, 2.20×10^{-13} S/m, and 7.31×10^{-14} S/m, respectively, under a field strength of 20 kV/mm. At the same field strength, the DC conductivity of 25LP is about 1/10 of that of LCBPP hot pressed film. This represents that under high temperature and strong field conditions, the biaxial stretching process not only does not damage the unique thermal stability of the long-branched chain structure in the thin film sample, but also further enhances the dielectric properties of the thin film by inducing the biaxial orientation of molecular chains and improving the crystal morphology. It is worth noting that the contribution of traps introduced after stretching to the decrease in conductivity cannot be ignored. Local deep traps can effectively limit the transport of charge carriers inside the thin film. The decrease in conductivity is the result of the combination of the thermal stability of long branched chain structures, the improvement of crystal morphology introduced by stretching, and the trap effect of deep traps. As shown in the figure, the DC conductivity of 15LP at 25 °C, 85 °C, 105 °C, and 125 °C under a field strength of 20 kV/mm is 7.54×10^{-16} S/m, 1.64×10^{-13} S/m, 3.32×10^{-13} S/m, and 5.01×10^{-13} S/m, respectively, which are the highest among the three stretching samples. The preceding analysis reveals that sample 15LP exhibits the highest deep trap energy level and density across all four temperatures assessed. This observation suggests that, in comparison to the enhancement of crystal morphology observed in thin film samples subjected to high stretching rates, the localized introduction of deep traps in thin film samples subjected to low stretching rates does not exert a predominant influence in suppressing the conductivity characteristics.

In order to investigate the influence of stretching ratio on the dielectric constant and dielectric loss spectrum of biaxially stretched long-chain branched polypropylene films, this section tested the dielectric constant and dielectric loss tangent of five samples, and the results are shown in Fig. 6.21. As shown in Fig. 6.21a, the dielectric constant of PP is around 2.21, while the dielectric constant of LCBPP increases to around 2.38. The increase in the dielectric constant of LCBPP is due to the increased polarity and asymmetry of the original molecular chain caused by the long-branched chain structure. The rearrangement of asymmetric molecular chains with branched chain structures under an electric field can introduce additional dipole moments and enhance polarization. The dielectric constant of the three stretching samples slightly decreased compared to the LCBPP hot pressed film, and with the increase of stretching rate, the dielectric constant showed a decreasing trend. The dielectric constant of a sample with a stretching ratio of 2.5 times is about 2.37. This phenomenon may occur due to biaxial stretching, which hinders the dipole from alternating with the electric field and leads to a decrease in dielectric constant. As shown in Fig. 6.21b, at low frequencies, it can be clearly observed that the dielectric loss of PP hot pressed film is the highest, followed by LP hot pressed film. As the stretching ratio increases, the dielectric loss of the sample decreases. This is because the dielectric loss at low frequencies is mainly determined by the conductivity loss. At high frequencies, the dielectric loss of all samples is almost the same. Compared with other samples, the dielectric loss of PP hot pressed film is slightly lower. This is because the dielectric loss at high frequencies is mainly determined by polarization characteristics, and the dielectric constant spectrum shows that the dielectric constant of PP is significantly lower than that of other samples.

The breakdown strength of polypropylene dielectric directly affects the energy storage density and service life of capacitors. This section investigates the DC

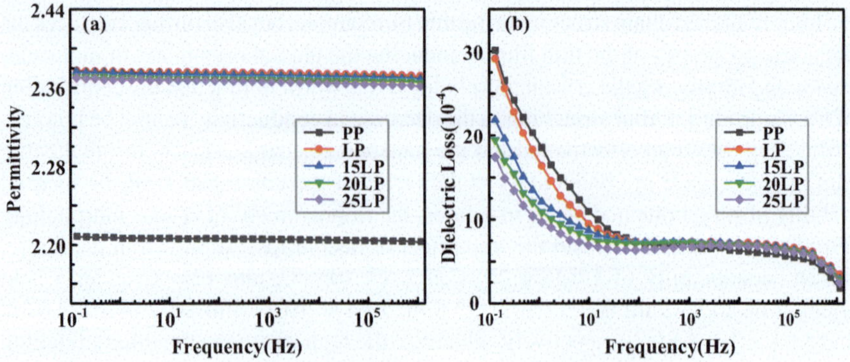

Fig. 6.21 Relationship between the **a** permittivity, **b** dielectric loss and the frequency

breakdown characteristics of biaxially stretched long-chain branched polypropylene films with different stretching ratios at ambient temperatures of 25, 85, 105, and 125 °C. The Weibull probability distribution of breakdown field strength is shown in Fig. 6.22. At 25 °C, the characteristic breakdown field strengths of PP and LCBPP hot pressed films are similar, with values of 580.4 kV/mm and 583.6 kV/mm, respectively. However, as the stretching ratio increases, the characteristic breakdown field strengths of the stretched samples increase. The characteristic breakdown field strengths of 15LP, 20LP, and 25LP are 616.3 kV/mm, 632.3 kV/mm, and 644.6 kV/mm, respectively. As the testing temperature increased, the breakdown strength of PP sharply decreased, decreasing by 12.3%, 14.9%, and 20.0% from 25 °C to 85 °C, 105 °C, and 125 °C, respectively. This is due to the increase in free volume and average free path of charge carriers caused by molecular chain slip and local chain segment relaxation in the amorphous region at high temperatures, resulting in a significant decrease in breakdown strength. At 85 °C, 105 °C, and 125 °C, the characteristic breakdown strength of LCBPP hot pressed film is much higher than that of PP. This is because the introduction of long branched chain structure can increase the entanglement and friction between molecular chains, suppress the slip between polypropylene molecular chains, enhance the intermolecular interaction force, and thereby improve the thermal stability of molecular chains. The breakdown performance of the three types of tensile samples at high temperatures follows the same pattern of increasing characteristic breakdown field strength with the increase of tensile ratio as at room temperature. At 125 °C, the characteristic breakdown field strengths of 15LP, 20LP, and 25LP were 556.7 kV/mm, 584.0 kV/mm, and 606.0 kV/mm, respectively, which were increased by 19.9%, 25.6%, and 30.5% compared to PP hot pressed film, and 8.01%, 13.3%, and 17.6% compared to LCBPP hot pressed film, respectively. On the one hand, based on the original thermal stability of LCBPP materials, stretching has refined the connections between crystal regions, and the dense crystal structure induced by stretching has eliminated the defects and electrical weaknesses of residual broken crystals; On the other hand, the deep traps introduced by stretching enhance the film's ability to capture charge carriers, thus synergistically improving the high-temperature breakdown performance of the new polypropylene film from two aspects and further reducing the thermal sensitivity of the film's dielectric properties. Furthermore, it is worth mentioning that the shape parameters of the thin film significantly improved after stretching. At 125 °C, the shape parameters of 15LP, 20LP, and 25LP were 18.27, 15.29, and 18.63, respectively, and the dispersibility of the breakdown field strength decreased. This may be due to the fact that long chain branching promotes more consistent deformation during the strain hardening process induced by biaxial tension, thereby improving the uniformity of the film.

In order to study the effect of stretching ratio on the energy storage density of biaxially stretched long-chain branched polypropylene films, the discharge energy density and charge discharge efficiency were calculated based on the hysteresis loops of the film samples under different polarization electric fields. The charge discharge efficiency and discharge energy density curves of the five samples at 25 °C, 85 °C, 105 °C, and 125 °C are shown in Fig. 6.23.

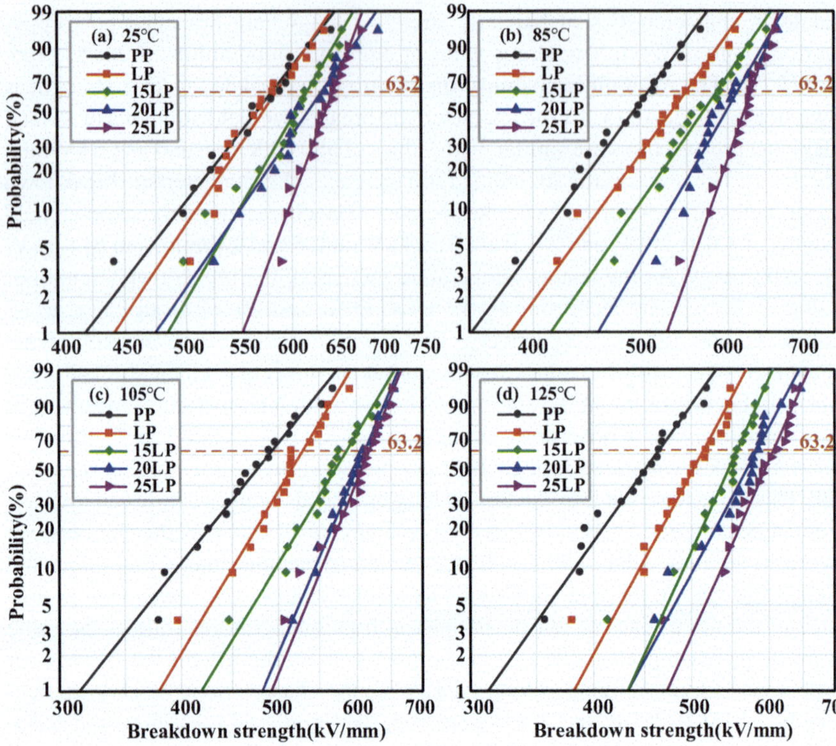

Fig. 6.22 Relationship between the breakdown strength and the temperature

As shown in the figure, at the same temperature and polarization electric field, compared to PP hot pressed film, the discharge energy density of all long-branched chain samples is slightly increased, which can be attributed to the higher dielectric permittivity of the long-branched chain samples, that is, the slightly improved polarization ability. The maximum discharge energy density of PP hot pressed film at 25 °C is 3.166 J/cm^3, and the energy storage density of LP hot pressed film is 3.383 J/cm^3. At the same temperature, the energy storage density of the thin film sample increases with the increase of stretching ratio. This is because the energy storage density is determined by the breakdown field strength of the sample when the difference in dielectric constant is not significant. The energy storage density of sample 25LP is the highest, at 4.569 J/cm^3, which is 44.3% and 35.1% higher than that of PP and LP hot pressed films, respectively. When the temperature increased to 85 °C and 105 °C, the energy storage density of PP hot pressed film decreased to 2.488 J/cm^3 and 1.904 J/cm^3, respectively, while the energy storage density of LP hot pressed film decreased to 3.198 J/cm^3 and 2.525 J/cm^3, respectively. At these two temperatures, the energy storage density of sample 25LP remained the highest, at 4.095 J/cm^3 and 3.831 J/cm^3, respectively. Compared with PP hot pressed film, it

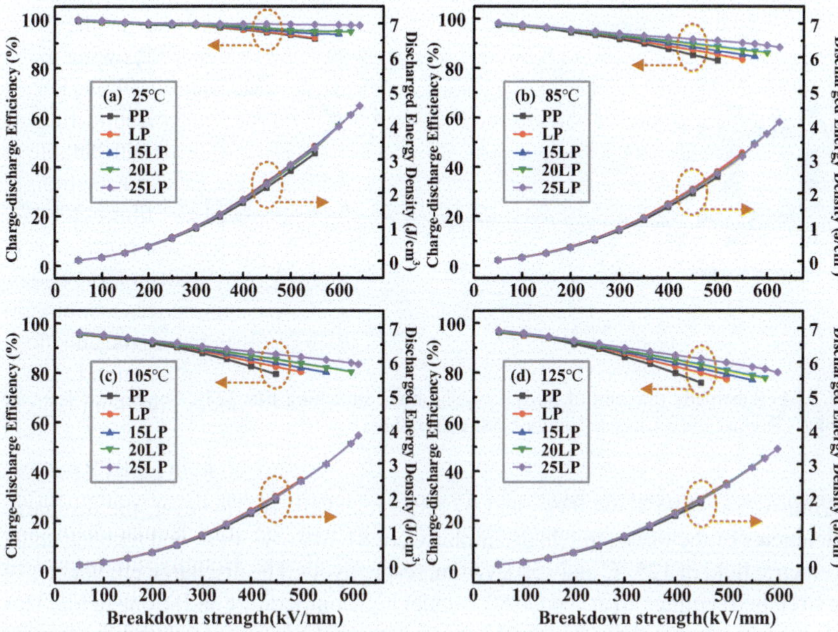

Fig. 6.23 Relationship between the energy storage characteristic and the temperature

increased by 64.6% and 101.2%, respectively. The synergistic effect of crystal structure optimization and deep trap introduction on the enhancement of breakdown field strength at high temperatures and the advantage of long branched chain structure with strong thermal stability at high temperatures began to emerge. At 125 °C, the energy storage density of PP and LP hot pressed films decreased to 1.637 J/cm^3 and 2.429 J/cm^3, respectively. The energy storage density of sample 25LP was 3.446 J/cm^3, even higher than that of PP and LP hot pressed films at 25 °C. At this time, compared to PP and LP hot pressed films, the energy storage density increased by 110.5% and 41.9%, respectively. This indicates that biaxially stretched long-chain branched polypropylene films can not only significantly improve the energy storage characteristics of traditional polypropylene films, but also enhance the long-term stability of energy storage characteristics in high-temperature and harsh environments.

As shown in the figure, with the increase of polarization electric field, the discharge efficiency of the thin film sample shows a decreasing trend. Although the polarization loss of PP hot pressed film itself is the lowest, as the field strength increases, the high conductivity loss leads to the fastest decrease in its discharge efficiency, which is 91.73% at 25 °C and 550 kV/mm electric fields, and only 75.83% at 125 °C and 450 kV/mm electric fields. The performance of LP hot pressed film at 25 °C is similar to that of PP hot pressed film, with 92.09% at an electric field of 550 kV/mm and 80.01% at a high temperature of 125 °C and 450 kV/mm electric field. The discharge efficiencies of 15LP, 20LP, and 25LP stretched samples at 25 °C and 550

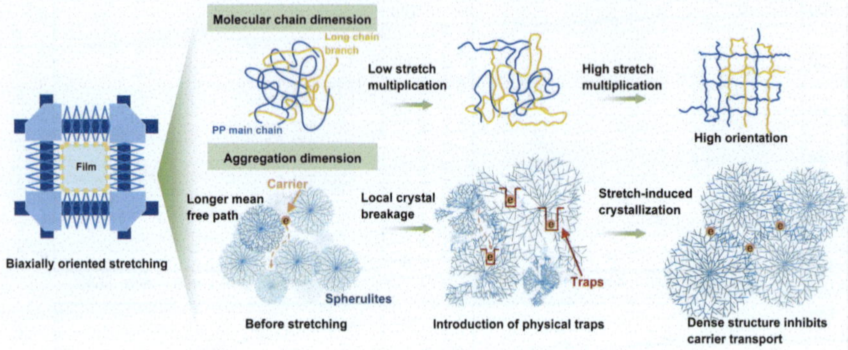

Fig. 6.24 Schematic diagram of biaxial synchronous stretching film [23]. Copyright Clearance Center's RightsLink© License Number: 5811400103321

kV/mm were 93.94%, 94.93%, and 97.88%, respectively. However, the discharge efficiencies of the three stretched samples were 81.74%, 83.49%, and 85.65% under an electric field of 125 °C and 450 kV/mm, respectively. The discharge efficiencies of the stretched samples all exceeded 80% under high temperature and strong fields. This indicates that the biaxially stretched long-chain branched polypropylene film not only significantly enhances the maximum discharge energy density, but also maintains a high level of discharge efficiency under high temperature and strong field conditions. The synergistic improvement of energy storage density, charge discharge efficiency, and service life of this new type of polypropylene film is of great significance for further improving the economy of high-voltage DC capacitors and promoting their volume miniaturization.

Biaxial stretching of PP films is a continuous process of rapid structural evolution of molecular chains, lamellar crystals, fibrous crystals, and other multi-scale structures under the simultaneous action of high speed, high temperature and stretching stress. Figure 6.24 illustrates the changes in the microstructure of LCBPP during the simultaneous biaxial stretching of the film.

In the molecular chain dimension, as the stretching ratio increases, the orientation of the molecular chain due to stretching subsequently grows. In the highly oriented state, the degree of ordering of the molecular chain arrangement rises significantly, resulting in an increment in the number of entanglement points. However, the long-chain branched structure itself already introduces many entanglement points within the dielectric, as mentioned above in the tensile viscosity test for strain hardening. The tensile resistance increases remarkably under the simultaneous action of both, which is what causes the film to break at higher stretching multiples. Yet, the long-branched structure has its own advantages in stretching. The extended relaxation time due to the high degree of entanglement reduces the spontaneous unorientation of the molecular chains, allowing the dielectric to remain more stably in a highly oriented state.

In the aggregated state dimension, the spherical crystals before LCBPP stretching are small in size and fragmented, which is due to the restriction of the growth of spherical crystals by the tangling of molecular chains. The amorphous region between the small spherical crystals provides a large area for carrier acceleration to gain energy. Therefore, carriers are easily transported in the dielectric, resulting in increased losses and reduced breakdown strength of the film. Biaxial stretching makes LCBPP produce ordered rearrangement of crystals under stress induction, and the spherical crystals tend to be large and dense. The tight arrangement of spherical crystals effectively suppresses the carrier migration process and reduces the leakage conductivity. It is also important to note that damage such as slip and minor rupture of the spherical crystal during stretching introduces physical traps to the dielectric. The initial spherocrystalline morphology is destroyed due to a sharp reduction of the local cross-section, which leads to high stress concentrations. The deep trap energy level has a strong attraction to carriers, which hinders carrier transmission.

6.4.2 Stretching Temperature Adjusting

The previous section indicated that an appropriate stretching ratio can improve the crystal structure of long-chain branched polypropylene and thereby enhance its high-temperature dielectric properties. The stretching temperature is also a key factor affecting the microstructure evolution of the film during the stretching process. However, the effect of stretching temperature on the dielectric properties of long-chain branched polypropylene films during synchronous biaxial stretching is still unclear. This section analyzes the crystallization characteristics of biaxially stretched long-chain branched polypropylene films prepared at different stretching temperatures, investigates the influence of stretching temperature on the DC conductivity, breakdown field strength, and energy storage density of long-chain branched polypropylene films, determines the optimal stretching temperature suitable for long-chain branched polypropylene, and explores the mechanism of the influence of stretching temperature on the high-temperature dielectric properties of long-chain branched polypropylene films.

The XRD spectra of long-chain branched polypropylene samples prepared at different stretching temperatures are shown in Fig. 6.25. It can be clearly seen from the figure that the crystal structure of the two hot pressed film samples and the three stretched samples is mainly composed of α crystals. The diffraction peaks all appear on the (110), (040), (130), (111), and (131) crystal planes, indicating that changes in stretching temperature will not cause a transformation of the α crystal structure. Compared with the control group samples that were not stretched, the intensities of all XX scattering peaks in the three samples after biaxial stretching were significantly reduced, which may be related to the oriented grain arrangement caused by stretching, indicating that stretching gradually transforms α crystals into oriented mesocrystals. Different intergranular phase transitions occur during the stretching process, accompanied by the destruction of the original crystal and the formation

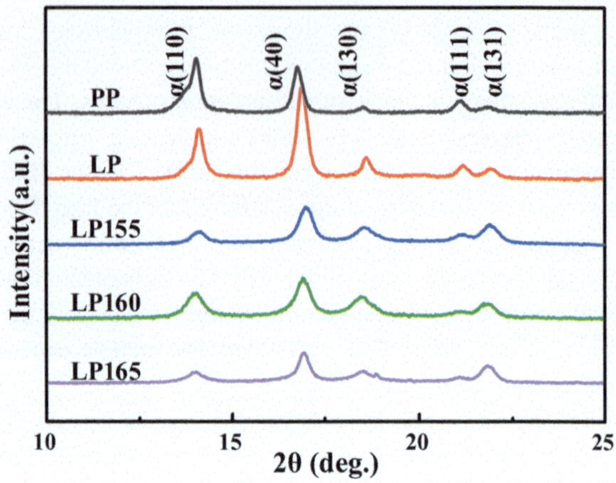

Fig. 6.25 XRD spectra with different stretching temperatures

of new fiber crystals. As a result, the transformation from α crystal to meso crystal implies the occurrence of crystal orientation, slip, and melting during the stretching process.

The melting curves of long-chain branched polypropylene samples at different stretching temperatures were tested using DSC differential scanning calorimetry. The peak melting temperatures of samples LP155 and LP160 are 166.69 °C and 167.32 °C, respectively, which are significantly lower than the peak melting temperature of the LP hot pressed film of the unstretched sample, which is 168.32 °C. This can be attributed to the fact that stretching reduces the conformational entropy of molecular chains during the process of generating biaxial orientation. The crystal slip and dislocation introduced by stretching can cause damage to the overall structure, thereby promoting the decrease in melting temperature. It is worth noting that for three different stretching temperatures of the stretching samples, the peak melting temperature shows an upward trend with the increase of stretching temperature. The peak melting temperature of sample LP165 is 168.66 °C, which is slightly higher than that of the LP hot pressed film. The melting point of semi crystalline polymers depends on the thickness of the flakes, and higher temperatures are more conducive to further crystallization and refinement of the flake layer. At a stretching temperature of 155 °C, due to the poor mobility and activity of molecular chains, the bound chain segments play a role in transmitting stress during the stretching process, which increases the possibility of crystal fracture during the stretching process. From the DSC curve, it can be seen that 165 °C is close to the melting temperature of the film, indicating that the stretching process at 165 °C is close to melt stretching. When the stretching temperature is relatively high and close to the melting temperature, the entanglement of molecular chain segments at high temperatures will lead to strain relaxation of the remaining chain segments, ultimately resulting in the production of more folded chain layer crystals. Therefore, sample LP165 can generate thicker flakes under temperature coupled tensile stress, leading to an increase in melting temperature.

The crystallinity of the stretched samples with three different stretching temperatures was 46.69%, 48.27%, and 50.44%, respectively. Not only did it increase with the increase of stretching temperature, but it was also higher than the crystallinity of LP hot pressed film by 43.58%. This can be explained by the theory of tensile stress induced crystallization. When subjected to external tensile stress, the interpenetrating network inside long-chain branched polypropylene causes the melt to exhibit similar characteristics to crosslinked elastomers, where local molecular chain segments are stretched and oriented, leading to melt crystallization. It can be seen that the crystallinity of the sample gradually increases with the increase of stretching temperature, indicating that the effect of stress induced crystallization can be enhanced at higher stretching temperatures.

In addition to observing the surface crystal morphology of the thin film samples, this section used a 3D optical profilometer to measure the surface 3D contours of four types of long-chain branched polypropylene samples. The pseudo color images of the surface 3D contours of LP hot pressed film and three different stretching temperature stretched sample films are shown in Fig. 6.26. Observation of the surface structure shows that compared to the hot-pressed film, the surfaces of the three stretched films exhibit varying degrees of coarsening. From the graph, it can be observed that as the stretching temperature increases, the roughness of the film gradually decreases. The crystal dislocations related to surface grain collapse during the stretching process may be the reason for the deterioration of the surface roughness of the thin film. Just as the increase in stretching temperature enhances the effect of stress induced crystallization, the increase in stretching temperature also improves the structure of surface crystals damaged by stretching stress, thereby reducing the deterioration of surface roughness during the stretching process. During the operation of metalized thin film capacitors, the increase in surface roughness at the interface between the metal electrode and the dielectric thin film may lead to the enhancement of the local electric field near the surface micro tip, which promotes charge carrier injection and local space charge accumulation, thereby having a subsequent impact on the breakdown process.

To investigate the effect of stretching temperature on the internal microstructure and charge transfer characteristics of biaxially stretched long-chain branched polypropylene films, this section conducted bulk conductivity tests on the film samples at different environmental temperatures and electric strengths, as shown in Fig. 6.27.

As shown in the figure, compared with the control groups PP and LP hot pressed films under the same testing conditions, the DC conductivity of biaxially stretched long-chain branched polypropylene films at three stretching temperatures significantly decreased. This is because biaxial stretching promotes the biaxial orientation and regular arrangement of molecular chains, and the dense microstructure after stretching effectively hinders the transport of charge carriers. At the same time, the local deep traps introduced by stretching can also effectively capture charge carriers inside the film. The synergistic effect of the two factors leads to a decrease in conductivity, as mentioned earlier. As the stretching temperature increases, the DC conductivity of the stretched film sample shows a trend of first decreasing and then

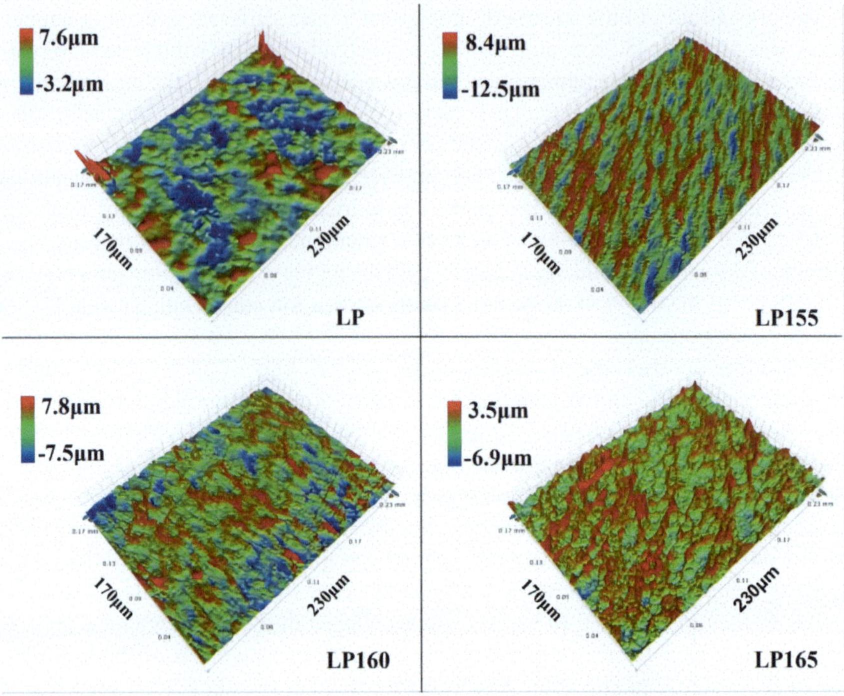

Fig. 6.26 Surface three-dimensional profiles with different stretching temperatures

increasing. At 25 °C, the DC conductivity of samples with stretching temperatures of 155 °C, 160 °C, and 165 °C under a field strength of 20 kV/mm were 3.53×10^{-16} S/m, 1.41×10^{-16} S/m, and 1.96×10^{-16} S/m, respectively. Among them, the DC conductivity of sample LP160 was the lowest. At high temperatures, the variation pattern of DC conductivity of thin film samples at different stretching temperatures is consistent with that at 25 °C. The DC conductivity of sample LP160 is the lowest, and at 125 °C, the DC conductivity at 10 kV/mm, 20 kV/mm, and 30 kV/mm field strengths is 2.10×10^{-14} S/m, 7.31×10^{-14} S/m, and 1.74×10^{-13} S/m, respectively. It can be inferred that the increase in stretching temperature enhances the stress induced crystallization effect, and the crystal morphology inside the stretched film is further improved through induced recrystallization. This tighter crystal structure further inhibits the migration and transmission of charge carriers, which is also the reason for the significant decrease in DC conductivity as the stretching temperature increases from 155 to 160 °C. However, as the stretching temperature increased from 160 to 165 °C, the DC conductivity of the thin film sample actually increased, and this pattern was reflected in all four testing temperatures. Based on the previous description, it can be seen that the higher stretching temperature enhances the effect of stress induced crystallization and reduces the defects and defects introduced into the aggregated structure by stretching. Therefore, compared to the other two stretching samples, sample LP165 exhibits the lowest deep trap energy level and density at

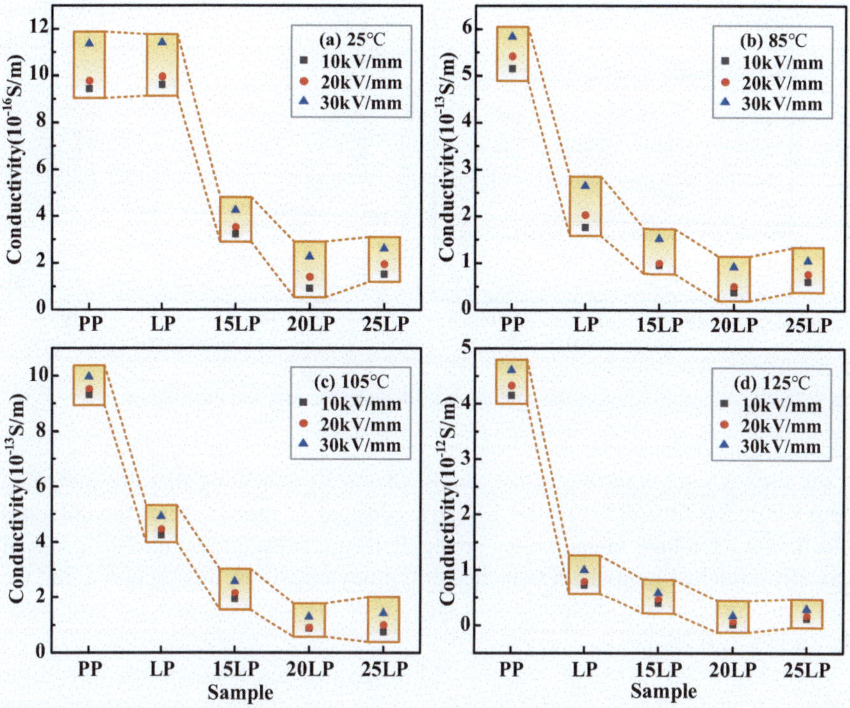

Fig. 6.27 Relationship between the conductivity and the temperature

all testing temperatures. Deep traps can capture the number and mobility of charge carriers inside polypropylene, enhance electron scattering effects, reduce high-energy electron energy, and ultimately lower DC conductivity. From this, it can be seen that although the stretching temperature of sample LP165 is the highest and the effect of tensile stress induced crystallization is the strongest, the decrease in deep trap energy level and density limits the further improvement of the dielectric properties of the sample under this process. As the stretching temperature increases from 160 to 165 °C, the DC conductivity of the sample does not decrease but instead increases.

In order to study the influence of different stretching temperatures on the dielectric constant and dielectric loss of biaxially stretched long-chain branched polypropylene films, this section tested the dielectric constant and dielectric loss tangent of the samples, and the results are shown in Fig. 6.28.

As shown in Fig. 6.28a, due to the introduction of long branched chain structure, the dielectric constant of LP hot pressed film is significantly higher than that of PP hot pressed film. The dielectric constant of the three stretched samples slightly decreased compared to the LP hot pressed film, all around 2.37. This is because the biaxial stretching orientation changes the distribution of intermolecular and main valence bond forces inside the film, hindering the dipole from alternating with the electric field, resulting in a slight decrease in dielectric constant. The dielectric constants

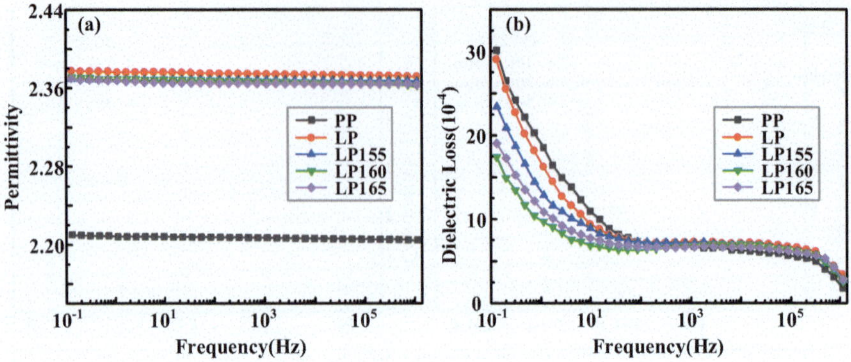

Fig. 6.28 Relationship between the **a** permittivity, **b** dielectric loss and the frequency

of the three stretching samples are almost identical, indicating that the stretching temperature has little effect on the dielectric constant. In order to study the influence of different stretching temperatures on the dielectric constant and dielectric loss of biaxially stretched long-chain branched polypropylene films, this section tested the dielectric constant and dielectric loss tangent of the samples, and the results are shown in Fig. 6.28.

From Fig. 6.28b, it can be seen that at low frequencies, the dielectric loss of the three stretching samples is lower than that of the control group, and first decreases and then increases with the increase of stretching temperature. The trend of change is consistent with the conductivity loss mentioned in Sect. 4.2.2 above. This can be attributed to the fact that the dielectric loss of the dielectric at low frequencies is mainly determined by the conductivity loss. It can be observed that at high frequencies, the dielectric loss of PP hot press film is the lowest, while that of LP hot press film is the highest. The dielectric loss of the three stretched samples is between the two, consistent with the distribution of dielectric constant. This is because the dielectric loss at high frequencies is mainly determined by the polarization characteristics of the dielectric. Overall, the dielectric properties of the stretched film sample are similar to those of the LP hot pressed film, indicating that the influence of stretching temperature on the dielectric constant and dielectric loss of the biaxially stretched long-chain branched polypropylene film sample is very limited.

This section investigates the DC breakdown characteristics of biaxially stretched long-chain branched polypropylene films prepared by different stretching temperatures at test temperatures of 25, 85, 105, and 125 °C. The Weibull probability distribution of breakdown field strength is shown in Fig. 6.29.

As shown in the figure, the characteristic breakdown field strengths of PP hot pressed film and LP hot pressed film are similar at 25 °C, indicating that the advantage of strong thermal stability of long branched chain structure at room temperature is not yet obvious. The characteristic breakdown field strength of the three stretching samples first increases and then decreases with the increase of stretching temperature, and the highest characteristic breakdown field strength of sample LP160 is

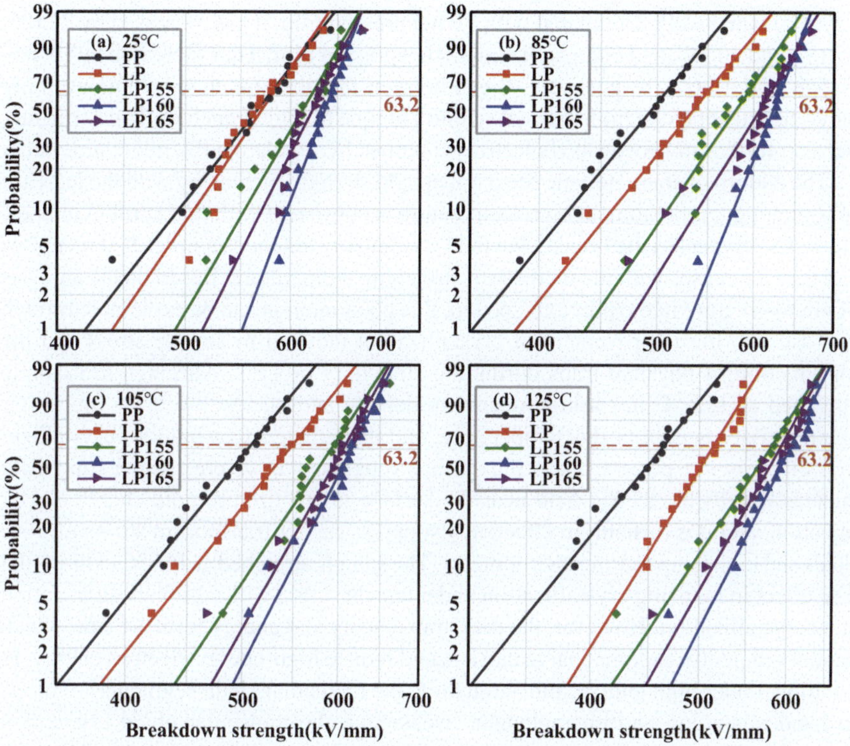

Fig. 6.29 Relationship between the breakdown strength and the temperature

644.6 kV/mm. As the testing temperature increases, the characteristic breakdown field strength of LP hot pressed film is significantly higher than that of PP hot pressed film. This is because the introduced long branched chain structure increases the entanglement between molecular chains, suppresses the slip of molecular chain segments, and thus reduces the influence of temperature on the free volume. The breakdown performance of the three tensile samples under high temperature has been further improved compared to the LP hot pressed film. The characteristic breakdown field strengths of samples LP155, LP160, and LP165 at 125 °C are 580.8 kV/mm, 606.0 kV/mm, and 592.2 kV/mm, respectively. This is because synchronous biaxial stretching transforms long-chain branched polypropylene from its original spherical crystal morphology to a densely stacked sheet-like fiber network structure. This dense crystal structure can effectively limit the transport of charge carriers inside polypropylene. At the same time, the introduction of sheet-like stacking faults and the band gap deep traps generated by dislocations during the stretching process further reduce the number and mobility of charge carriers inside polypropylene. The synergistic effect of the two aspects significantly increases the breakdown strength of the thin film sample. The characteristic breakdown field strength of sample LP160 is the highest, reaching 631.3 kV/mm, 618.2 kV/mm, and 606.0 kV/mm at 85 °C, 105 °C, and

125 °C, respectively. Compared with PP hot pressed film, it has increased by 24.0%, 25.1%, and 30.5%, and compared with LP hot pressed film, it has increased by 14.7%, 16.5%, and 17.6%, respectively. It can be seen that the stretched sample not only plays the advantage of high thermal stability unique to long branched chain structure, but also further improves its dielectric properties by regulating the microstructure.

The pattern of characteristic breakdown field strength increasing with the increase of stretching temperature at high temperature is consistent with that at room temperature. The characteristic breakdown field strength of stretched samples first increases and then decreases with the increase of stretching temperature. Based on the analysis of microstructure in the previous section, this phenomenon can be explained through two aspects: the improvement of crystallization induced by tensile stress and the introduction of physical traps during the stretching process. Although the sample stretched at 155 °C has a high deep trap density and energy level, the molecular chain activity is poor at this temperature, and the tensile stress induced crystallization effect is not perfect. Even the surface roughness is severe, which can easily enhance the local electric field near the surface micro tip, promote charge carrier injection, and it is difficult to effectively suppress carrier transport in strong electric fields and high temperature environments. The sample stretched at 165 °C exhibits the best effect in inducing crystallization under tensile stress, with complete aggregation state crystallization. However, the deep trap density and energy level are low, which cannot efficiently capture charge carriers and reduce high-energy electron energy. In addition, under high temperature stretching, the molecular chain segments are prone to local relaxation and entanglement, weakening the molecular chain orientation arrangement generated by biaxial stretching, making it difficult for the molecular chains to withstand high-speed carrier impacts. Therefore, compared to the sample LP160, the breakdown field strength decreases.

The temperature parameter settings during synchronous biaxial stretching directly affect the microstructure of long-chain branched polypropylene films, thereby affecting their dielectric properties. The microstructural differences between biaxial oriented long-chain branched polypropylene films prepared at different stretching temperatures and their impact mechanism on dielectric properties can be analyzed from both the molecular chain level and the aggregated state level. At a stretching temperature of 155 °C, the mobility and activity of the film's molecular chains are poor, which hinders the stretching orientation and increases the possibility of molecular chain and folded chain crystal breakage during the stretching process. Defects and defects such as stacking faults and dislocations introduced in the aggregated state increase, ultimately introducing a large number of local physical deep traps in the bandgap, enhancing the ability to capture charge carriers. However, the effect of tensile stress induced recrystallization at this temperature is not perfect, making it difficult to form a complete and dense ripple slip crystal structure to efficiently hinder carrier transport. At the same time, the deterioration of surface roughness enhances the local electric field nearby, promoting charge injection. At a stretching temperature of 165 °C, the stretching temperature is close to the melting temperature. At high temperatures, the unwinding and reorientation of molecular chain segments can cause strain relaxation in the surrounding chain segments, and also affect the optimal arrangement of molecular chain segments along the stretching direction. The advantage of stretching at this temperature is that the thickness and crystallinity of the crystal are greatly increased, and the effect of tensile stress induced crystallization is the best. The recrystallization caused by stretching refines the connections between crystal regions, and the dense crystal structure induced by stretching eliminates the original crystal fragmentation defects and electrical weaknesses. However, this also leads to low trap density and energy levels, which cannot efficiently capture charge carriers and reduce the energy of high-energy electrons. At a stretching temperature of 160 °C, on the basis of the original high thermal stability of long-chain branched polypropylene materials, synchronous biaxial stretching can transform the long-chain branched polypropylene matrix from the original spherical crystal morphology to a compact and dense network structure. The dense crystal structure induced by tensile stress can effectively limit the internal carrier transport of polypropylene. At the same time, the appropriate number of band gap deep traps caused by the introduction of stacking faults and dislocations during the stretching process further reduce the number and migration rate of carriers inside the thin film sample. The synergistic effect of two aspects significantly enhances the breakdown strength and energy storage characteristics of the thin film sample, which can achieve a synergistic improvement in the energy storage density, charge and discharge efficiency, and service life of polypropylene thin film for capacitors.

6.5 Summary

Aiming at the problems of increasing conductivity loss and decreasing breakdown strength of PP films at high temperature, this chapter introduces the method of improving dielectric properties of PP films based on long chain branching modification. The effects of the introduction of long branched chain structure, the grafting of functional monomers and the regulation of crystallization on the dielectric properties of PP films are described in detail. Compared with the current PP film, the dielectric properties of LCBPP have been greatly improved, but there are still some problems in theoretical research and practical application, and the following studies need to be carried out in the future:

Development of high-performance PP based on long chain branched modification. Long chain branched polypropylene with different molecular structures can be synthesized by changing the preparation technology. The modified PP with high dielectric properties was obtained by preparing LCBPP with different chain length and branched density.

Large-scale preparation of LCBPP films. PP films in metallized film capacitors are prepared by bidirectional stretching, so it is necessary to develop LCBPP films with excellent mechanical and processability to achieve batch preparation of LCBPP films.

Study on dielectric properties of LCBPP films under complex working conditions. In fact, metallized film capacitors are faced with complex operating conditions such as DC voltage, pulse voltage and harmonic voltage superimposed on each other. Therefore, it is urgent to investigate the dielectric properties of modified films under complex working conditions.

References

1. J. B. P. Soares, "Mathematical modeling of the long-chain branch structure of polyolefins made with two metallocene catalysts: An algebraic solution," *Macromol. Theor. Simul.*, vol. 11, no. 2, pp. 184-198, 2002.
2. W. Q. Weng, W. G. Hu, A. H. Dekmezian and C. J. Ruff, "Long chain branched isotactic polypropylene," *Macromolecules*, vol. 35, no. 10, pp. 3838–3843, 2002.
3. E. Kolodka, W. J. Wang, S. P. Zhu and A. E. Hamielec, "Copolymerization of propylene with poly(ethylene-co-propylene) macromonomer and branch chain-length dependence of rheological properties," *Macromolecules*, vol. 35, no. 27, pp. 10062–10070, 2002.
4. P. K. Agarwal et al, "Shear-induced crystallization in novel long chain branched polypropylenes by in situ rheo-SAXS and -WAXD," Macromolecules, vol. 36, no. 14, pp. 5226–5235, 2003.
5. Z. B. Ye, F. AlObaidi and S. P. zhu, "Synthesis and rheological properties of long-chain-branched isotactic polypropylenes prepared by copolymerization of propylene and nonconjugated dienes," *Ind. Eng. Chem. Res.*, vol. 43, no. 11, pp. 2860–2870, 2004.
6. M. Rätzsch, M. Arnold, E.Borsig, H. Bucka and N. Reichelt, "Radical reactions on polypropylene in the solid state," *Prog. Polym. Sci.*, vol. 27, pp. 1195–1282, 2002.
7. D. Graebling, "Synthesis of branched polypropylene by a reactive extrusion process," *Macromolecules*, vol. 35, no. 12, pp. 4602–4610, 2002.
8. F. Yoshii, K. Makuuchi, K. Keizo, T. Tanaka, J. Saitoh and K. Koyamaet, "High-melt-strength polypropylene with electron beam irradiation in the presence of polyfunctional monomers," *J. Appl. Polym. Sci.*, vol. 60, no. 4, pp. 317–622, 1996.
9. M. Sugimoto, T. Tanaka, Y. Masubuchi, J. Takimoto and K. Koyama, "Effect of chain structure on the melt rheology of modified polypropylene," *J. Appl. Polym. Sci.*, vol. 73, no. 8, pp. 1493–1500, 1999.
10. J. H. Tian, W. Yu and C. X. Zhou, "The preparation and rheology characterization of long chain branching polypropylene," *Polymer*, vol. 48, no. 7, pp. 7962–7969, 2007.
11. R. Ohnishi, T. Fujimura, R. Tsunori and Y. Sugita, "A new method for producing high melt strength poly(propylene) with a reactive extrusion," *Macromol. Mater. Eng.*, vol. 290, no. 12, pp. 1227–1234, 2005.
12. Y. J. An, H. P. Xing, Y. H. Wang and T. Tang, "Characterization of high melt strength propylene/1-butene copolymer synthesized by in situ heat induction melt reaction," *J. Appl. Polym. Sci.*, vol. 125, no. 4, pp. 2724–2731, 2012
13. K. Li, H. S. Zhou, Y. W. Qin, Y. Zhao, D. J. Wang and J. Y. Dong, "ω-Alkenylmethyldichlorosilane-assisted propylene polymerization with Ziegler-Natta catalyst to long chain-branched polypropylene," *Polymer*, vol. 202, 2020, Art. no. 122737.
14. K. Li, Y. W. Qin, S. M. Zhao and J. Y. Dong, "Blending behavior of high-degree long-chain-branched polypropylene prepared by Ziegler–Natta catalysis with common polypropylene," *Ind. Eng. Chem. Res.*, vol. 60, pp. 13614–13626, 2021.
15. M. Xiao, M. D. Zhang, B. X. Du, Z. Y. Ran, H. L. Liu and Y. W. Qin, "Breakdown strength improvement of polypropylene by introducing long-chain branched structures for film capacitors," *J. Phys. D: Appl. Phys.*, vol. 55, 2022. Art. no. 245501.
16. J. A. Langston, R. H. Colby and T. C. Chung, "Synthesis and characterization of long chain branched isotactic polypropylene via metallocene catalyst and T-reagent," *Macromolecules*, vol. 40, no. 8, pp. 2712–2720, 2007.

17. L. Wang, S. Ishihara, M. Ando, A. Minato, Y. Hikima and M. Ohshima, "Fabrication of high expansion microcellular injection-molded polypropylene foams by adding long-chain branches. *Ind. Eng. Chem. Res.*, vol. 55, no. 46, pp. 11970–11982, 2016.
18. L. Wang, D. Wang, J. Qiu and T. Tang, "Effects of long chain branches on the crystallization and foaming behaviors of polypropylene-g-poly(ethylene-co-1-butene) graft copolymers with well-defined molecular structures," *Polymer*, vol. 53, no. 21, pp. 4737–4757, 2012.
19. T. C. Chung, "Functionalization of polypropylene with high dielectric properties: applications in electric energy storage," *Green Sustain. Chem.*, vol. 2, pp. 29–37, 2012.
20. H. L. Liu, B. X. Du, Xiao M, Z. Y. Ran and Y. W Ma, "Aromatic compound improving polypropylene film performance for power capacitor in magnetic field," *IEEE Trans. Dielect Elect. Insul.*, vol. 29, no. 5, pp. 1754–1762, 2022.
21. M. Xiao, Y. S. Zhao and B. X. Du, "Dielectric performance of furan-functionalized polypropylene film for HVDC capacitors," *IEEE Trans. Dielect Elect. Insul.*, vol. 31, no. 1, pp. 160–167, 2024.
22. M. Xiao, M. D. Zhang, Liu H L, B. X. Du and Y. W. Qin, "Dielectric property and breakdown strength performance of long-chain branched polypropylene for metallized film capacitors," *Materials*, vol. 15, 2022, Art. no. 3071.
23. M. Xiao, Y. S. Zhao, K. X. Wang and B. X. Du, "Effect of biaxial stretching on the dielectric performance of long-chain branched polypropylene for metallized film capacitors," *IEEE Trans. Dielect Elect. Insul,* 2024.

Chapter 7
Irradiation Crosslinking Modification

7.1 Introduction

Currently, conventional polypropylene films encounter performance bottlenecks, such as poor dielectric properties and insufficient heat resistance, in devices such as high-voltage DC transmission converters, DC/DC converters for electric vehicles, and pulsed power supplies, which constrain their application in a number of critical areas. These issues have driven scientists and engineers to search for new methods and technologies to improve the performance of polypropylene films [1–3]. Irradiation cross-linking, as a modification technology, has been proven to have great potential in improving the properties of polypropylene films after years of research and practice. The development and optimization of irradiation cross-linking technology has made it one of the main means to improve the properties of polypropylene films [4]. In recent years, an increasing number of studies have focused on the optimization of irradiation cross-linking techniques, cross-linker selection, irradiation dose control, and in-depth studies of the effects on film properties [5]. These research efforts are geared towards achieving more efficient and controllable performance enhancement of polypropylene films, paving the way for their wider industrial applications. In summary, the enhancement of the dielectric properties of polypropylene films is crucial for their development in specific applications, and irradiation cross-linking shows great potential as an effective means of modification.

In this chapter, we will comprehensively explore the effects, mechanisms and application prospects of irradiation cross-linking on the enhancement of dielectric properties of polypropylene films. We will give an overview of different types of irradiation crosslinking techniques, including γ-ray, electron beam and UV crosslinking. Then, we will discuss in detail the effects of different irradiation methods on the dielectric properties of polypropylene films and their improvement effects. Finally, we will explore how to optimize the irradiation cross-linking conditions to improve the film properties and give an outlook on the future development direction. By exploring these topics in depth, this chapter will provide researchers and practitioners

B. Du and M. Xiao, *Polypropylene Film for HVDC Capacitors*,
https://doi.org/10.1007/978-981-96-3029-5_7

with an in-depth understanding of the role of irradiation cross-linking modification in enhancing the dielectric properties of polypropylene films, as well as guidance and insights for future developments.

7.2 Irradiation Crosslinking Method

7.2.1 Basic Principle

Crosslinking modification refers to the process in which polymer linear macromolecules are chemically bonded under the action of radiation (γ-rays, high-energy electron beams, ultraviolet light, etc.) or crosslinking agents to form a three-dimensional crosslinked network. At present, the main means of crosslinking modification are chemical crosslinking method and irradiation crosslinking method. Chemical crosslinking requires high-temperature melting conditions, after crosslinking polypropylene from thermoplastic to thermosetting materials, it is difficult to realize the subsequent stretching of polypropylene processing [6, 7]. Therefore, the irradiation crosslinking method has better application prospects for modified polypropylene films. The method carries out a solid-state crosslinking reaction at room temperature after the film is bi-directionally stretched, which has the advantages of simple preparation process and good applicability [8]. The crosslinking structure enhances the thermal stability of molecular chains in the amorphous region of polypropylene and inhibits the movement of molecular chain segments, which greatly reduces the influence of strong electric field and high temperature environment on the carrier transport and breakdown characteristics. The weak interaction force between the molecular chains in the amorphous region of polypropylene under extreme conditions is an important reason for the decrease of its dielectric properties, irradiation crosslinking modification is mainly used in the amorphous region of the molecular chains, and the improvement of crosslinking in the amorphous region can greatly improve the dielectric properties of polypropylene films (Fig. 7.1).

7.2.2 Irradiation Crosslinking Technology

(1) **γ-Ray Crosslinking**

γ-ray crosslinking is a key method widely used to enhance the properties of materials. γ-ray crosslinking is a technique that utilizes γ-rays as a source of radiation for crosslinking polymers for modification. This type of crosslinking plays an important role in the modification of polypropylene films and has significant potential to improve the properties and application range of the films. During γ-ray crosslinking, γ-rays, which are high-energy radiation generated by radioactive isotopes (e.g., cobalt 60), penetrate the polypropylene film and interact with carbon–carbon or

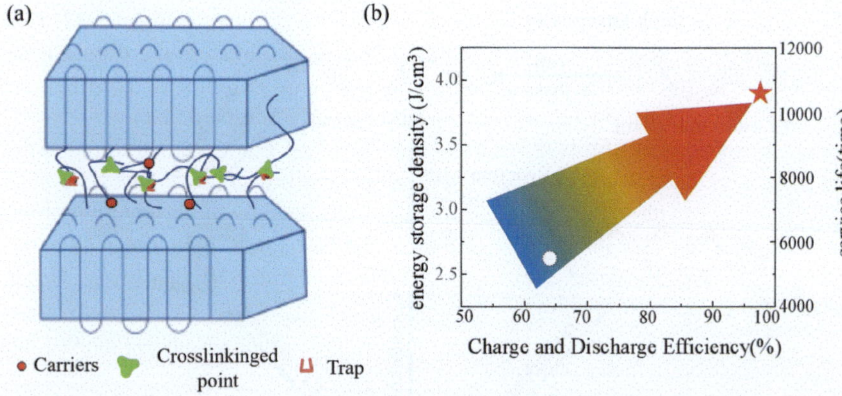

Fig. 7.1 **a** Microstructure of crosslinked PP, **b** comparison of dielectric performance between the linear PP and the crosslinked PP [9]. Copyright Clearance Center's RightsLink© License Number: 5811941028925

carbon-hydrogen bonds within the polymer molecule, triggering the formation of free radicals. These free radicals then initiate crosslinking of the polymer chains to form a more stable three-dimensional network structure, which improves the physical and chemical properties of the polypropylene film. The reaction mechanism of γ-irradiation crosslinking of polypropylene films is shown in Fig. 7.2.

The advantages of γ-ray crosslinking are characterized by its easy operation, no need to add additional chemicals, fast reaction speed and controllable crosslinking depth. This modification method is widely used in medical devices, electronic devices, packaging materials and other fields. In polymer film modification, γ-ray crosslinking technology is particularly notable because it can significantly improve the mechanical properties, heat resistance, chemical stability, and dielectric properties of polypropylene films, making them more suitable for applications in the field of electrical appliances. However, the effect of γ-ray crosslinking is affected by factors such as irradiation dose, temperature, and atmosphere. The choice of irradiation dose directly affects the degree of crosslinking and film properties, and therefore needs to be precisely controlled. In addition, when γ-ray crosslinking is used, the irradiated product should be treated appropriately to remove radiation-induced adverse effects.

(2) **Electron-Beam Crosslinking**

Electron beam crosslinking is a technique for crosslinking polymers by electron beam irradiation. In polypropylene film modification, E-beam crosslinking has attracted much attention for its high efficiency, precision and controllability, and has a significant effect on improving the performance of polypropylene films.

E-beam irradiation crosslinking polypropylene reaction mechanism and γ-ray irradiation is similar, are the use of high-energy rays and polypropylene molecules of chemical bonding to generate free radicals, these free radicals and then trigger the polymer chain crosslinking reaction. The difference is that the high-energy rays

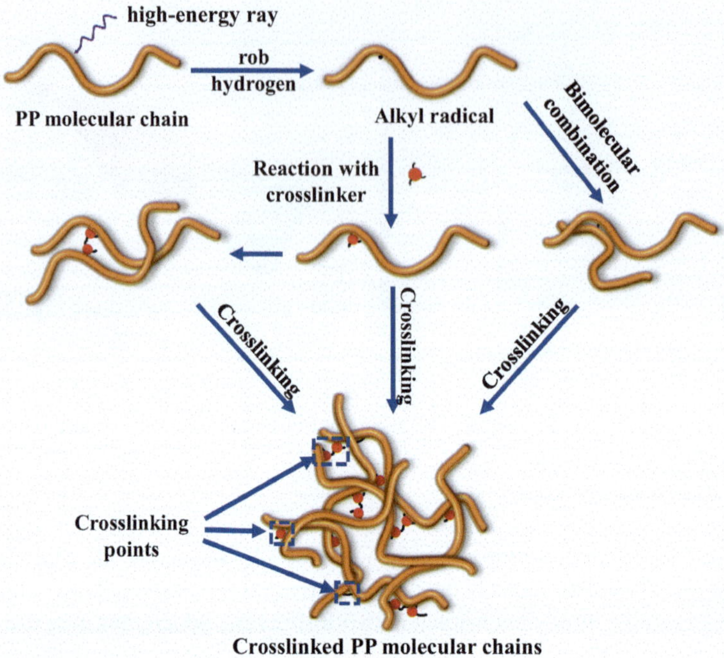

Fig. 7.2 Crosslinked PP formed by γ-irradiation [9]. Copyright Clearance Center's RightsLink© License Number: 5811941028925

of the E-beam crosslinking technology are accelerator-generated high-speed electron beams. Compared to γ-ray crosslinking, E-beam crosslinking does not require radioactive isotopes, is safer to operate, and allows for more precise control of crosslinking depth and degree of crosslinking. This makes E-beam crosslinking technology widely used in medical, food packaging, electronic devices and other fields. E-beam crosslinking improves the performance of polypropylene films significantly. First, the crosslinked polypropylene films have higher strength and heat resistance, making them more suitable for applications in high-temperature environments. Second, the crosslinking-modified films have better chemical stability and aging resistance, extending the service life of the material. Most importantly, electron beam crosslinking can significantly improve the dielectric properties of polypropylene films, including dielectric constant and dielectric loss, which is of great significance for its application in the field of electronic devices.

The effect of electron beam crosslinking can be realized by regulating parameters such as irradiation dose, energy, crosslinking temperature and rate. In the crosslinking process, the irradiation dose and energy directly affect the degree of crosslinking, higher dose and energy can improve the crosslinking effect, but need to be optimized and controlled according to the specific situation, in order to avoid excessive crosslinking and lead to embrittlement or other adverse effects. In addition, the crosslinking temperature is also an important factor affecting the crosslinking effect, the appropriate temperature can promote the crosslinking reaction, but too high a temperature may cause thermal degradation of the material.

(3) **UV Crosslinking**

Ultraviolet (UV) crosslinking is an important technology widely used in the field of polymer modification, and it also plays an important role in the crosslinking modification of polypropylene films. UV crosslinking is the use of ultraviolet light as a source of irradiation to crosslink polymers. In polypropylene film modification, UV crosslinking technology has attracted much attention because of its high efficiency, low temperature and high speed. The reaction mechanism of UV irradiation crosslinking of polypropylene films is shown in Fig. 7.3. The UV irradiation crosslinking reaction consists of two stages: radical generation and crosslinking. In the radical generation stage, the photoinitiator absorbs UV light, leaps to a single excited state, and crosses over to a triply excited state through the intersystem, thus seizing the hydrogen atoms on the tertiary atoms and forming the corresponding molecular radicals. In the free radical crosslinking stage, the active alkyl group combines with the crosslinking agent through the double base coupling reaction, and finally forms the crosslinking network structure. Compared to other irradiation crosslinking methods, UV crosslinking produces no radioactive waste, requires no high-temperature treatment, is easy to operate, enables localized or total crosslinking, and has a low thermal impact on the film.

UV crosslinking is effective in improving the properties of polypropylene films. First, UV crosslinking improves the mechanical strength and abrasion resistance of the film, and enhances its tensile and tear resistance. Secondly, the crosslinked film has better chemical resistance and aging resistance, and is more stable to environmental factors. The applicable conditions for UV crosslinking include the intensity of the light source, the irradiation time, and the concentration of the photoinitiator, etc. The appropriate selection of the photoinitiator and the suitable conditions for crosslinking are as follows. Appropriate selection of photoinitiators and irradiation parameters is the key to effective UV crosslinking. In addition, the absorption characteristics of the material, the surface state and other factors should also be taken into account to achieve the best crosslinking effect.

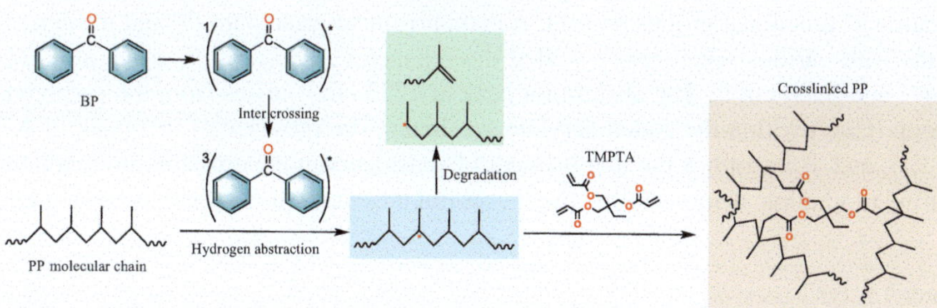

Fig. 7.3 Crosslinked PP formed by UV-irradiation [10]. Copyright Clearance Center's RightsLink© License Number: 5811941254988

7.3 γ-Rays

7.3.1 *Effect on the Microstructure*

Trimethylolpropane triacrylate (TMPTA), trimethylolpropane trimethacrylate (TMPT), and pentaerythritol triacrylate (PETA) were chosen as sensitizers. The experimental samples maintained a sensitizer ratio of 1.0 wt%. The samples were named A-5, T-5 and E-5, respectively.

The gel contents of samples containing various sensitizers are presented in Table 7.1. When compared to P-0, the introduction of sensitizers prior to irradiation led to a significant increase in the crosslinking of PP films. This resulted in a maximum crosslinking of 25.68% for E-5, with smaller differences of 22.54% and 21.75% for A-5 and T-5, respectively. The underlying principle behind the promotion of the crosslinking reaction by the sensitizer is the abundant generation of radicals post-irradiation, where the unsaturated double bond in the sensitizer takes precedence in the addition reaction. Ultimately, radical coupling terminates to form a three-dimensional network structure. It is evident that, at the same irradiation dose, the double bond content of the sensitizer plays a crucial role in influencing the crosslinking degree of the irradiated films. Upon analyzing the molecular structure of the three sensitizers under study, it was found that all of them contain three terminal unsaturated carbon–carbon double bonds, signifying an equal number of functional groups involved in the crosslinking reaction. In comparison to TMPTA, TMPT features a CH_3 group attached to the double bond's carbon, where the carbon–carbon covalent bond hinders the participation of the double bond in the crosslinking reaction, leading to a reduced crosslinking degree for T-5 when compared to A-5. On the other hand, PETA possesses a hydroxyl group (OH) attached to the terminal carbon, and the highly polar OH group is conducive to enhancing the activity of double bonds in the crosslinking reaction, thereby resulting in the highest crosslinking degree for E-5. It is evident that, aside from the double bond content, the functional groups in different types of sensitizers also impact the activity of the double bonds, ultimately leading to variations in the crosslinking outcomes.

Fourier transform infrared spectroscopy (FTIR) is commonly utilized to examine the structure of polymers, enabling the reflection of molecular structure and identification of functional groups present. A comparison between pure PP and irradiated films with various sensitizers revealed discrepancies in absorption peaks at 1723 cm^{-1} and 3300 cm^{-1}. The absorption peak at 1723 cm^{-1} arises from the carbonyl group (C=O) within the added acrylate sensitizer. The presence of this peak in A-5, E-5, and T-5 signifies the involvement of sensitizers in the crosslinking reaction during irradiation, demonstrating the successful introduction of sensitizer molecules

Table 7.1 Gel content of testing samples

Sample	A-5	T-5	E-5
Gel content (%)	22.54	21.75	25.68

into the PP matrix. The peak at 3300 cm^{-1} corresponds to the stretching vibration of OH, with the broadening of the OH peak attributed to hydrogen bonds. As only the PETA molecule contains the OH group, the corresponding absorption peak is exclusively observed in the infrared spectral curve of E-5.

The melting temperature (Tm) of E-5 is 165.1 °C, representing the most significant enhancement among the three sensitizers utilized, with a 2.4 °C increase compared to P-0. The melting peak temperatures of A-5 and T-5 are 164.9 °C and 164.8 °C, respectively. The considerably higher Tm of E-5 can be attributed to the heightened activity of the sensitizer, involving double bonds in the crosslinking reaction during irradiation. The formation of a three-dimensional network structure between the molecular chains through chemical bonding increases the intermolecular chain forces, leading to enhanced thermal stability of the films and requiring more energy during melting. The crystallinity of A-5 and T-5 decreased by 3.7% and 3.5%, respectively, compared to P-0, while E-5 exhibited the lowest crystallinity index (Xe) at 40.5%. The type of sensitizer impacts the degree of crosslinking, and combined analysis with the crosslinking degree data indicates that the activity between molecular chains is weakened due to crosslinking. Consequently, the participation of the chain segments in the crystallization process becomes slow and challenging, resulting in decreased crystallinity with an increase in the degree of crosslinking.

Differences in the enhancement of crystallization peak temperature by various sensitizers are observed. Following irradiation, E-5 shows an increase in Tc by 1.8 °C, whereas A-5 and T-5 exhibit improvements of 1.5 °C and 1.4 °C, respectively. The sequential increase in the crosslinking degree of T-5, A-5, and E-5 is attributed to variations in double bond activity, leading to a decrease in the free energy of nucleation during crystallization, consequently favoring the rise in crystallization peak temperature. Analyzing the data in the table provides insight into the sensitizer's impact on the crystalline half-peak width. E-5 demonstrates a small ΔW and narrow crystalline peak width, indicating uniform crystal size, thereby contributing to improved thermal stability.

7.3.2 *Effect on the Dielectric Properties*

(1) **Dielectric Constant and Dielectric Loss Factor**

The polarization of a dielectric refers to the phenomenon of elastic displacement and dipole orientation of bound charges under the action of an applied electric field. Dielectric constant is a measure of the degree of polarization of the dielectric under the action of the electric field of the macroscopic physical quantity, can reflect the ability of the dielectric storage charge, and the relative dielectric constant refers to the capacitance of the dielectric capacitor and vacuum capacitor ratio, the strong degree of polarization of the dielectric means that the electrode plate charge density is high, and therefore the value of the relative dielectric constant is also the greater. Due to the dielectric conductivity and dielectric polarization of the hysteresis effect, insulating materials in the electric field under the action of a certain energy loss, mainly by the conductivity and part of the lossy polarization caused by the total called dielectric loss, characterized by the dielectric loss factor, is an important evaluation of insulating materials parameters.

Polypropylene is a non-polar molecular polymer, and its internal polarization is dominated by electronic polarization, with a small number of other polarization forms originating from trace impurities such as oxidative decomposition products and antioxidant residues during processing. Electronic polarization establishment time is fast, dipole polarization by the polar molecules with permanent dipole moment along the external electric field orderly arrangement, the time required is longer. Under the action of the external electric field, the accumulation of electrons or ions on the surface of impurities in polypropylene or at the interface between the crystalline and amorphous regions produces interfacial polarization, which takes longer and therefore exists only in the low-frequency range.

Figure 7.4a shows the variation of relative dielectric constant with frequency for irradiated polypropylene films. As can be seen from the figure, the dielectric constant of polypropylene before and after modification are in the range of 2.20–2.25, the dielectric constant in the process of frequency increase from 100 to 105 Hz changes in the magnitude of the smaller, select the frequency of 50 Hz when the value of the longitudinal coordinates for analysis. After γ-irradiation, the dielectric constant of polypropylene increased; after adding sensitizer and then γ-irradiation, its relative dielectric constant was higher. The irradiation promotes the crosslinking degree, and the formation of three-dimensional mesh structure contributes to the improvement of film dielectric properties. On the one hand, the three-dimensional mesh structure can inhibit the dipole steering and reduce the dipole polarization; on the other hand, after the increase of crosslinking degree, the number of spherical crystals increases, the structure is dense, and the interfacial polarization between crystalline and amorphous regions is enhanced, and due to the latter's more significant effect, the relative permittivity of polypropylene films shows an increasing trend with the increase of crosslinking degree.

Figure 7.4b shows the variation of dielectric loss factor with frequency for irradiated crosslinking polypropylene. Due to the nonpolar nature of polypropylene, the

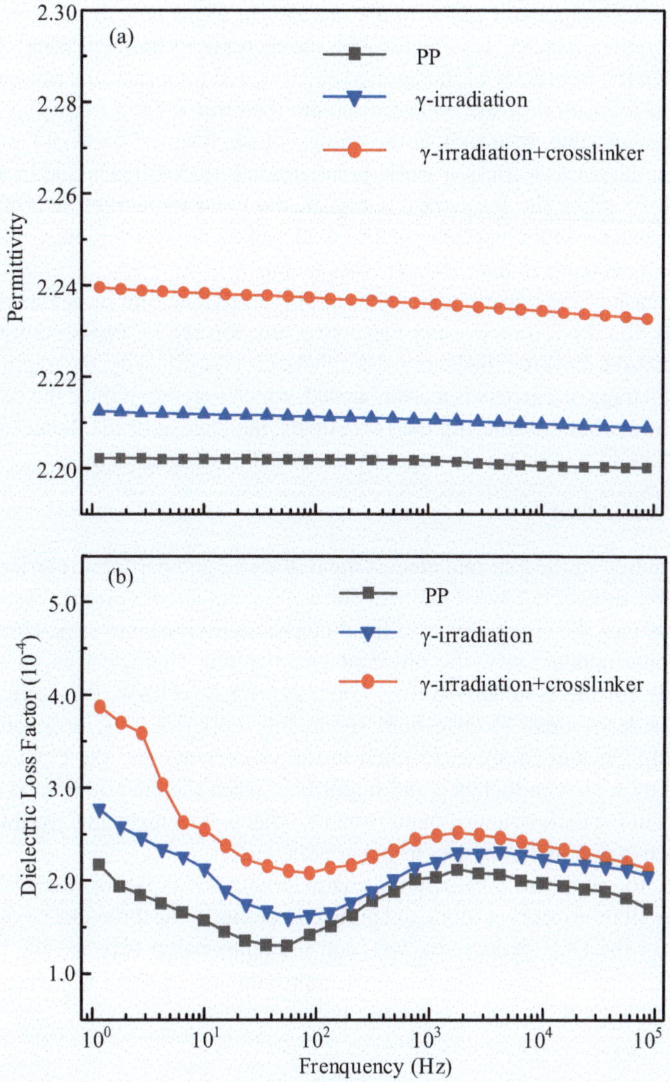

Fig. 7.4 Relationship between the **a** permittivity, **b** dielectric loss factor and the temperature

value of dielectric loss factor is low and varies between 0.0001 and 0.0005. As can be seen from the figure, when the frequency of the applied electric field increases, the dielectric loss factor of the sample first decreases and then increases, and then slightly decreases. There are two main factors affecting the dielectric loss, which are the conductivity loss due to leakage current and the polarization loss due to polarization. In the low-frequency range, polar molecules to polarization can be consistent with the frequency of the applied electric field, polypropylene film loss changes depend on the conductive loss, with the increase in frequency, the conductive loss decreases, so the dielectric loss factor with the increase in frequency and gradually reduce; when the frequency of the applied electric field increases, the polypropylene film internal relaxation of the polarization and the electric field frequency changes gradually out of sync, resulting in the energy in the form of frictional work Loss of energy in the form of friction work, polarization loss dominates and increase the dielectric loss; when the frequency is higher, the polar molecules steering cannot follow the changes in the applied electric field, polarization completely out of sync resulting in a decrease in the dielectric loss factor.

The increase of the dielectric constant of polypropylene film increases the polarization loss; the three-dimensional mesh structure formed by crosslinking inhibits carrier migration, and after the increase of spherical crystal density, the trapping effect of interfacial traps on carriers is also enhanced, which together reduce the conductive loss; therefore, after considering the two effects, the change of the dielectric loss in polypropylene film is relatively small by the effect of irradiation dose.

(2) **DC Conductivity**

Under the action of the external electric field, the charged particles (carriers) inside the dielectric directional movement to form the conductive current, based on the different carriers, the conductivity of the dielectric is divided into ionic conductivity and electronic conductivity, the physical quantity that characterizes its electrical conductivity for the conductivity (σ). For polypropylene films, the carriers in the dielectric under a weak electric field are mainly impurity ions, originating from various additives and catalysts residual in the processing, and the conductance is dominated by ionic conductance and impurities; when the electric field is stronger, the carriers in the dielectric originate from the injection of metal electrodes, and the conductance is dominated by electronic conductance.

In order to study the conductivity characteristics of polypropylene films with different irradiation doses at room and high temperatures, the three-electrode method was used for the DC conductivity test, and the relationship between the dielectric conductivity and the temperature can be approximated by the following equation when the strength of the applied electric field is certain:

$$\sigma = \mathrm{A}e^{-\frac{B}{T}} \tag{7.1}$$

where, is the dielectric conductivity (S/m), A and B are constants associated with the dielectric, and T is the absolute temperature (K). The equation shows that the conductivity of the dielectric increases with temperature in an exponential law.

The conductivity of polypropylene increases significantly under the influence of high temperature, and the order of magnitude of the DC conductivity of the film increases from 10^{-16} S/m to about 10^{-13} S/m, when the temperature is increased from 25 to 115 °C. On the one hand, the temperature increases, the thermal electron emission effect of the electrode is enhanced, a large number of electrons escape from the metal, and the carrier concentration inside the injected medium increases; on

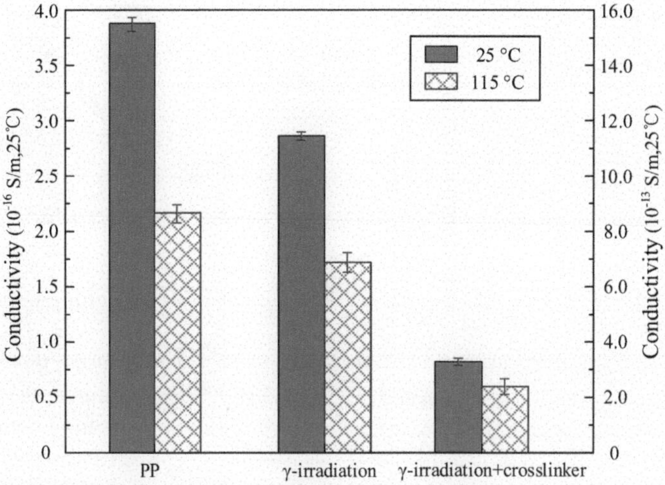

Fig. 7.5 Relationship between the conductivity and the temperature

the other hand, the molecular chain segments slip and relaxation intensify at high temperatures, which provide more space for the acceleration of the carriers, and contribute to the migration of carriers inside the thin film, and thus the conductive current increases, and the conductivity rises. The conductivity of polypropylene before and after irradiation is shown in Fig. 7.5.

At room temperature, direct irradiation of polypropylene films contributes to the reduction of conductivity, on the basis of which the conductivity properties are further improved by the addition of sensitizers. The three-dimensional mesh structure formed by crosslinking restricts the slip and relaxation of the polypropylene molecular chain segments, which effectively inhibits the carrier transport, and this inhibition is more significant as the crosslinking degree of the samples increases, resulting in the decrease of the conductivity of the samples; in addition, the increase of the density of spherical crystals introduces more interfacial traps, which improves the probability of the carriers to be trapped during transport, and contributes to a further decrease of the conductivity of the samples.

At high temperature, the rule of change of conductivity of polypropylene film is basically the same as that at room temperature, and the crosslinking film has the lowest DC conductivity. The crosslinking degree of the sample increases, on the one hand, the movement of the molecular chain at high temperature is inhibited, and the carrier migration is limited; on the other hand, the interface trap on the carrier trapping effect is enhanced, and the two together make the conductivity of the sample significantly reduced.

(3) **DC Breakdown Strength**

When the applied voltage is increased, the dielectric insulation performance is reduced due to breakdown, and the ratio of the breakdown voltage to the thickness of

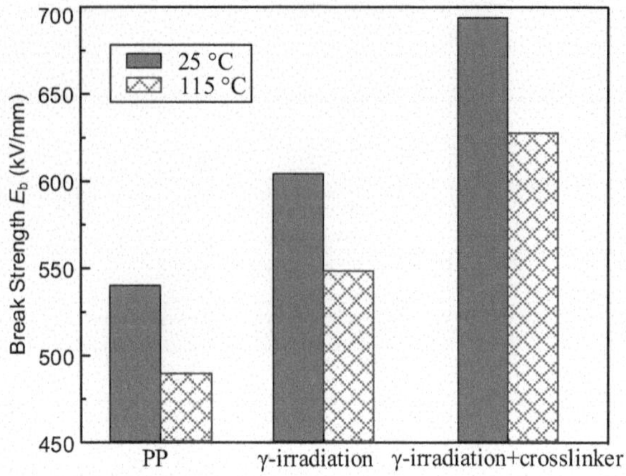

Fig. 7.6 Relationship between the breakdown strength and the temperature

the dielectric is the breakdown field strength at the temperature. At 25 °C, compared with the pure PP film, the crosslinking film has different degrees of improvement in the breakdown performance, the sensitizer-free samples irradiated with the increase in the breakdown field strength; the addition of crosslinking agent and irradiated with the increase in the breakdown field strength. Figure 7.6 shows the DC breakdown field strengths of polypropylene films before and after irradiation at 25 °C and 115 °C. At 115 °C, the breakdown field strengths of the irradiated polypropylene films follow the same trend as that at 25 °C, and the crosslinking reaction forms a three-dimensional mesh structure of the molecular chains through chemical bonding, which inhibits inter-chain slippage and localized relaxation of the chain segments, and enhances the thermal stability, resulting in a decrease in the mean free range of carriers, which is conducive to the breakdown of films at room temperature and elevated temperature. The increase of breakdown strength of the film at room temperature and high temperature.

When the temperature increases from 25 to 115 °C, the breakdown field strength decreases. This is because the thermal movement of polypropylene molecular chain segments at high temperatures is intensified, the free volume of the amorphous region increases, the carriers gain more energy in the migration process, which increases the chances of the molecular chain being broken by impact, and the breakdown of the film is more likely to occur at high temperatures. As the temperature increases from 25 to 115 °C, the breakdown strength of crosslinking polypropylene decreases to a lesser extent than that of pure polypropylene films. Therefore, crosslinking not only enhances the breakdown field strength of polypropylene films at ambient and elevated temperatures, but also plays a certain inhibitory effect on the weakening of the breakdown field strength at elevated temperatures.

(4) **Energy Storage Density**

The energy storage density of the dielectric material determines the volume occupied by the film capacitor in the power equipment. When using dielectrics with excellent

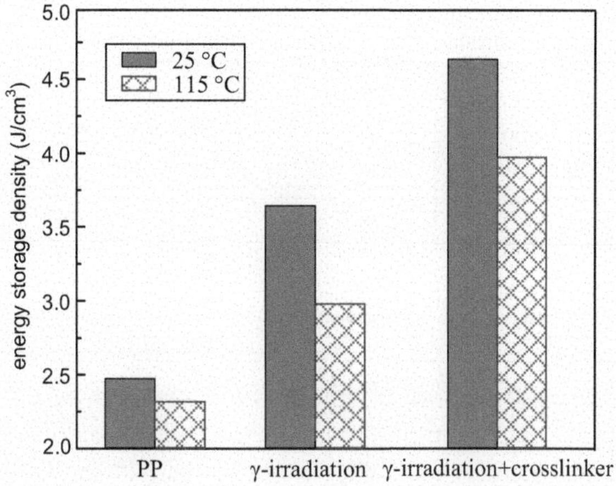

Fig. 7.7 Relationship between the theoretical energy storage density and temperature

energy storage characteristics, the capacitor requires less volume to store the same amount of energy, which is in line with the trend of miniaturization of the power equipment, so it is important to improve the energy storage density of the dielectric material. In this section, the theoretical energy storage density of polypropylene films with different irradiation doses is analyzed, and based on the relative permittivity of the film samples at the industrial frequency (50 Hz) and the DC breakdown field strengths at different temperatures, the maximum theoretical energy storage density at the corresponding temperature of the samples is obtained, as shown in Fig. 7.7.

As can be seen from the energy storage density calculation formula, the energy storage density results are mainly determined by the DC breakdown field strength due to the small difference in the relative dielectric constant values before and after irradiation. The polypropylene breakdown field strength decreases significantly during the temperature from 25 to 115 °C, and the theoretical energy storage density also decreases significantly, and the crosslinking structure enhances the thermal stability of the films. The crosslinking modification of the film has a considerable degree of enhancement. Irradiation promotes the generation of free radicals in the molecular chains, which provides conditions for the crosslinking reaction, and the selection of irradiation dosage and the addition of sensitizers contribute to the enhancement of the degree of crosslinking, which significantly enhances the breakdown field strength of polypropylene films at ambient and high temperatures through the enhancement of the intermolecular chain force and the improvement of the crystalline properties, thus increasing the energy storage density of the films.

Based on the above microstructure characterization and dielectric properties test results analysis, γ irradiation and added sensitizers make polypropylene film performance has been significantly improved, γ irradiation on polypropylene film microstructure and dielectric properties of the mechanism shown in Fig. 7.8.

Irradiation for polypropylene molecular chain breaking covalent bonds to provide energy, inducing the generation of free radicals, and then by the polymer free radicals to produce subsequent reactions, is a necessary condition to cause the molecular chain

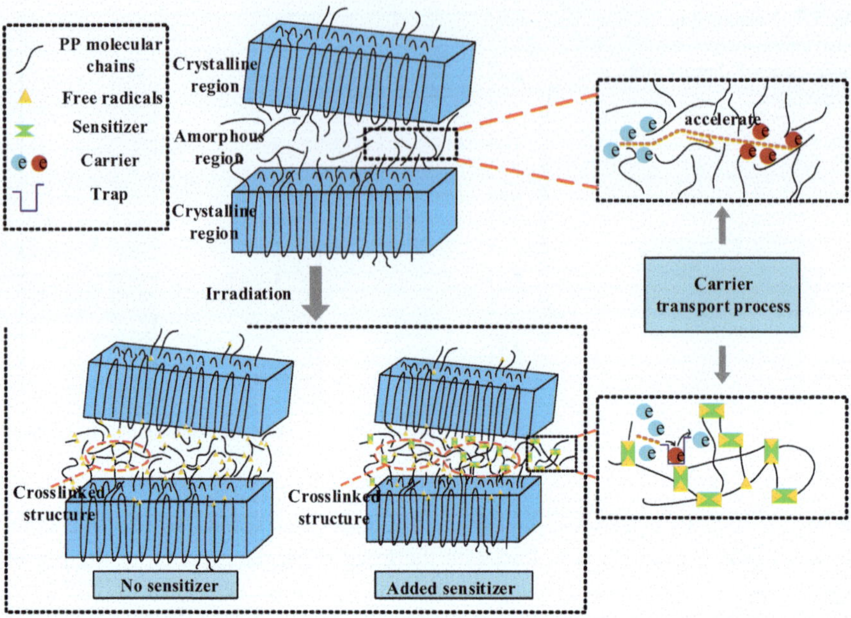

Fig. 7.8 Schematic diagram of the γ irradiation process

crosslinking. In the irradiation process, degradation and crosslinking reactions are in a competitive state, and the higher dose increases the speed and number of free radicals, the molecular chain free radicals have a small probability of collision and bonding to form a three-dimensional mesh structure, and when the irradiation dose is too high most of the tendency is to degradation. Sensitizers play a bridging role in the irradiation process of polypropylene films, helping to enhance the degree of crosslinking between molecular chains to form a large number of three-dimensional mesh structures.

After γ irradiation, the degree of crosslinking to form a three-dimensional mesh structure increase, and the irradiated DC conductivity is low, the breakdown field strength and energy storage density are high. On the one hand, the three-dimensional mesh structure can enhance the interactions between molecular chains, and the chemical bonding between the chains inhibits the slip and relaxation of the chain segments at room and high temperatures, reduces the average free range of the carriers, and thus reduces the probability of their impact on the molecular chains, and enhances the molecular chain thermal stability; on the other hand, the formation of the crosslinking point as a heterogeneous nucleation point, which significantly increases the density of the nucleation point, the number of spherical crystals, and the number of spherical crystals increases, and due to the crosslinking of molecular chains after the folding into a The process of layered structure after crosslinking of molecular chains is significantly inhibited, resulting in the reduction of spherical crystal size and the enhancement of carrier trapping by the interfacial traps between the crystalline and amorphous regions, and therefore, the conductivity of polypropylene films with high degree of crosslinking is significantly reduced, and the breakdown field strength and the maximum energy storage density are substantially increased. In addition, the dense spherical crystal structure of the films with high crosslinking degree and the enhanced interfacial polarization between the crystalline and amorphous regions are conducive to the enhancement of the relative dielectric constant.

7.4 UV Irradiation

7.4.1 Effect on the Microstructure

As an important medium for absorbing ultraviolet rays and participating in subsequent hydrogen abstraction reactions, the content of photoinitiators directly affects the efficiency of photopolymerization reactions, which in turn affects the amount of free radicals generated in polypropylene molecules, thereby affecting the overall reaction efficiency of UV crosslinking. In this chapter, the microstructure of crosslinked polypropylene film under UV irradiation was investigated with different content of photoinitiator.

The absorption peaks of the untreated polypropylene film did not show any changes, mainly concentrated between the 2830–3000 cm^{-1} wavelength range and the lower wavelength range of 500–1500 cm^{-1}. These absorption peaks correspond to the chemical bond vibrations inherent in the polypropylene molecules themselves. Compared with polypropylene, a new absorption peak appeared at 1740 cm^{-1} in the infrared spectrum of polypropylene film after UV irradiation, mainly attributed to the C=O group contained in the crosslinking agent monomer TMPTA. After UV irradiation, the crosslinking agent monomer binds with polypropylene molecules through chemical bonds, resulting in the formation of new polar groups in the chemical structure of polypropylene. However, there are significant differences in the intensity of absorption peaks among samples with different levels of photoinitiators added. The samples treated only by irradiation without the addition of photoinitiators showed weaker absorption intensity, which not only proves the occurrence of crosslinking reaction in the samples, but also reflects a smaller degree of reaction. As the content of photoinitiator increases, the crosslinking reaction rate increases and the degree of crosslinking increases, which is reflected in the intensity of new absorption peaks in different samples.

Under UV irradiation, the spherulite size of cross-linked polypropylene samples decreases and the structure becomes denser. The degree of improvement varies depending on the amount of photoinitiator added. With the increase of the content of photoinitiators added, the number of photoinitiators that can participate in photoinitiated reactions increases, and the degree of cross-linking of the generated macromolecular free radicals significantly improves. The cross-linking point increases the initial nucleus density through heterogeneous nucleation, which limits the growth of spherulites by adjacent crystals. In addition, cross-linking networks limit the degree of freedom of chain segments, causing more spatial obstacles for polypropylene molecules during folding and arrangement, making it difficult to form large-sized spherical crystals. After crystallization, the spherical morphology of modified polypropylene is compact and small. When the content of photoinitiators increases to a certain extent, excessive photoinitiators cause the formation of small molecular chain segments in the matrix, further increasing the crystallization ability of polypropylene.

7.4.2 *Effect on the Dielectric Properties*

(1) **Dielectric Constant and Dielectric Loss Factor**

Following exposure to UV irradiation, the infusion of polar groups into the crosslinking sites within the crosslinked polypropylene (PP) matrix serves to heighten the polarity of the molecular chains. Nevertheless, the presence of the crosslinking structure poses a barrier to the mobility of chain segments, thereby impacting the alignment of the molecular chains and subsequently influencing the polarization of the modified polypropylene. The relative permittivity of the irradiated films exhibits a marginally lower value in comparison to unmodified PP. Figure 7.9 provides a depiction of the relative permittivity and dielectric loss across varying frequencies for the samples. Moreover, the dielectric loss subsequent to irradiation is noted to be lower than that of pure polypropylene. This occurrence can be attributed to the establishment of a crosslinking network, which hampers the flow of charge carriers and consequently reduces conduction losses. Consequently, the irradiated samples demonstrate heightened dielectric stability within the tested frequency range, with further enhancements in dielectric stability discernible with prolonged irradiation durations.

(2) **DC Conductivity**

When exploring the mechanism through which UV irradiation influences electrical conductivity, our initial investigation focused on the manipulation of material properties by means of crosslinking structures. The creation of a three-dimensional network through crosslinking has a substantial impact on conductivity. This stable crosslinking framework acts as a center for capturing molecules, effectively restricting the movement of charge carriers. As a result, the irradiated samples demonstrate reduced conductivity at 25 °C, particularly pronounced in the modified film exposed to 90 s of irradiation, which exhibited only 55.8% of the conductivity observed in the pure film. This finding highlights the advantageous adjustment of conductivity properties facilitated by the presence of crosslinking structures.

In addition, the irradiated modified samples maintained relatively low conductivity levels when the ambient temperature was increased from 25 to 115 °C, with the modified films irradiated for 90 s being only 2.5% of the pure films. The main reason

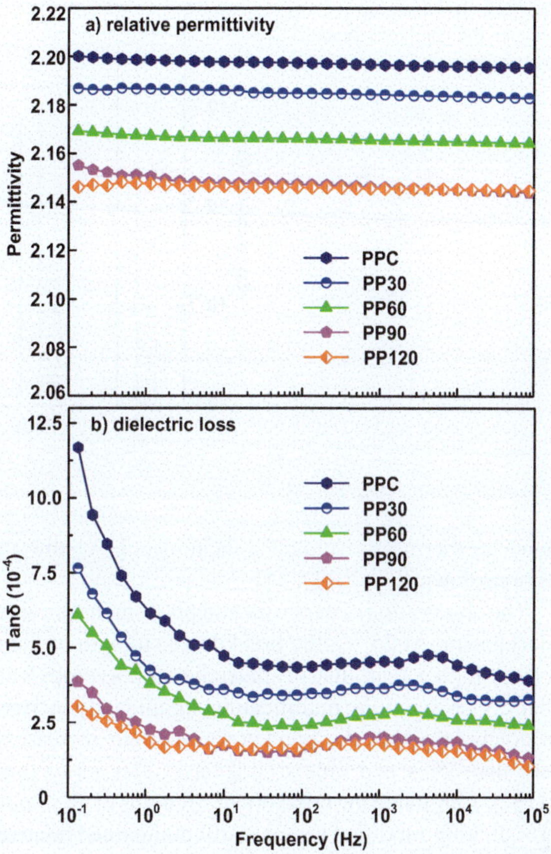

Fig. 7.9 Relationship between the permittivity, the dielectric loss and the frequency [11]. Copyright Clearance Center's RightsLink© License Number: 5811950157963

for this difference is that the crosslinking structure induced by the irradiation treatment forms a barrier in the carrier transport path, which restricts the free migration of charge and thus effectively reduces the leakage current. The crosslinking reaction in the polypropylene matrix gradually weakened with the increase of irradiation time, leading to the stabilization of the crosslinking structure ratio (Fig. 7.10).

(3) **DC Breakdown Strength**

In understanding the internal mechanism of the effect of crosslinking on breakdown strength, we focus on the structural changes induced by crosslinking at the molecular level. UV crosslinking modification triggers the formation of covalent bonds between molecular chains, building a three-dimensional network structure. This structural modification not only effectively reduces the distance between molecular chains to form a more compact network, but is also able to limit the migration of electrons or holes in the presence of an electric field, thus increasing the breakdown strength. At the same time, crosslinking also reduces the free volume of the material and slows

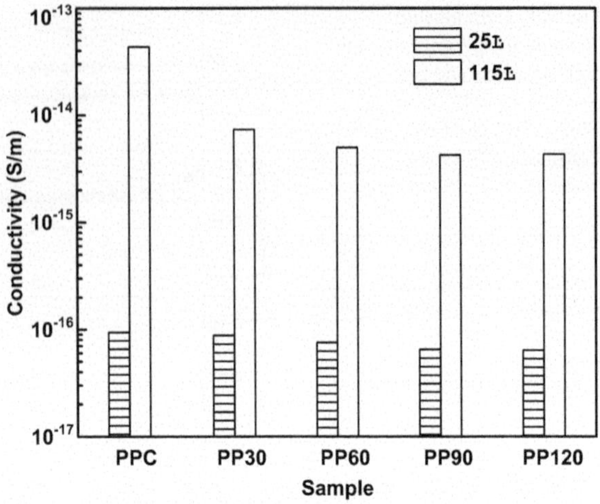

Fig. 7.10 Relationship between the conductivity and the temperature

down the thermal movement of the molecular chains, thus suppressing the possibility of breakdown (Fig. 7.11).

The observation of experimental phenomena provides support for this crosslinking mechanism. At 25 °C, the breakdown strength of the modified samples was significantly higher than that of unirradiated polypropylene (PP). This clearly indicates that UV crosslinking modification significantly increases the breakdown strength. It is noteworthy that the breakdown strength increases rapidly for irradiation times less than 90 s and almost stagnates after more than 90 s. At high temperature (115 °C), although the breakdown strengths of all samples decreased significantly, the crosslinking-modified samples still maintained relatively high breakdown strengths. This verifies that the crosslinking reaction significantly improves the thermal stability

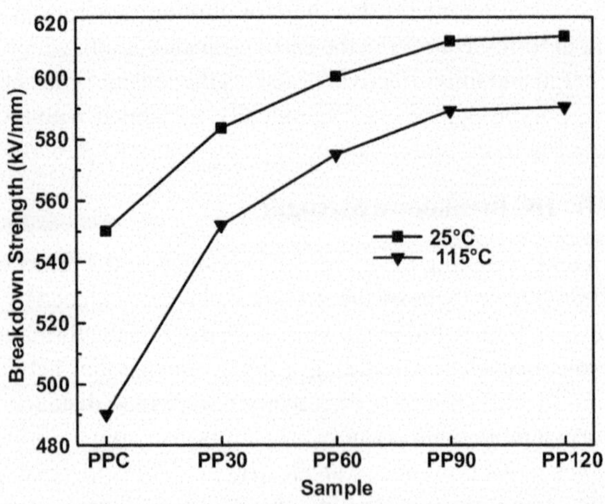

Fig. 7.11 Relationship between the breakdown strength and the temperature

of the material, mainly originating from the formation of a three-dimensional network structure that reduces the distance between molecular chains and improves the thermal resistance of the material. There is no tendency for crosslinking to have a further positive effect on improving the stability of the dielectric strength of the modified PP after irradiation times exceeding 90 s. This experimental phenomenon is consistent with the internal mechanism of crosslinking and provides a key clue for a deeper understanding of the effect of crosslinking on breakdown strength.

(4) **Energy Storage Density**

The internal mechanism of UV crosslinking on energy storage density involves the influence of several aspects, mainly focusing on the adjustment of molecular structure and the optimization of electrical properties. First, the crosslinking network allows the formation of tighter connections between the molecular chains of the material. The formation of this structure effectively reduces the distance between molecular chains and improves the mechanical strength of the material. The crosslinking structure also restricts the migration of electrons or holes under the action of an electric field, which improves the breakdown strength and plays a key role in the increase of energy storage density. At the same time, the additional crosslinking points introduced by the crosslinking process play an important role in the enhancement of energy storage density. These crosslinking points become fixed points in the material, restricting the movement of molecular chains. This restriction effect helps to inhibit charge migration and maintain the stability of the material under electric field, which is especially significant under high temperature conditions (Fig. 7.12).

Observing the experimental phenomena, the energy density of the film gradually increases and tends to be constant with the extension of UV irradiation time. At 25 °C, the energy density of the modified samples was significantly higher than that of unirradiated polypropylene (PP). This indicates that the UV crosslinking modification effectively increased the energy storage density. However, the energy density of

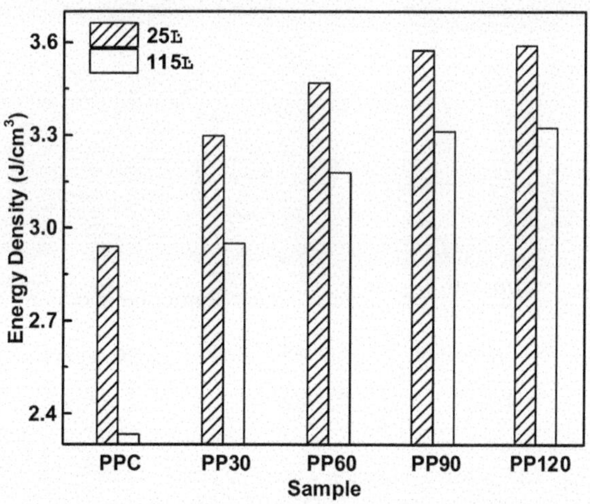

Fig. 7.12 Relationship between the energy density and the temperature

the unmodified samples decreased significantly with increasing temperature, while the decrease of the modified samples was relatively small. This reflects the critical role of the crosslinking network structure in maintaining the electrical and energy storage properties of the material under high temperature conditions. The fact that the modified samples still maintain relatively high energy density at high temperatures indicates that the formation of crosslinking structure improves the thermal stability of the material and enhances breakdown strength, thus increasing the energy storage density.

The modification through crosslinking influences the migration of charges by regulating the molecular structure, as depicted in Fig. 7.13. Owing to the dense structure of polypropylene (PP) spherulites, the migration of carriers primarily takes place in the amorphous region under the influence of an electric field. At elevated temperatures, a greater number of charges are injected from the electrodes into the medium, leading to increased charge mobility. The UV crosslinking modification method enhances the density of chemical traps in the matrix by introducing new polar groups. The deeper traps have the ability to capture carriers and immobilize them in localized areas, thus impeding charge conduction and improving charge transport characteristics. Furthermore, the crosslinking network constrains the movement of molecular chains, impeding the expansion of free volume at high temperatures, thereby diminishing the escalation of carrier energy. Consequently, the sharp decline in dielectric properties observed in original PP with increasing temperature is not observed in the crosslinked modified samples.

The duration of irradiation is a critical factor governing the crosslinking reaction within the medium during UV exposure. When the irradiation period is less than 90 s, the heightened concentration of photoinitiator facilitates the generation

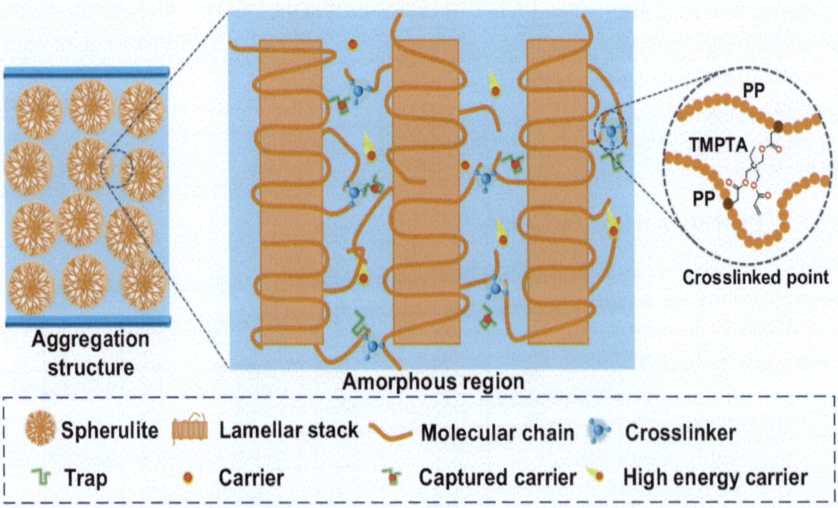

Fig. 7.13 Charge migration within the crosslinked PP [11]. Copyright Clearance Center's RightsLink© License Number: 5811950157963

of an ample number of macromolecular radicals. These radicals actively participate in the crosslinking process, leading to a swift and pronounced increase in the crosslinking degree of the modified samples. However, as the duration of irradiation extends, the concentration of the photoinitiator gradually diminishes. Consequently, the crosslinking reaction decelerates, and there is a gradual augmentation in the gel content within the medium. Once the irradiation time surpasses the 90-s threshold, the consumption of the photoinitiator becomes significant, ultimately halting the crosslinking reaction altogether. It is noteworthy that prolonged irradiation does not yield proportional improvements in the dielectric properties of the medium. This fact is substantiated by the outcomes of dielectric assessments conducted on samples PP90 and PP120. Thus, it is prudent to limit the irradiation time to 90 s for UV crosslinked modified polypropylene to minimize energy wastage and optimize the crosslinking process (Fig. 7.14).

7.5 EB Irradiation

7.5.1 Effect on the Microstructure

In the irradiation cross-linking modification of polypropylene films, the irradiation dose is an important parameter of the irradiation treatment, which can affect the generation of free radicals and thus change the degree of cross-linking of the films. In this chapter, the microstructure of polypropylene films crosslinked under electron beam irradiation at different irradiation doses is investigated.

Gel content was absent in the initial PP samples, signifying that cross-linking had not taken place. As irradiation dose escalates, it brings about an increase in the free radicals engaged in crosslinking, which also escalates the gel content to its highest recorded value of 17.35% at an irradiation dose of 15 kGy. Upon the application of PETA as a crosslinking agent and subsequent exposure to high-energy rays, PETA catalyzed the free radical interactions, thereby promoting PP cross-linking and causing a substantial elevation in the gel content of the treated samples. However, at an irradiation dose over 15 kGy, the overabundance of tertiary carbon radicals might trigger oxidative degradation of the polypropylene matrix.

The infrared spectral curve band position and absorbance reflect the type and content of the characteristic groups of the sample, respectively. During the irradiation process, the free radicals generated on the polypropylene molecular chain easily react with oxygen to form peroxides, which in turn generate oxygen-containing macromolecules, and this process ultimately leads to the degradation of the molecular chain, and the degree of degradation is reflected by the absorbance of the corresponding bands. Comparing the infrared spectral curves of the modified polypropylene films at different irradiation doses, the four absorption peaks between 2830 cm^{-1} and 2950 cm^{-1} are the saturated C–H stretching vibration peaks in $-CH_3$ and $-CH_2-$. Compared with pure PP, the changes in the modified samples are mainly reflected

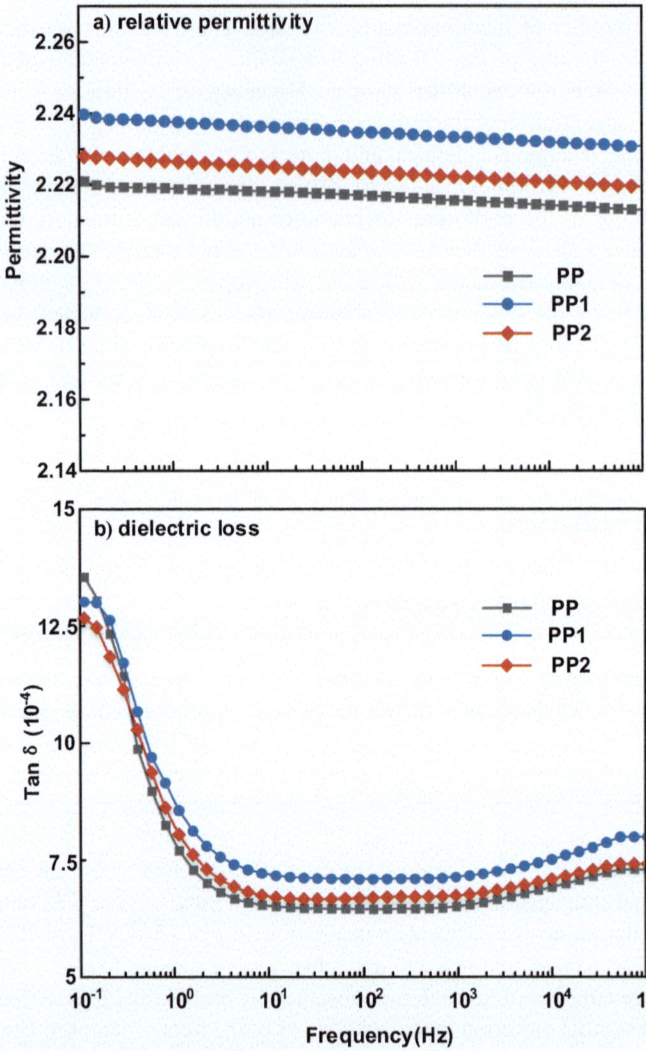

Fig. 7.14 Relationship between the permittivity, the dielectric loss and the frequency

in the strength of the absorption peak at 1728 cm^{-1}, which is caused by the C=O stretching vibration resulting from the cleavage and oxidation of the PP main chain. Compared with PP, the difference in absorption peaks of cross-linked PP is also mainly reflected at 3300 cm^{-1}, corresponding to the hydroxyl group (OH). This is related to the introduction of functional groups into the matrix in PETA.

As a result, the crystallisation temperature of the modified samples was generally lower than that of pure PP, especially for the sample with an irradiation dose of 15 kGy, where the crystallisation temperature decreased by 3%. In the matrix of cross-linked polypropylene, the stable network structure acts as an obstacle to the orderly arrangement of the molecular chains and has a negative effect on the crystallisation behaviour. And the melting temperature of crosslinking PP increased to 165.3 °C, which was higher than that of pure PP. The network structure formed by cross-linking strengthens the intertwining of molecular chains and slows down the slip of molecular chains, which improves the heat resistance.

As the dose increases, the size of the spherical crystal decreases, the boundary is clear, the crystallisation is dense, the irradiation dose of 20kGy, the density of the nucleus is slightly reduced. The change of irradiation dose affects the degree of crosslinking of the sample, polypropylene molecules crosslinked to a high degree, the formation of three-dimensional mesh structure so that the intermolecular chain force to enhance the combination of close, reducing the activity of the macromolecular chain, the crystallisation process, the chain segments of the wafer to the front edge of the migration and diffusion rate is significantly slowed down, the process of folding into a layered structure is inhibited, resulting in the formation of a larger size of the ball crystal is difficult to form. The cross-linking point as a heterogeneous nucleation centre is conducive to increasing the number and density of the initial nuclei, and thus the ball crystals are more likely to come into contact with each other during the growth process, which ultimately leads to a significant reduction in the size of the ball crystals.

Analysing the above experimental results, it is found that E-beam irradiation is a necessary condition to cause cross-linking of polypropylene molecular chains, and the irradiation promotes the generation of free radicals in polypropylene, and the rate and quantity of free radicals generated increase with the higher dose; the addition of cross-linking agent helps to form a three-dimensional mesh structure by cross-linking the free radicals of the molecular chains through the chemical bonding, and the enhancement of the cross-linking degree contributes to the improvement of microstructure of the polypropylene films.

7.5.2 Effect on the Dielectric Properties

(1) **Dielectric Constant and Dielectric Loss Factor**

In the experimental setup, the crosslinking agent PETA, comprising polar groups, plays a pivotal role. Its incorporation into the films induces a notable augmentation in dipole polarization. This augmentation can be attributed to the formation of stable crosslinking points by the polar groups of PETA during the crosslinking process. Consequently, there is an effective elevation in the dipole moment within the film matrix. This heightened dipole polarization manifests in the UV irradiation crosslinked modified PP films exhibiting a marginally elevated relative dielectric constant compared to pure PP films. The presence of PETA within the film structure thus engenders subtle yet discernible alterations in the dielectric properties, underscoring its significance in the modification process.

Furthermore, it is imperative to underscore the profound influence of the degree of crosslinking on the relative dielectric constant of the modified PP films. As the degree of crosslinking diminishes, there is a notable escalation in the relative dielectric constant. This phenomenon can be primarily attributed to the highly stable network structure established within the crosslinked PP matrix. The stability inherent in this crosslinking structure imposes constraints on dipole polarization, thereby impeding fluctuations in the dipole moment. Consequently, the extent of the increase in the relative dielectric constant is directly impacted by the degree of crosslinking. This elucidates the intricate interplay between the structural integrity of the crosslinked network and its ramifications on the dielectric properties of the modified PP films.

The rise in the dielectric constant of crosslinked PP films leads to an increase in their polarization loss. However, the crosslinking structure hinders the migration of carriers and the trapping effect of carriers by interfacial traps is enhanced. This leads to a decrease in the conductivity loss. Therefore, the results indicate that the effect of electron beam irradiation dose on the dielectric loss of PP films is relatively small.

(2) **DC Conductivity**

The irradiated samples had lower conductivity at lower temperatures. The irradiation-induced crosslinking structure has an inhibitory effect on the conductivity. This is consistent with the mechanism that the crosslinking structure restricts the molecular chain movement of PP. The conductivity of the films increased significantly with increasing temperature, especially at 125 °C, by three orders of magnitude. This can be explained by the enhancement of Schottky emission and more drastic thermal movement of the molecular chains at high temperatures leading to the increase in conductivity. However, it is worth noting that the crosslinked structure reduces the free volume of PP, and therefore the decrease in conductivity of the crosslinked PP films is more pronounced at high temperatures. This experimental phenomenon provides a substantial observation of the temperature dependence of crosslinking on conductivity (Fig. 7.15).

(3) **Dynamic Charge**

The Q_0 is defined as the initial charge generated at the plate interface, and its relationship with the relative permittivity is shown in (7.2).

$$Q_0 = \varepsilon_0 \varepsilon_r \frac{S}{d} V_{app} \tag{7.2}$$

where V_{app} is the test voltage, ε_r is the relative permittivity of the polymer, S and d the area and thickness of the polymer, respectively.

The charge accumulation rate K is identified as the ratio of the accumulated charge in the polymer Q within 500 s to the initial charge Q_0 stored after the application of high voltage.

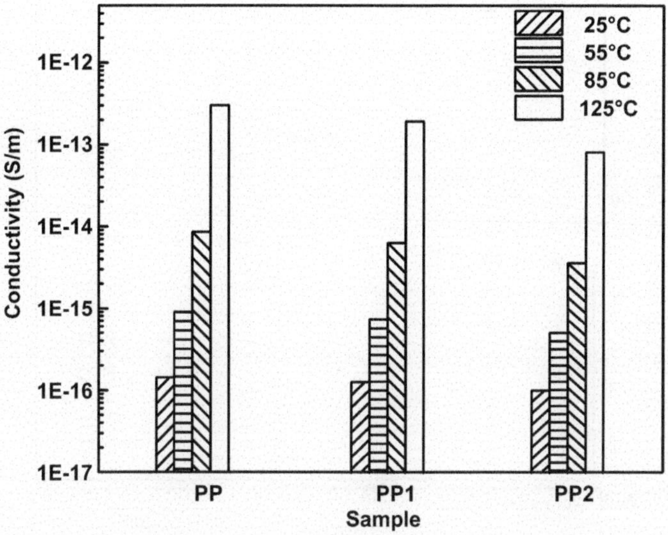

Fig. 7.15 Relationship between the conductivity and the temperature

$$K = \frac{Q(t) - Q_0}{Q_0} \tag{7.3}$$

The relationship between K and the irradiation dose at different temperatures is shown in Fig. 7.17. The charge accumulation rate of the modified samples is significantly lower than that of the unirradiated samples at 25 °C. With the increase of temperature, the K increases continuously (Fig. 7.16).

At the high temperature of 125 °C, the charge accumulation rate of the modified films still dominates, and its decreasing pattern is consistent with the trend of the crosslinking degree. This change reflects that the introduction of crosslinking structure and polar groups forms a deep trap inside PP, which in turn increases the trap probability during charge injection. The introduction of crosslinking structures provides more scattering centers for charge injection and the presence of polar groups increases the deep traps, which together lead to a significant decrease in the charge accumulation rate.

(4) **DC Breakdown Strength**

Breakdown strength, as one of the key parameters of insulation and energy storage properties, is concerned with the stability of materials under extreme conditions. The breakdown strength of the film decreases as the temperature rises, due to the intense thermal movement of PP molecular chains caused by the temperature increase, which increases the possibility of collisions and breakage of the molecular chains. The strong crosslinking formed after modification increases the intermolecular interaction force, which helps to further maintain the insulating properties of PP films at high temperatures. The results show that the trend of increase in breakdown strength is

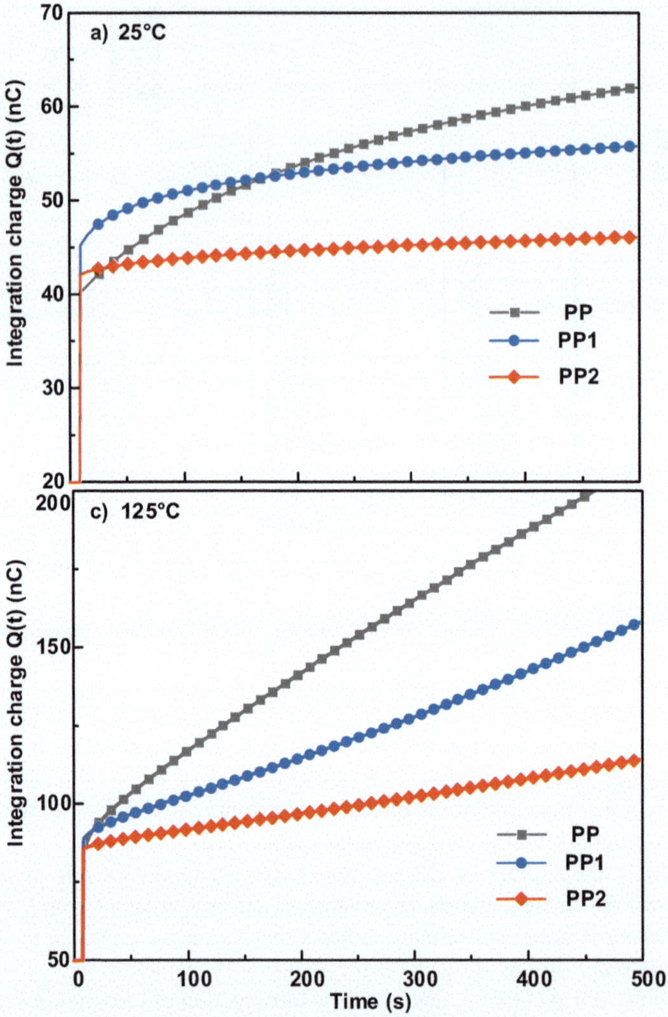

Fig. 7.16 Relationship between the integration charge and the temperature

consistent with the trend of change in crosslinking degree, indicating that crosslinking modification by EB irradiation can significantly enhance the breakdown strength of PP.

(5) **Energy Storage Performance**

At elevated temperatures, the charge/discharge efficiency (η) of polypropylene (PP) experiences a decline, concomitant with a decrease in energy density (Ue), leading to a notable reduction in energy storage performance. This diminution is primarily attributed to limitations imposed by the breakdown strength of the material. However,

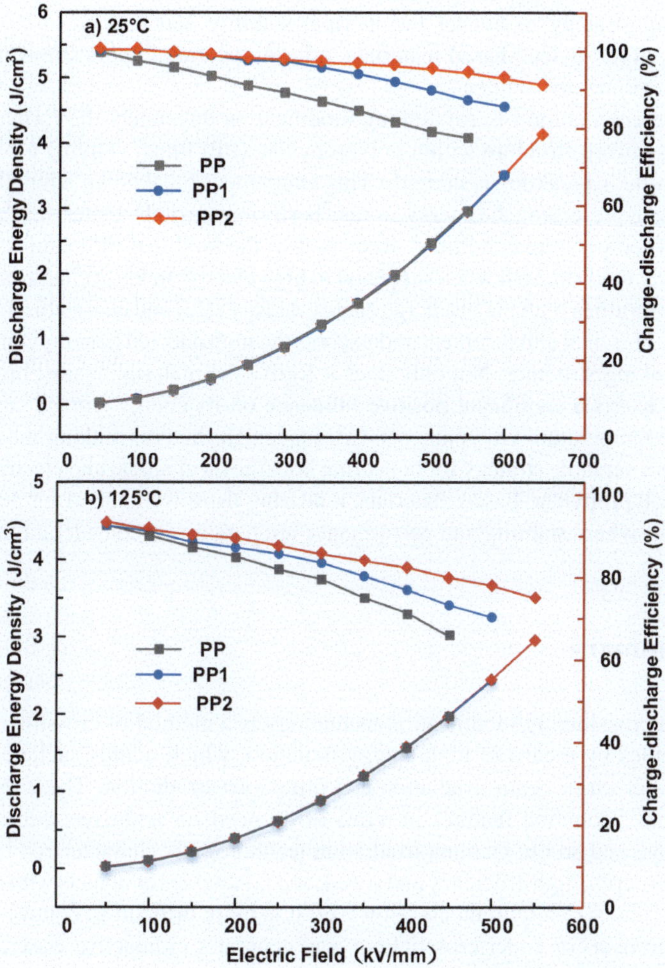

Fig. 7.17 Relationship between the discharged energy density, the charge–discharge efficiency and the temperature

when comparing modified PP with a higher degree of crosslinking to its pure counterpart, there emerges a noteworthy enhancement in both Ue and η at elevated temperatures. This observation underscores the pivotal role played by crosslinking in mitigating the degradation of energy storage performance under high-temperature conditions. The substantial improvement in breakdown strength post-crosslinking elucidates this phenomenon, highlighting the effectiveness of crosslinking in fortifying the material against thermal degradation and thereby preserving its energy storage capabilities. Hence, the incorporation of crosslinking agents emerges as a promising strategy to bolster the thermal stability and energy storage performance of polypropylene-based materials, offering potential applications in diverse high-temperature environments.

The implementation of crosslinking modification engenders the formation of a robust crosslinked structure within polypropylene (PP) films, yielding a substantial increase in their breakdown strength. This augmented breakdown strength plays a pivotal role in preserving the energy storage performance of the material, particularly under high-temperature conditions. Remarkably, the modified PP manifests heightened electric field strength when subjected to high electric fields, a characteristic that markedly amplifies both the discharge energy density (Ue) and charge/discharge efficiency (η). This dual enhancement underscores the multifaceted benefits conferred by the crosslinking structure. Not only does it fortify the material against breakdown, but it also exerts a significant positive influence on its energy storage capabilities in extreme temperature environments. Consequently, the crosslinking modification emerges as a versatile strategy for bolstering both the mechanical and electrical properties of polypropylene-based materials, rendering them well-suited for demanding applications where stability and performance are paramount considerations.

7.6 Summary

Irradiation crosslinking modification technology is a method of crosslinking material molecules by means of high-energy radiation, which mainly includes gamma irradiation, electron beam irradiation and ultraviolet irradiation. These three technologies have common features in terms of no need for additives, wide range of applications, and so on. Gamma irradiation technology is characterized by a large penetration depth, high efficiency and speed, and is suitable for objects with less stringent shape and size requirements. However, it is more difficult to control the dose, may lead to over- or under-crosslinking, and generates radioactive waste, which is more complicated to dispose of environmentally. On the other hand, E-beam irradiation technology is more outstanding in terms of precise dose control and environmental protection. Through the electron beam, precise adjustment of the amount of crosslinking agent can be realized, and no waste is generated, which is more friendly to the environment. However, the relatively shallow penetration depth of the electron beam is not suitable for materials that do not require a high radiation dose, are thick or dense, and the equipment cost is relatively high. Another technology is UV irradiation, which offers the advantages of greater selectivity and real-time control. UV irradiation can be carried out at specific wavelengths to achieve selective crosslinking of specific materials, and the irradiation process can be controlled in real time with a high degree of flexibility. However, due to the limited penetration depth of UV light, it is suitable for thin films and surface materials, and some materials may require the addition of curing agents, which has a certain impact on the material performance.

Taken together, these three irradiation crosslinking modification techniques have their own advantages and disadvantages. In selecting the appropriate technology, it is necessary to take into account the characteristics of the material, the needs of production and the impact on the environment. This selective consideration is crucial for the material modification process in different applications.

In practice, the choice of the appropriate irradiation crosslinking modification technique depends on the specific material and process requirements. Gamma irradiation is suitable for applications with less stringent shape and size requirements and where efficient production is required. E-beam irradiation excels where precise irradiation dosage is required for environmentally friendly production. UV irradiation technology, due to its strong selectivity and real-time control advantages, is more suitable for the application of fine control of specific areas. With the continuous development of science and technology, irradiation crosslinking modification technology is also evolving. In the future, the emergence of more efficient, precise and environmentally friendly irradiation technologies can be expected. More advanced control systems may emerge to make the irradiation process more flexible and better meet the needs of different application scenarios. In addition, the combination of research on new materials and irradiation crosslinking modification technology will bring broader application prospects for materials science.

References

1. B. Du, J. Zhang, M. Xiao, H. Liu, and Z. Ran, "Dielectric breakdown performance of deashed polypropylene for HVDC film capacitors," *IEEE Trans. Dielectr. Electr. Insul.*, vol. 29, no. 6, pp. 2209–2217, 2022.
2. M. Xiao et al, "Breakdown strength improvement of polypropylene by introducing long-chain branched structures for film capacitors," *J. Phys. D: Appl. Phys.*, vol. 55, no. 24, pp. 5501–5508, 2022.
3. H. Li et al, "Insulation resistance of metallized polypropylene film capacitors under different electric fields," *IEEE Trans. Dielect Elect. Insul.*, vol. 20, no. 4, pp. 1315–1320, 2013.
4. G. Zhang, C. Nam, T. C. M. Chung, L. Petersson, and H. Hillborg, "Polypropylene copolymer containing cross-linkable antioxidant moieties with long-term stability under elevated temperature conditions," *Macromolecules*, vol. 50, no. 18, pp. 7041–7051, 2017.
5. Y. Tang, W. Xu, S. Niu, Z. Zhang, Y. Zhang, and Z. Jiang, "Crosslinked dielectric materials for high-temperature capacitive energy storage," *J. Mater. Chem. A*, vol. 9, no. 16, pp. 10000–10011, 2021.
6. Y.-W. Fu, W.-F. Sun, and X. Wang, "UV-initiated crosslinking reaction mechanism and electrical breakdown performance of crosslinked polyethylene," *Polymers*, vol. 12, no. 2, pp. 420, 2020.
7. H. Liu et al., "Improving foam ability of polypropylene by crosslinking," *J. Appl. Polym. Sci.*, vol. 122, no. 2, pp. 973–980, 2011.
8. J. Y. Lin and H. J. Zhang, "The effect of TAIC-assisted crosslinking on the foaming performance of polypropylene," *Polym. Mater. Sci. Eng.*, vol. 26, no. 9, pp. 68–71, 2010.
9. M. Xiao, K. Wang, Y. Song and B. Du, "Dielectric performance of polypropylene films modified by UV irradiation for HVDC capacitors," *IEEE Trans. Dielect Elect. Insul.*, vol. 31, no. 1, pp. 176–183, 2024.

10. M. Xiao, K. X. Wang, K. Chen and B. X. Du, "Effect of photoinitiator on dielectric properties of polypropylene films modified by UV irradiation," *IEEE Trans. Dielect Elect. Insul.*, vol. 31, no. 1, pp. 168–175, 2024.
11. M. Xiao, K. X. Wang, Y. S. Zhao and B. X. Du, "Effect of UV irradiation time on dielectric performance of PP film for HVDC capacitor," *IEEE Trans. Dielect Elect. Insul.*, vol. 31, no. 3, pp. 1268–1275, 2024.

Chapter 8
Aggregated Structure Regulation

8.1 Introduction

PP as a polymer consists of long molecular chains with a complex tertiary structure. Among them, according to the chemical structure of molecular chains, linkage and spatial configuration, etc. are divided into the primary structure; molecular chains can change their conformation through the internal rotation of the performance is called the chain flexibility, usually the greater the chain flexibility the easier it is to crystallize, according to the molecular chain flexibility of the differences, such as straight chains, folded chains and helical chains and other morphology composed of the polymer's secondary structure; furthermore, in the secondary structure of the different forms of the molecular chain Further, the combination of different forms of molecular chains in the secondary structure forms an aggregated structure, which is usually divided into a disordered amorphous region and a dense and ordered crystalline region, called the tertiary structure (Aggregate structure). Among them, the molecular chains in the crystalline structure are arranged in a dense and regular manner, while the loose amorphous region is mainly composed of the ends of some molecular chains. The amorphous region consists mainly of the ends of some molecular chains and the molecular chain segments that are not discharged into the crystal lattice [1].

PP crystalline morphology has a significant impact on its space charge, breakdown and mechanical properties, and the modulation of crystallization parameters is a key link in the formation of multilayered aggregated state structures. Therefore, the PP substrate as capacitor film needs to pay more attention to the control of the aggregation structure, to prevent uneven crystallization of the inside and outside of the insulation to produce physical defects, which in turn cause the capacitor local insulation weakness. This puts higher requirements on the manufacturing of metallized film capacitors. Therefore, the study of modification means to improve the PP space charge, breakdown characteristics and mechanical toughness must focus on the

B. Du and M. Xiao, *Polypropylene Film for HVDC Capacitors*,
https://doi.org/10.1007/978-981-96-3029-5_8

PP aggregation state structure, to achieve the goal of inhibiting charge accumulation and improving the breakdown characteristics [2].

8.2 Aggregated Structure Regulation Method

The crystallization process of PP is mainly regulated by two means: non-isothermal crystallization and isothermal crystallization treatment. In non-isothermal crystallization, the faster cooling rate will limit the relaxation of molecular chain segments, resulting in the formation of smaller and more imperfect spherical crystal structures. Under micro stress or static conditions, PP usually starts from the nucleus of the crystal and grows radially around it at the same rate, eventually forming spherical aggregated structures [3]. Spherical crystals will appear at the interface of aspherical crystals after the growth and expansion. The structure at the interface plays a crucial role in the electrical and mechanical properties of the polymer (Fig. 8.1).

The addition of nucleating agents to PP increases its crystallization temperature, crystallinity, and melting point and decreases the size of the spherical crystals. The addition of nucleating agents to PP increases the crystallization temperature, crystallinity and melting point, and reduces the size of the spherical crystals [4]. The amount of nucleating agents and the cooling rate of non-isothermal crystallization have little effect on the DC conductivity characteristics of PP insulation [5]. The addition of α-nucleating agent to PP after a slower cooling rate of non-isothermal crystallization process will form a relatively perfect crystalline morphology, the crystallinity of the specimen will be increased, the DC breakdown field strength tested at 25 °C compared with the pure PP increased by 25.7%; and too fast cooling rate may lead to the surface of the PP film macroscopic regularity is damaged [6, 7].

By regulating the isothermal crystallization time and temperature can also improve the PP lamellar parameters (such as the long period and thickness of lamellar crystals, etc.) and the distribution structure of the spherical crystals, which will improve the mechanical properties of PP such as Young's modulus and elongation at break,

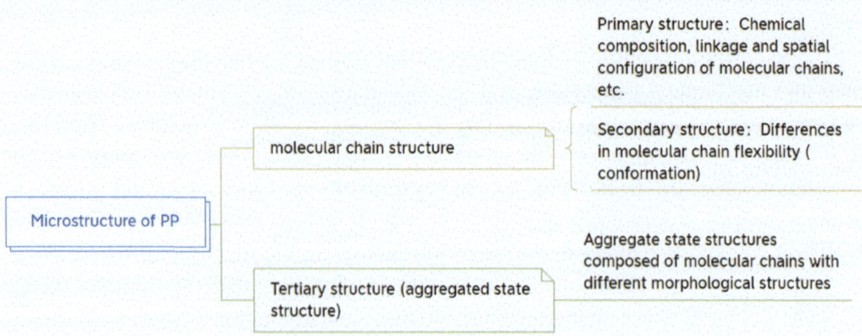

Fig. 8.1 Microstructural hierarchy of PP

etc. The crystallization process of PP will have a large impact on the electrical properties [8]. The crystallization process of PP will have a greater impact on the electrical properties, usually the higher the isothermal crystallization temperature and the longer the crystallization time, the larger the size of the spherical crystals and the narrower the size distribution, the interaction force at the interface of the spherical crystals will be reduced, lowering the energy level and density of the deep traps, and the binding capacity of the carriers is weakened, which accelerates the transport behavior of carriers at the interfaces of the spherical crystals and, as a result, the electrical properties of the PP insulation will be greatly reduced [9].

In essence, controlling the crystallization process in the production of PP insulation significantly impacts its overall performance. By precisely regulating this process, it becomes feasible to enhance the structural integrity of insulation aggregates, thereby effectively improving both electrical and mechanical properties [10]. Nevertheless, current research by domestic and international scholars has yet to fully comprehend the evolution of PP aggregate structures and corresponding control methodologies. There is a pressing need to establish a clear correlation between the structural states of PP aggregates and their macroscopic conductivity and breakdown characteristics. Building upon this foundation, a synergistic approach can be proposed to optimize the electrical and mechanical properties of PP insulation based on aggregate structures. Such advancements are crucial for advancing the theoretical framework necessary for optimizing processes in the development of high-performance, long-lasting metallized thin-film capacitors.

This chapter focuses on the study of PP aggregate structure on the electrical and mechanical properties of the influence of the law and the mechanism of the molecular chain has a variety of conformation, a variety of molecular chains folded to form a lamellar crystal layer, lamellar crystal layer further stacking growth to form a spherical crystal structure, and ultimately by the stacking of spherical crystal structure to form a multi-layered PP aggregate structure. The structure of aggregated state plays an important role in the electrical properties of PP, and the crystallization process is an important factor affecting the structure of PP aggregated state. In this chapter, polypropylene composite films were obtained by adding a small amount of nucleating agent and organic crystallization accelerator, respectively, aiming to regulate the size and shape of spherulites of the film sample.

8.3 Organic Nucleating Agent Doping

In this study, PP film samples doped with an organic phosphorus nucleating agent ((sodium 2, 2'-methylenebis (4, 6-di-tert-butylphenyl) phosphate) under three cooling processes were prepared to evaluate the effects of regulating the crystallization [11]. The micromorphology and crystal structure were observed. The influences of the nucleating agent content and the cooling rate were analyzed. In addition, the conductivity of the film sample was tested at different temperatures to compare the conduction loss [12]. The DC breakdown strength was investigated, and the mechanism was discussed. Based on the above experimental results, the relationship between the crystallization parameters and the electrical properties was established [13, 14]. A comprehensive control method for the conduction loss and breakdown performance of the PP film was proposed, showing the potential for application in metallized film capacitors.

8.3.1 Effect on the Microstructure

Figure 8.2 shows the crystallization processes of the pure PP and the PP composite samples, from which the doping of the nucleating agent does not induce a new crystal form. Following the addition of the nucleating agent, there is a noticeable increase in the number of nucleation sites observed at the initiation of crystallization, as depicted in Fig. 8.2(a1) and (b1). This indicates a significant enhancement in nucleation density upon doping with the nucleating agent, facilitating an accelerated onset of crystallization compared to undoped conditions. The dotted circles in Fig. 8.2(a_4) and (b_4) represent the spherulites with elatively complete structure when the crystallization ends. The entire crystallization processes of both (a) and (b) PP composite samples containing 0.05 wt% nucleating agent were observed. In the case of PP composite samples with 0.05% nucleating agent, the average diameter measures approximately 48.61 μm, indicating a reduction of approximately 30.20% compared to pure PP. Moreover, the crystallization onset in samples with nucleating agent precedes that of pure PP [15]. These findings highlight that the nucleating agent accelerates crystallization and effectively reduces the average diameter of the composite samples by about one-third when compared to pure PP. The crystallization temperature is elevated from 124 °C to approximately 128 °C, showing the impact of the nucleating agent. The V-shaped configuration of the two benzene rings in the nucleating agent serves to confine the helical structure of PP molecules, enhancing their stability significantly. As crystallization proceeds, the nucleating agent, functioning as a heterogeneous crystal nucleus, initiates crystallization prior to the PP melt. PP molecules subsequently epitaxially crystallize on the surface of the nucleating agent crystal, gradually forming a fully developed spherulite [16]. This process ensures robust stabilization of PP molecules throughout crystallization.

For samples doped with 0.07% nucleating agent and cooled at room temperature, the possesses the highest Xc of 54.35%, which is increased by 67.59% compared with Xc of the undoped nucleating agent samples and by 34.96% compared with Xc of samples doped with 0.07% nucleating agent and prepared by quench cooling. It can be concluded that the value of Xc has a positive correlation with the content of the nucleating agent. Moreover, the relatively slow cooling rate promotes the crystallization process. After adding the nucleating agent, additional molecular chains and segments tend to be arranged in an orderly manner, which increases the density and crystallinity. For all quench-prepared samples, crystallization occurs under extreme non-isothermal conditions [17, 18]. The macromolecular segments do not have enough time to fold and form a wafer. At this time, the polymer relaxation process cannot

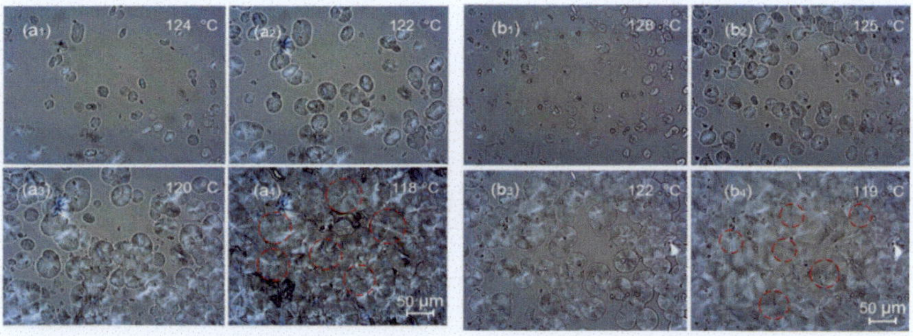

Fig. 8.2 Whole crystallization processes of **a** the pure PP and **b** the doped PP with 0.05 wt% nucleating agent

keep up with the change of temperature. Thus, the molecular chains form a bulky amorphous region because they could not crystallize during the rapid cooling process.

8.3.2 Effect on the Dielectric Properties

(1) **DC Conductivity**

Figure 8.3 shows the conductivity of the samples B (The samples were cooled under a pressure of 35 MPa from 190 to 120 °C and then cooled to a room temperature of 25 °C in air) with different doping contents. At 25 °C, there is a slight decrease and then an increase with the nucleating agent doping content. The conductivity of the sample with a nucleating agent content of 0.01 wt% is 8.35×10^{-15} S/m, which is 40.78% lower than that of pure PP (1.41×10^{-14} S/m). The DC conductivity of polymers is determined by the carrier mobility, which is negatively correlated with the migration barrier [19]. According to the microscopic experimental results, it can be seen that a proper addition of nucleating agent can improve the crystallinity of PP films. The conduction process mainly occurs in the amorphous region. For the modified films, the amorphous region is very small and only exists at the local irregularities inside or at the boundary of spherulites. Thus, the carrier mobility is limited due to the effects of crystal interfaces and relatively high conduction barrier. However, the conductivity is also related to the carrier concentration [20]. Since the nucleating agent itself may induce some carriers, with the addition of excess nucleating agent, the carrier concentration increases, resulting in higher carrier mobility and conductivity. Considering the effects of the crystallization rate, the conductivity of three series of testing samples is shown in Fig. 8.3. It is worth noting that the conductivity of PP film sample A (quench-prepared samples) may be the highest relative to the other two series of samples B. and C (he samples were cooled under a pressure of 35 MPa from 190 to 150 °C and then cooled to a room temperature of 25 °C in air.). Because of the slow cooling rate, there is sufficient time for the spherulites inside the

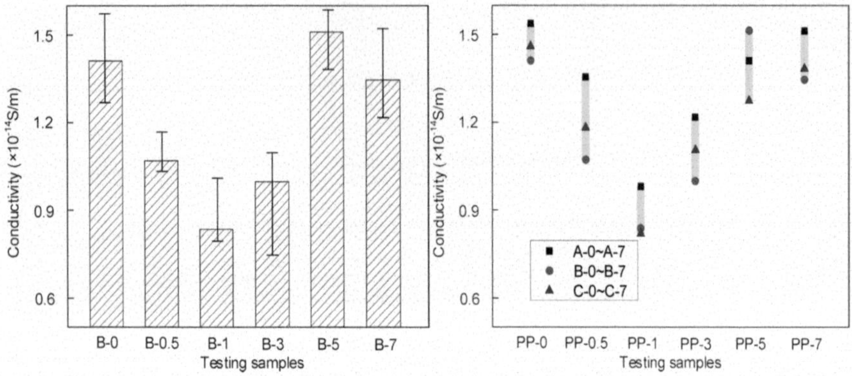

Fig. 8.3 Conductivity of the PP films with **a** the different content and **b** the cooling methods

film to be formed, which leads to high integrity of the spherulites and carrier migration barrier. On the contrary, with fast crystallization rate, the time for spherulites to grow is insufficient. Many broken crystals are formed and the amorphous region is enlarged, increasing the DC conductivity. Based on the discussion above, the introduction of the nucleating agent does not increase the conduction loss in the PP film. In addition, it can be speculated that the fast rate of cooling has a negative impact on the conduction loss of PP film [21].

(2) **Permittivity**

As for the dielectric parameters and surface potential test data, because the three series of samples with different cooling rates have the same trend, only the test and analysis data of samples B-0–B-7 are shown here. Figure 8.4 shows the permittivity of the testing samples under different frequencies. With the doping of the nucleating agent, the dielectric constant of the modified PP film changes slightly, with an increase of 0.91–3.18%. The results indicate that a micro amount of nucleating agent does not cause much change in dielectric properties. The tanδ of film samples is shown in Fig. 8.4. When the doping content is 0.005 wt%, compared with pure PP, the loss of the modified film decreases slightly at low frequency, and is basically unchanged above 50 Hz. When the content increases to 0.01 wt%, the tanδ of the sample decreases to below 5×10^{-4}. This can be attributed to the limited carrier migration process and conductivity reduction of the film. The area of the crystalline region increases significantly, and it is difficult to move the regular structure. On the other hand, the addition of nucleating agent may introduce heterogeneous interface. The crystalline/amorphous interfaces also increase due to the increase of the number of spherulites. As the addition content increases to 0.05 or 0.07 wt%, the tanδ is very close to that of the pure PP and may even increase above 20 Hz. In summary, the proper regulation of the microscopic morphology can play a positive role in the suppression of dielectric loss [22].

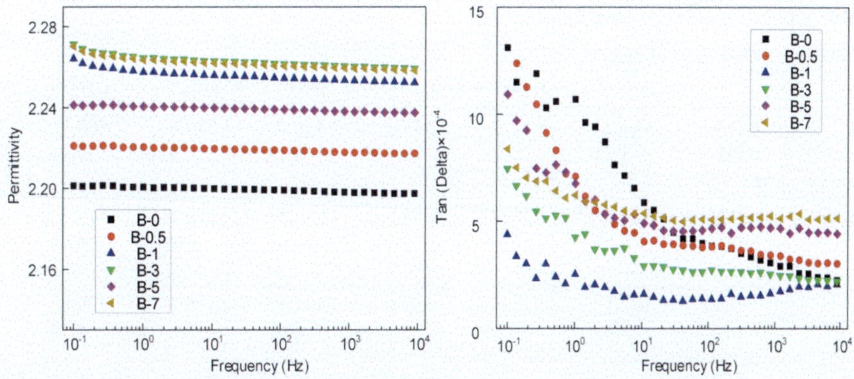

Fig. 8.4 Relationship between the permittivity, the dielectric loss and the frequency

(3) DC Breakdown Strength

The breakdown strength of all samples was tested in the same way at 25 °C. Table 8.1 shows the scale and shape parameters of samples. The variation law of breakdown strength of the three series of PP films is basically the same, thus, B-0–B-7 are taken as an example to analyze the influence of nucleating agent content. As shown in Table 8.1, with the addition of the nucleating agent, the breakdown strength of the film sample first increases and then decreases. Sample B-1 possesses the highest scale parameter of 565.9 kV/mm as well as the highest shape parameter of 7.19 compared with the scale parameter of 511.5 kV/mm and the shape parameter of 5.56 for B-0, indicating a reliable improvement in the DC breakdown characteristics. When the content reaches 0.07 wt%, the breakdown value of the film sample is basically equivalent to that of the pure PP sample. An appropriate doping of the nucleating agent helps to improve the microstructures of the PP film. After the nucleating agent is added, the crystalline region in the film increases and the amorphous region decreases [23]. The electron acceleration process is limited; thus, the breakdown strength increases. Although the crystallinity increases with the content, the agglomeration caused by the poor dispersion of the excessive nucleating agent cannot be ignored because it decreases the electrical characteristics to an extent. It can be seen clearly that the breakdown strength of sample B is the highest among the three series of data, showing a stable trend. Based on the results, the cooling treatment method plays a substantial role in the regulation of the breakdown characteristics. The slow rate of the cooling process is conducive to the formation of crystal regions, which may have the potential to improve the breakdown performance of the PP film sample. However, when the cooling rate is too fast, the quality of the film formation is affected. For example, the uniformity of the film thickness is reduced, and wrinkles are likely to occur.

Figure 8.5 shows the model for the electron transport in the PP film. It can be seen in area 1 that the spherulites consist of radially grown lamellas and epitaxial

Table 8.1 The scale and shape parameters at 25 °C

Samp.	Para.	Cont.					
		0	0.005	0.01	0.03	0.05	0.07
A	α (kV/mm) β	484.2 5.23	517.2 8.13	547.1 9.43	526.4 8.13	510.6 7.93	508.7 7.47
B	α (kV/mm) β	511.5 5.56	538.9 6.95	565.9 7.19	543.5 8.01	529.2 6.79	517.3 6.57
C	α (kV/mm) β	507.1 7.35	528.6 8.35	559.6 6.92	531.5 6.96	516.3 7.63	479.8 6.76

tangential lamellas. Some of the molecular chains in the amorphous region are bound between the lamellas; thus, the arrangement is relatively tight. In area 2, which is the interface between the spherulites, the lamellas are not tightly connected due to the high packing density of the lamellas in single spherulites. Molecular chains at the boundaries of the spherulites are tightly entangled, increasing the difficulty of electron transport. The closer the amorphous region is to the crystal region, the greater the molecular chains are constrained by the lamellas, and the more ordered the distribution is [24].

When a DC electric field is applied to pure PP film sample B-0, the electrons are preferentially transported in the amorphous regions, where the molecular chains are loosely arranged in a disordered manner, such as path 1 and path 2. After the crystallization is improved by doping with the nucleating agent and cooling at a slow rate, the microstructure of B-1 is improved, and the density of the spherulites increases and they slightly change their shape. The decreased number of amorphous

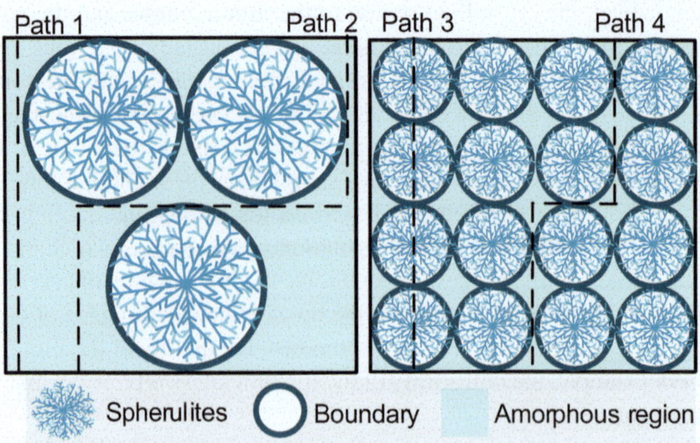

Fig. 8.5 Model for the electron transport in the PP film

regions in the film likely force the electrons to move around the spherulites. Spherulite is a system in which crystalline and amorphous phases coexist. There are very few amorphous parts inside the spherulites. If the electrons choose to transport in the amorphous region of the crystal, the route is simplified as path 3 consisting of area 1. If the electrons avoid moving in the spherulites as much as possible, they can only choose to transport along the grain boundaries as path 4 consisting of area 2. Whether the electrons are transported from inside the spherulites or from the interface between the spherulites, their movement is greatly restricted, which not only reduces the conduction loss but also inhibits the destruction of the molecular chains and increases the breakdown strength of the PP film. In view of the above analysis, the crystalline structure determines the transport of the electrons, which affects the electrical properties of the PP film. The control of the electronic transport process connects the microstructures and macro electrical characteristics, making it possible to improve the breakdown strength by adjusting the crystallization characteristics.

8.4 Crystallization Accelerator Doping

The processing of materials must be carried out at a temperature higher than or close to their melting point. Usually, the melting point of the nucleating agent is believed to be higher than that of pure PP. However, when the melting point is too high, part of the nucleating agent will cause defects, or degradation of the PP may occur. In this study, a novel organic crystallization accelerator with a low melting point and stable molecular structure was investigated to avoid decomposition. A method of microstructure optimization was proposed based on the micro doping of the organic crystallization accelerator. The crystallization process and thermal properties of PP films were studied. The effects of the organic crystallization accelerator on the breakdown strength were tested under DC, DC superimposed harmonic and DC superimposed pulse voltages [25]. In addition, the DC conductivity of PP samples was measured to compare the conduction loss. Trap characteristics were studied, and the mechanism is discussed. This PP film modification method showed good results and theoretical and practical significance in the field of PP films for power capacitors.

8.4.1 Effect on the Microstructure

The microstructures of the melted samples are shown in Fig. 8.6, and the distribution of the nucleus can be seen clearly. At approximately 190 °C, the lamellae structures almost disappear, and the crystals do not possess a complete spherical form. Figure 8.6a shows the crystal nucleus in PP-0 (PP-X stands for 0.0X% doping content). There are far smaller dots in PP-3 that have obvious advantages in quantity, as shown in Fig. 8.6b. The small nuclei are well dispersed and considered to be caused by the doped organic crystallization accelerator. Because of its low melting

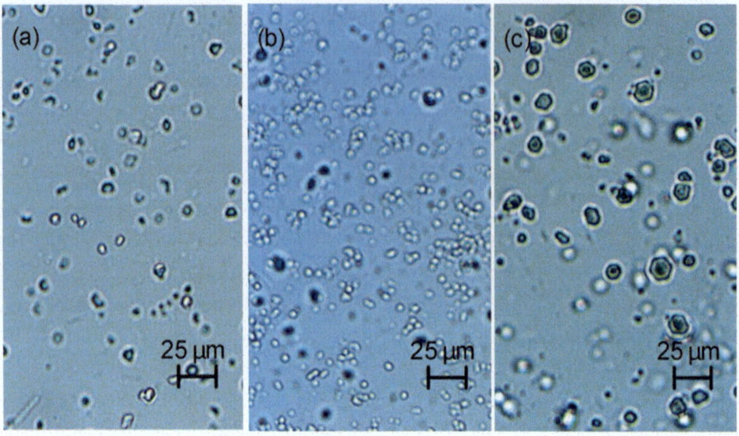

Fig. 8.6 Microstructure of **a** PP-0, **b** PP-3, and **c** PP-7 [26]. Copyright Clearance Center's RightsLink© License Number: 5812520397887

point and high decomposition temperature, the organic crystallization accelerator exists as a stable liquid form during the blending process, thus promoting uniform distribution. However, as shown in Fig. 8.6c, there are no similar small dots in PP-7 but circles with a different micro morphology. This indicates that the organic crystallization accelerator may have agglomerated, which deserves attention, and the doping content should not be more than 0.07 wt% in follow-up studies [26, 27].

The results of DSC are calculated and listed in Table 8.2. From the melting curves, it can be seen that the melting temperature shows an increasing trend with the doping content from 161.33 to 167.35 °C. This indicates relatively complete micro structures and strong forces between molecular chains in the sample. Although the melting point of the organic crystallization accelerator is low, the negative effect can be ignored due to the minimal addition. In addition, the area enclosed by the melting curve represents the melting enthalpy, which obviously increases. With the increase in crystal nucleus, the growth of lamellas is gradually improved, and the crystal area significantly increases, resulting in more energy required for melting the sample. In the cooling process, PP-7 has a lower crystallization temperature of 114.26 °C than that of PP-0 (119.55 °C). The introduction enhances the crystallization process, resulting in a reduction of the amorphous regions within the material. This acceleration effectively decreases the extent of amorphous phases present in the structure.

Table 8.2 Parameters of the samples with different doping contents of nucleating agent

Parameter	Sample					
	0%	0.005%	0.01%	0.03%	0.05%	0.07%
Tm (°C)	161.33	162.95	162.34	161.50	163.33	167.35
ΔH (J/g)	56.32	72.24	74.33	58.72	109.11	107.03
Tc (°C)	131.85	130.07	131.38	130.96	129.83	129.54
Tp (°C)	119.55	117.85	118.97	119.26	116.57	114.26
ΔT (°C)	12.30	12.22	12.41	11.70	13.26	15.51
ΔW	11.34	13.70	13.06	11.35	14.03	15.65
Xc (%)	39.72	54.44	56.87	66.03	86.56	86.33

8.4.2 *Effect on the Dielectric Properties*

(1) **Conductivity and Trap Distributions**

To prevent the increase in dielectric loss caused by additives, the conductivity was tested to compare the conduction loss, as shown in Fig. 8.7a. The minor addition of the organic crystallization accelerator helps to reduce the conductivity, which can be attributed to the restriction of the improved crystal region structures on the carrier migration. However, as the additive content increases, the agglomeration phenomenon or crystal overlaps may cause a slight increase in the conductivity. The DC conductivity of samples with different doping contents ranges from 7.3–14.1 × 10^{-15} S/m, indicating that a proper addition of the organic crystallization accelerator will have a reduction effect on the conduction loss. The DC conductivity value of PP-3 is lower than that of PP-0, with a decrease of nearly 50%; therefore, this method can be considered for the modification of PP films without increasing leakage current.

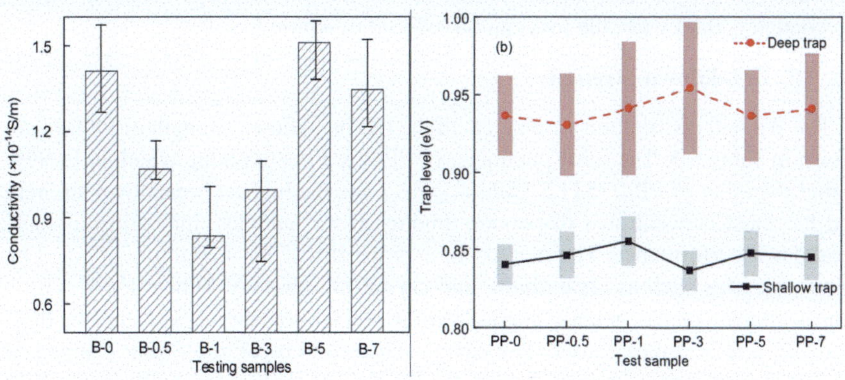

Fig. 8.7 Conductivity of the PP films with **a** the different content and **b** the cooling methods [26]. Copyright Clearance Center's RightsLink© License Number: 5812520397887

The trap distribution of the thin film sample is calculated, as shown in Fig. 8.7b. After the doping of the crystallization accelerator, the energy level of the shallow traps in the films increases from 0.93 to0.95–0.97 eV, and the deep trap level increases from 1.00 to 1.01–1.04 eV. Among them, PP-3 has the deep trap with the highest energy level. The energy level basically shows a trend of first rising and then falling. As for the trap density, the density of deep traps increases for all modified films, while the shallow trap decreases. Compared with PP-0, the deep trap density of PP-3 increases to 1.48 × 1022 eV^{-1} m^{-3}, while the shallow trap density decreases to 4.76 × 1021 eV^{-1} m^{-3}. This may be because the area of the amorphous region is reduced, and the gaps between the spherulites are filled. It can be seen that the change of micro structures affects the defect distribution, thereby changing the trap site distribution of the sample. The increase of deep traps in the crystal region helps to shorten the free path of carriers and reduces the charge mobility.

(2) **Permittivity**

As shown in Fig. 8.8a, the permittivity of the modified films with different contents of crystallization accelerators changes from approximately 2.20–2.25. Because PP is a kind of non-polar materials, the additives may introduce polarization effect. Therefore, the permittivity of the samples shows a weak upward trend. Considering the test error due to the environment, the amount of crystallization accelerator has little effect on the dielectric properties of the samples. The tanδ of the film sample is shown in Fig. 8.8b. Although the polarization of the additive itself is stronger than that of PP. However, its content is so small that the influence is small. Also, as the microstructure is refined and dense, the conductivity of the samples decreases, indicating a reduction in conduction losses. With the effect of the crystallization accelerators, the number of crystals increases and the size decreases, and the interaction between molecular chains is enhanced, thus increasing the energy required for turning or moving the groups inside the films to a certain extent. Considering the effects of both polarization loss and conduction loss, the addition of 0.005–0.05 wt% nucleating agent can slightly reduce tanδ of the film samples. Especially when the content is 0.03 wt%, the loss can be reduced to within 5×10^{-4}.

(3) **DC Breakdown Strength**

The Weibull distribution parameters of the DC breakdown strength of PP film are shown in Table 8.3. The scale parameter of PP-3 is 598.4 kV/mm, which is 16.99% higher than that of PP-0 (511.5 kV/mm). PP-3 has the highest breakdown strength and the most concentrated dispersion among all the samples, indicating a stable breakdown performance. The macroscopic properties of PP are largely affected by microstructures such as crystallinity and crystal morphology. When doped with a proper content of the organic crystallization accelerator, the crystal area in PP-3 increases and becomes denser. The energy required to break molecular chains and destroy local structures increases; thus, the PP film can withstand a high electric field strength.

When the temperature increases, the size of defects inside the film and the injection and migration of charges increase, which further aggravates the scission of the

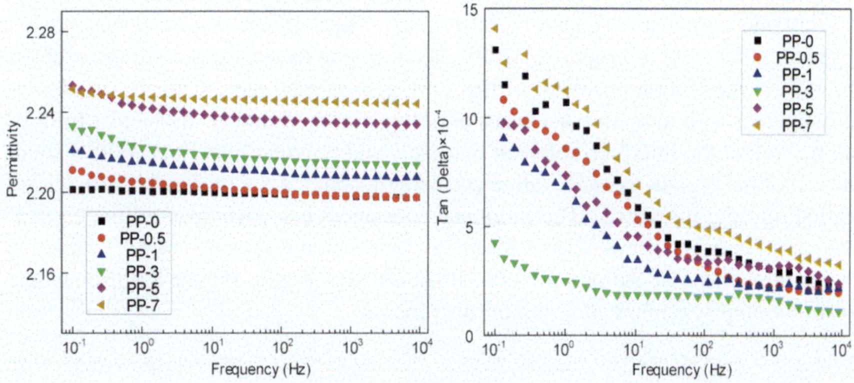

Fig. 8.8 Relationship between the permittivity, the dielectric loss and the frequency

Table 8.3 Scale and shape parameters at 25, 55 and 85 °C

Samp. (°C)	Para.	Cont.					
		PP-0	PP-0.5	PP-1	PP-3	PP-5	PP-7
25	α (kV/mm)	511.5	553.7	572.9	598.4	559.6	527.7
	β	5.56	8.88	8.90	8.35	9.03	7.85
55	α (kV/mm)	455.4	520.9	535.6	563.8	515.0	495.1
	β	4.36	7.02	8.78	7.36	9.88	9.05
85	α (kV/mm)	397.1	481.8	505.1	522.4	491.8	473.1
	β	4.98	8.52	8.34	7.03	7.73	6.16

polymer molecular chain, resulting in a decrease in the breakdown strength of all samples. The breakdown strength of PP-0 at 55 and 85 °C is 10.97–22.37% lower than that at 25 °C, respectively. While for PP-3, the performance degradation due to high temperature is about 5.78 and 12.70% at 55 and 85 °C, respectively.

Based on the above studies, the mechanism of controlling crystallization by the organic crystallization accelerator can be analyzed as follows. First, the organic crystallization accelerator itself acts as a nucleating agent, providing nucleation centers for the spherulites. PP is a semicrystalline and nonpolar polymer. Its molecular chains do not possess highly reactive polar groups, while the organic crystallization accelerator contains organophilic groups. The COR' group of the organic crystallization accelerator shown in Fig. 8.9 can interact with the PP matrix. After adding the organic crystallization accelerator, the molecular chains become entangled. The introduced interaction limited the movements of macro molecular chains, thereby effectively improving the structural stability and promoting the crystallization process of PP. The corresponding schematic diagram of this process is shown in Fig. 8.9a. This process can be considered a "self-nucleating" effect of the organic crystallization accelerator. In PP particles, there may be original impurities generated during the manufacturing procedure. In the blending process of PP composites, and an extremely small number of inorganic particles may also be introduced. These inorganic substances have a negative impact on the properties of PP. According to the molecular structure of the organic crystallization accelerator, there are groups that can physically or chemically interact with inorganic materials or polymers. Therefore, a "molecular bridge" can be built at the interface between inorganic and polymer materials. With a small amount of the organic crystallization accelerator, some of the inorganic particles are grafted into the PP matrix. The inorganic substances can participate in the crystallization process and act as centers for nucleation, which is regarded as the "inductive nucleating" effect of the organic crystallization accelerator. Figure 8.9b shows the corresponding schematic diagram of this process.

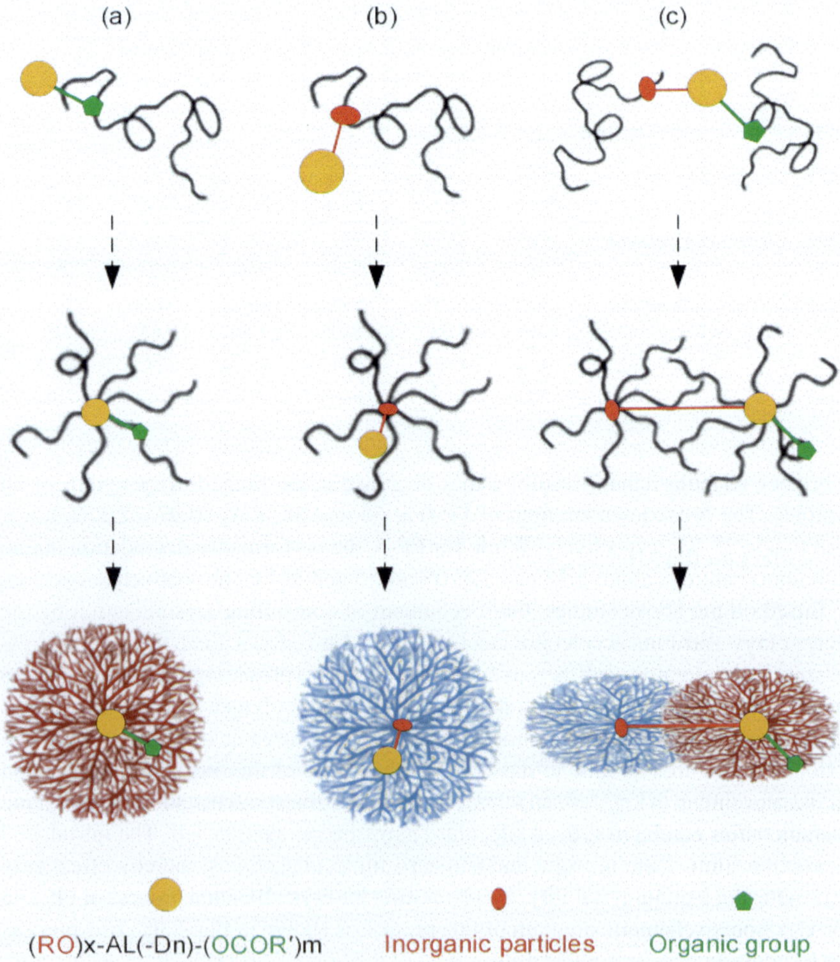

Fig. 8.9 Crystallization process with organic crystallization accelerator [26]. Copyright Clearance Center's RightsLink© License Number: 5812520397887

As shown in Fig. 8.9c, considering the different effects of molecular chains at both ends of the additive, the spherulites are stretched into ellipsoids; thus, the change in the crystalline morphology can be explained. Moreover, under the effects of the organic crystallization accelerator, PP molecules are more accessible to the surface of inorganic particles than without the accelerator, the interfacial gap is reduced, and the force between the two is strengthened so that air and other small molecules are driven out, making the structures denser and more complete.

Because of the above nucleation effects, more electrons in the deep traps need to overcome a huge potential barrier, and it takes a relatively long time for the de-trapping process in PP-3. Low carrier mobility leads to a relatively low conduction loss of the PP film. The high level and density of traps can also explain the excellent breakdown performance of PP-3. Due to a proper content of the organic crystallization accelerator, the crystal regions increase, effectively inhibiting the migration of carriers and avoiding damage to molecular chains. PP-7 possesses relatively deep trap sites, but microstructure destruction due to agglomeration results in similar breakdown properties as PP-0.

8.5 Aggregated Structure Regulation of LCBPP

The industrialization of long-chain-branched polypropylene (LCBPP) represents a significant milestone, facilitated by its commendable dispersion characteristics of organic nucleating agents within polypropylene matrices. Leveraging this advancement, this study endeavors to enhance the dielectric properties of polypropylene by modifying its microstructure through the incorporation of long-chain-branched polypropylene and nucleating agent (NA-21). Within this chapter, an exhaustive examination of the dielectric properties of LCBPP, modified with varying concentrations of nucleating agents, is meticulously conducted. Additionally, the intricate mechanisms underlying the interplay between crystallization properties and dielectric behavior in LCBPP are thoroughly investigated. By elucidating these mechanisms, this research aims to provide valuable insights into the rational design and optimization of polypropylene-based materials for enhanced dielectric performance. Through a systematic exploration of the relationship between microstructure modification, crystallization behavior, and dielectric properties, this study endeavors to contribute to the advancement of materials engineering and pave the way for the development of high-performance dielectric materials with tailored properties and applications.

8.5.1 Effect on the Microstructure

The DSC curves are shown in Fig. 8.10. An increase in sample number represents an increase in nucleating agent content. Figure 8.10a presents the crystallization

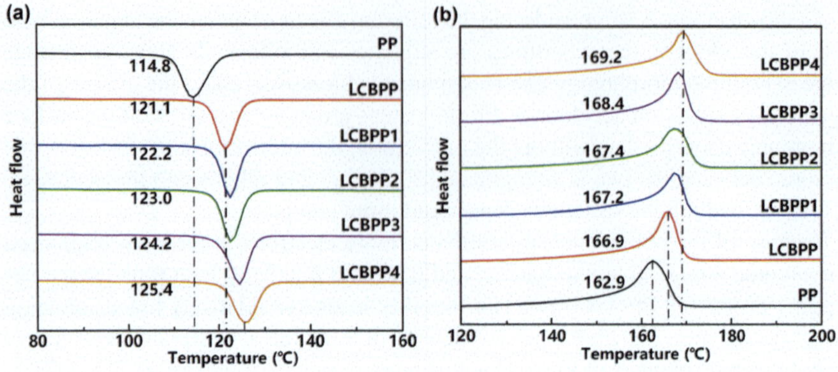

Fig. 8.10 **a** Crystallization process and **b** melting process of LCBPP [27]. Copyright Clearance Center's RightsLink© License Number: 5812520905738

curve during the cooling process. The crystallization temperature (Tc) of the films after modification with long-chain branching increases because it can promote heterogenous nucleation. In addition, the NA-21 can promote the crystallization of LCBPP. The free energy of nucleation of the system decreases and the crystallization temperature in-creases as the amount of NA-21 added increases. Compared with pure LCBPP, the crystallization temperature of LCBPP4 was increased by 4.3 °C.

Figure 8.10b presents the melting curve during the second heating process. The melting temperature (Tm) of LCBPP is increased by 4.0 °C compared to that of PP due to the increase in entanglement. The melting peak of the modified films with nucleating agent added shifted to the right. According to the melting enthalpy (Hm), it can be calculated that the crystallinities (Xc) of LCBPP and PP are 34.1% and 44.3% respectively. In this work, long-chain branches significantly reduced the crystallinity of the films. The enhancement of molecular chain entanglement and the increase of viscosity inhibits the growth of spherulites. The addition of nucleating agents improves the crystallinity of the films. The crystallinity of the modified films with the addition of 0.07 wt% of NA-21increased by 5.2%. The influence of nucleating agents on crystallization mainly depends on the lattice matching and structure matching between NA-21 and LCBPP. The NA-21 can induce epitaxial crystallization of LCBPP and lower the free energy barrier of nucleation, thereby promoting crystallization. The nucleation ability of NA-21 correlates with its effective nucleation concentration in LCBPP. The promoting effect of NA-21 on crystallization is proportional to the amount added within a certain range.

The NA-21 initiates the epiphytic crystallization of the LCBPP segments. The segments of LCBPP grow along the surface perpendicular to the edge of NA-21. The growth of spherulites is the process of the ordering and stabilization of the LCBPP molecular chains on the surface of the crystal nucleus and the continuous expansion in the radial direction until they encounter another spherulite. The presence of heterogeneous surfaces can reduce the critical size of crystal nuclei suitable for crystal growth. With the increase of NA-21 addition, the nucleation sites increase and the spherulite size decreases.

8.5.2 *Effect on the Dielectric Properties*

(1) **DC Conductivity**

Figure 8.11 shows the conductivity of the films at different ambient temperatures. At 25 °C, the conductance loss was increased for the long-chain branched-modified films, which may be caused by the decrease in the crystallinity of the films. The conductivity of LCBPP at 125 °C is 2.2×10^{-14} S/m, which is lower than that of PP (8.1×10^{-14} S/m). The increase of the interface between the spherulite and the amorphous region in LCBPP affects the trap energy level, which can be proved by analyzing the trap energy level distribution later. In addition, the entanglement of chains restricts the slippage of the molecular chains at high temperatures, enhancing thermal stability. Therefore, long-chain branching limits the transport of carriers in PP films at high temperatures.

The incorporation of nucleating agents plays a pivotal role in regulating the crystallization process within long-chain-branched polypropylene (LCBPP), thereby mitigating conductivity losses. Notably, the conductivity profile of the modified films exhibits a nuanced trend with varying NA-21 doping content across different ambient temperatures. At 25 °C, the conductivity of LCBPP3 demonstrates a remarkable reduction by an order of magnitude compared to both pure polypropylene (PP) and unmodified LCBPP, underscoring the efficacy of nucleating agent addition in enhancing the material's insulating properties. Furthermore, with increasing temperature, the conductivity loss in the modified films experiences a notable attenuation relative to the pristine dielectric material. Particularly at 125 °C, the conductivity of

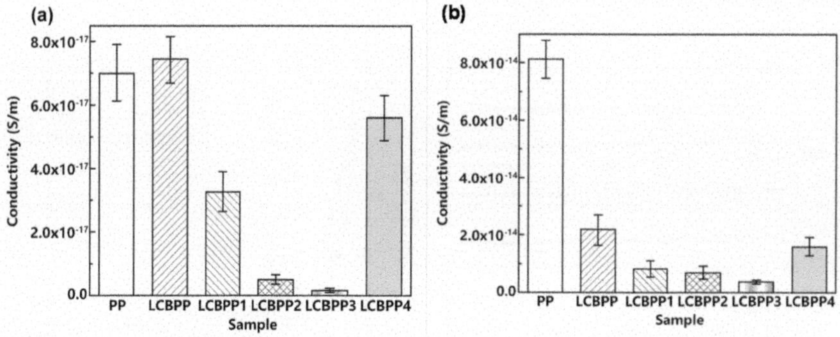

Fig. 8.11 Relationship between the conductivity and the temperature [27]. Copyright Clearance Center's RightsLink© License Number: 5812520905738

LCBPP3 exhibits a further reduction by an order of magnitude compared to unmodified LCBPP and two orders of magnitude compared to PP. This observed improvement can be attributed to the nucleating agents' role in promoting crystallization and augmenting the crystallinity of LCBPP. Consequently, the resultant compact crystal structure within the modified films impedes carrier mobility, thereby contributing to a decrease in conductivity loss.

However, it is noteworthy that the increase in conductivity observed in LCBPP4 may be attributed to the uneven dispersion of a relatively larger quantity of NA-21 within the films. This non-uniform distribution of nucleating agents may disrupt the crystalline structure, potentially creating localized regions with altered electrical properties. Thus, while nucleating agent addition generally enhances the crystallinity and insulating properties of LCBPP, the precise balance of doping content and dispersion uniformity warrants careful consideration to maximize the desired electrical performance enhancements. This nuanced understanding underscores the importance of meticulous optimization in the formulation of modified polypropylene materials for tailored dielectric applications [28].

(2) **Permittivity**

The relative permittivity and dissipation factor of the samples are shown in Fig. 8.12. The relative permittivity of LCBPP is greater than that of PP. The crystallinity of LCBPP is greatly reduced, and the size of spherulites is smaller. The increase of the interface between spherulites increases the polarization. With the increase of the nucleating agent content, the spherulites in the modified films become more compact and the movement of the segments is hindered.

It can be seen from Fig. 8.12b that the dissipation factor of the long-chain branched modified films increases. The crystallinity of LCBPP decreases, resulting in increased polarization loss. Moreover, doping a small amount of NA-21 can reduce the dielectric loss of LCBPP. The crystallinity of the modified LCBPP is increased and the space between the spherulites is denser, resulting in a decrease in polarization loss. Compared with LCBPP3, the dissipation factor of LCBPP4 is increased due to the

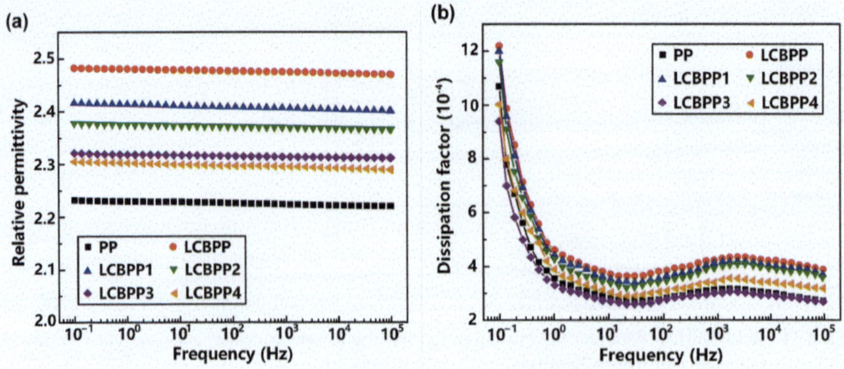

Fig. 8.12 Relationship between the permittivity, dielectric loss tangent and the frequency

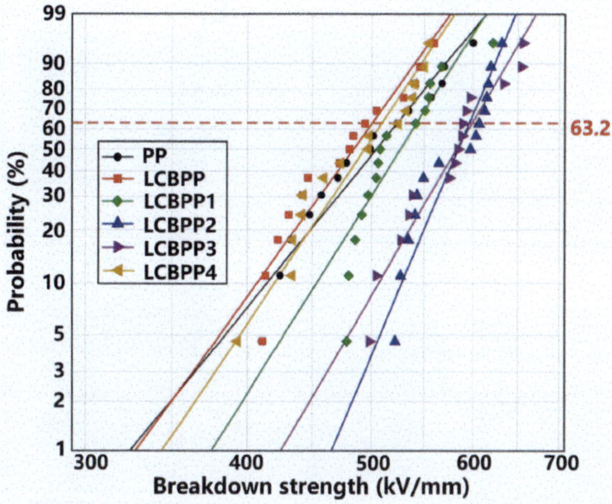

Fig. 8.13 Relationship between the breakdown strength and the temperature [27]. Copyright Clearance Center's RightsLink© License Number: 5812520905738

agglomeration of NA-21. Overall, the permittivity and dielectric loss of the modified LCBPP with the addition of nucleating agents varied little with the doping amount of NA-21.

(3) **DC Breakdown Strength**

Figure 8.13 shows the breakdown parameters of the films. At room temperature, the breakdown strength of the modified films with long chain branching decrease due to the lower crystallinity. The shape parameter of LCBPP films (10.9) is higher than that of PP films (9.6) because LCBPP with higher MS can form high quality films. As the temperature increases, the breakdown strength of LCBPP changes less. At 125 °C, the breakdown strength of the long-chain branched modified films is 442.3 kV/mm, which is 7.1% higher than that of PP. The increased number of spherulites in LCBPP affects the trap energy level and density. In addition, the introduction of long-chain branches leads to increased entanglement, hindering the movement of molecules and enhancing thermal stability.

The breakdown strength can be improved by adjusting the crystalline properties of LCBPP. As the content of the nucleating agent increases, the crystallinity increases and the crystal structure becomes tighter, which limits the transport of carriers in the films. Therefore, the breakdown strength of modified films with a lower nucleating agent content is improved. At 25 °C, the breakdown strength of LCBPP3 increases by 19.7%. The interaction force between the molecular chains in the modified films with higher crystallinity is stronger, and the thermal movement of the molecular chains is reduced at high temperatures. The breakdown strength of LCBPP3 is 521.5 kV/mm at 125 °C, which is 79.2 kV/mm higher than LCBPP. The breakdown strength of modified films with the addition of 0.07 wt% NA-21 was lower due to the agglomeration of the nucleating agent in the films. The shape parameter of the sample LCBPP3 is lower than that of the sample LCBPP2, which proves that the higher nucleating agent content will cause the problem of uneven dispersion [29].

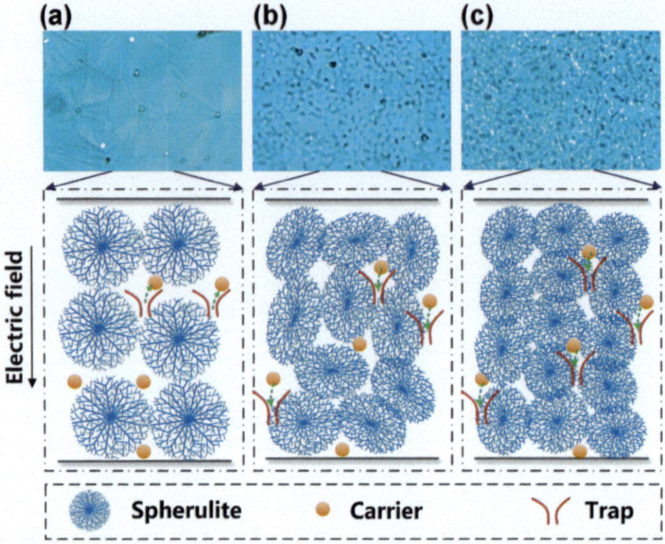

Fig. 8.14 Carrier transport in **a** PP, **b** LCBPP and **c** LCBPP + NA-21

The schematic diagram of the effects of crystalline properties on carrier transport in PP and LCBPP is shown in Fig. 8.14a, b respectively. Long-chain branches increase the nucleation density, resulting in an increase in the interface between the crystalline and amorphous zone. The spherulite boundary affects the trap level distribution, making the deep trap level and deep trap density of LCBPP relatively higher. The higher energy levels and density of traps limit the number and mobility of carriers. The combined effect of molecular structure and aggregated structure lead to the improvement of the high-temperature breakdown strength of films.

The increase in crystallinity leads to a denser internal structure of the modified films, which restricts the transport of carriers [30]. The spaces between the spherulites in the modified films doped with NA-21 are denser, reducing weak areas in the films. The distribution of nuclei in LCBPP4 is not uniform affecting the distribution of the electric field because the nucleating agent agglomerates. Therefore, as the increase of the amount of NA-21 is added within a certain range, the breakdown strength of modified LCBPP films increases. When the content of the NA-21 is higher, the breakdown strength of modified films decreases due to uneven dispersion. At high temperatures, the modified films with NA-21 have a more compact structure and higher deep trap energy level [31], so the breakdown strength is higher than that of pure LCBPP.

8.6 Summary

In this chapter, polypropylene composite films were obtained by adding a small amount of nucleating agents and organic crystallization promoters to PP or LCBPP species with the aim of regulating the size and shape of spherical crystals in the film samples. Through the crystallization control, the macroscopic electrical properties of the polypropylene films were effectively optimized due to the improvement of the microscopic morphology, which is expected to be applied in the field of thin-film capacitors. Based on the experimental data, the following conclusions can be drawn:

The nucleating agent inhibits the fluidity of the molecular chains, and there is an epitaxial crystallization effect between the nucleating agent and the polymer. Due to a reduction in the free energy barrier of the nucleating agent, the nucleation density increases, and the spherulite size is significantly reduced. With the doping of the nucleating agent, the crystallization temperature, the melting temperature and the melting enthalpy are elevated.

Both the content of the nucleating agent and the cooling rate do not obviously affect the conductivity and permittivity of the sample. The reduced amorphous regions and free volume result in a lower dielectric loss. By restricting the electron transport, a positive correlation is established between the micro structures and the dielectric properties of the PP film. An appropriate nucleating agent content (0.01 wt%) and slow cooling rate can improve the average breakdown strength by above 10% at 25–85 °C.

The organic crystallization accelerator participates in the crystallization process, which is embodied in increasing the spherulite density, significantly reducing the size and changing the shape. With an increase in nucleation sites and improvement of micro structures, the melting point and the regularity of the polypropylene films are elevated. However, excessive addition may have adverse effects.

By "self-nucleating" and "inductive-nucleating" effects, the doping of the organic crystallization accelerator causes an essential change of trap site distribution in PP films, introducing deep traps and restricting the carrier migration process. Without sacrificing the dielectric loss, the breakdown strength of the modified film samples under DC voltage is significantly improved by ~ 16.99%, showing stable superior electrical properties.

The introduction of the long-branched structure and the addition of NA-21 nucleating agent reduced the DC conductivity and increased the breakdown field strength and energy storage density of the polypropylene films; the breakdown field strength of the modified long-chain branched polypropylene film with 0.05 wt% NA-21 nucleating agent at 125 °C was increased by 108.0 kV/mm compared with that of the polypropylene.

References

1. Z. Y. Ran, B. X. Du, M. Xiao, H. L. Liu and J. W. Xing, "Dielectric property improvement of polypropylene films doped with POSS for HVDC polymer film capacitors," *IEEE Trans. Dielect Elect. Insul.*, vol. 28, no. 4, pp. 1308–1316, 2021.
2. Z. Y. Ran, B. X. Du, and M. Xiao, "High-temperature breakdown performance improvement of polypropylene films based on micromorphology control," *IEEE Trans. Dielect Elect. Insul.*, vol. 28, no. 4, pp. 1547–1554, 2021.
3. M. Xiao, M. D. Zhang and B. X. Du, "Crystallisation regulation of long-chain branched polypropylene on dielectric performance and energy density for metallised film capacitors," *High Volt.*, vol. 8, no. 5, pp. 921–930, 2023.
4. M. Xiao, K. L. Fan, M. D. Zhang, B. X. Du, Z. Y. Ran and Z. L. Xing, "Effect of cyclic olefin copolymer on dielectric performance of polypropylene films for capacitors," *IEEE Trans. Dielect Elect. Insul.*, vol. 29, no. 6, pp. 2266–2273, 2022.
5. C. Guillermin et al, "Metallized film power capacitors end-of-life study through monitored destruction tests," *IEEE Trans. Power Deliv.*, vol. 28, no. 1, pp. 368–375, 2013.
6. H. Wang and F. Blaabjerg, "Reliability of capacitors for dc-link applications in power electronic converters—an overview," *IEEE Trans. Ind. Appl.*, vol. 50, no. 5, pp. 368–375, 2014.
7. R. W. Brown, "Reliability Linking corrosion and catastrophic failure in low-power metallized polypropylene capacitors," *IEEE Trans. Device Mat Reliab.*, vol. 6, no. 2, pp. 326–333, 2006.
8. M. Horak and P. Mach, "Study of electrical ageing of polypropylene film capacitors using third harmonics measurement," 2014 IEEE 20th International Symposium for Design and Technology in Electronic Packaging, Bucharest, Romania, 2014, pp. 55–59.
9. Z. Chen et al, "Experimental study of the breakdown characteristics of polypropylene films under nanosecond voltage pulses," *IEEE Trans. Plasma Sci.*, vol. 46, no. 11, pp. 4010–4016, 2018.
10. M. Xiao, K. X. Wang, K. Chen and B. X. Du, "Effect of photoinitiator on dielectric properties of polypropylene films modified by UV irradiation," *IEEE Trans. Dielect Elect. Insul.*, vol. 31, no. 1, pp. 168–175, 2024.
11. C. Marco et al, "Comparative study of the nucleation activity of third-generation sorbitol-based nucleating agents for isotactic polypropylene," *J. Appl. Polym. Sci.*, vol. 84, no. 13, pp. 2440–2450, 2002.
12. Y. Feng et al, "Effect of nucleating agent addition on crystallization of isotactic polypropylene," *J. Appl. Polym. Sci.*, vol. 69, no. 10, pp. 2089–2095, 2015.
13. C. Li et al, "Effect of nucleating agent supported on zeolite via the impregnation on the crystallization ability of isotactic polypropylene and its mechanism," *Polym. Adv. Technol.*, vol. 30, no. 11, pp. 2674–2685, 2019.
14. Y. H. Wu, Y. H. Wu, J. W Zha, W. K. Li, S. J. Wang and Z. M. Dang, "A remarkable suppression on space charge in isotatic polypropylene by inducing the β-crystal formation," *Appl. Phys. Lett.*, vol. 107, no. 11, 2015, Art. no. 112901.
15. B. X. Du, Z. L. Li and Z. R. Yang, "Field-dependent conductivity and space charge behavior of silicone rubber/SiC composites," *IEEE Trans. Dielect Elect. Insul.*, vol. 23, no. 5, pp. 3108–3116, 2016.
16. B. X. Du, Z. Y. Ran, J. Li, H. C. Liang and H. Yao, "Fluorinated epoxy insulator with interfacial conductivity graded material for HVDC gaseous insulated pipeline," *IEEE Trans. Dielect Elect. Insul.*, vol. 27, no. 4, pp. 1287–1294, 2020.
17. W. Zhang et al, "Effect of β-crystals on the mechanical and electrica properties of β-nucleated isotactic polypropylene," *IEEE Trans. Dielect Elect. Insul.*, vol. 26, no. 3, pp. 714–721, 2019.
18. Z. S. Zhang, C. Y. Chen, C. G. Wang, J. Q. Guo and K. C. Mai, "Nonisothermal crystallization kinetics of isotactic polypropylene nucleated with a novel supported β-nucleating agent," *J. Therm. Anal. Calorim.*, vol. 103, pp. 311–318, 2011.
19. S. H. Kung and G. J. Kish, "Multiport modular multilevel converter for DC Systems," *IEEE Trans. Power Deliv.*, vol. 34, no. 1, pp. 73–83, 2019.

20. H. Y. Ye et al, "Review on HVDC cable terminations," *High Volt.*, vol. 3, no. 2, pp. 79–89, 2018.
21. P. Guido and R. Maurizio, "Pulse handling capability of energy storage metallized film capacitors," *IEEE Trans. Plasma Sci.*, vol. 28, no, 5, pp. 1603–1606, 2000.
22. M. H. El-Husseini et al, "Thermal simulation for geometric optimization of metallized polypropylene film capacitors," *IEEE Trans. Ind. Appl.*, vol. 38, no, 3, pp. 713–718, 2002.
23. E. J. Barshaw et al, "High energy density (HED) biaxially-oriented poly-propylene (BOPP) capacitors for pulse power applications," *IEEE T. Magn.*, vol. 43, no, 1, pp. 223–225, 2007.
24. Z. Y. Ran, B. X. Du, M. Xiao, H. L. Liu and J. W. Xing, "Effect of crystallization regulation on the breakdown strength of metallized polypropylene film capacitors," *IEEE Trans. Dielect Elect. Insul.*, vol. 28, no. 1, pp. 175–182, 2021.
25. H. Li et al, "The capacitance loss mechanism of metallized film capacitor under pulsed discharge condition," *IEEE Trans. Dielect Elect. Insul.*, vol. 18, no. 6, pp. 2089–2094, 2011.
26. Z. Y. Ran, B. X. Du, M. Xiao and J. Li, "Crystallization morphology-dependent breakdown strength of polypropylene films for converter valve capacitor," *IEEE Trans. Dielect Elect. Insul.*, vol. 28, no. 3, pp. 964–971,2021.
27. Z. Xing et al, "Improvement of breakdown performance of long-chain branched polypropylene for film capacitors based on crystallization Control," 2022 IEEE Conference on Electrical Insulation and Dielectric Phenomena (CEIDP), Denver, CO, USA, 2022, pp. 183–186.
28. H. Li, G. Zhang, Y. Wang and F. Lin, "Dependence of polarization behaviors on electrical and thermal aging for pulse metallized polypropylene film capacitor," *IEEE Trans. Dielect Elect. Insul.*, vol. 29, no. 4, pp. 1553–1562, 2022.
29. H. Li et al, "Electric field and temperature dependence of electrical conductivity in biaxially oriented polypropylene films," *IEEE Trans. Plasma Sci.*, vol. 42, no. 11, pp. 3585–3591, 2014.
30. W. Sima, J. Hua, P. Sun and Q. Chen, "Accumulative failure characteristic of polypropylene film in surge capacitor under repeated impulse voltage," *High. Vol. Eng.*, vol. 44, no. 2, pp. 358–365, 2018.
31. Y. H. Wu, J. W. Zha, W. K. Li, S. J. Wang and Z. M. Dang, "A remarkable suppression on space charge in isotatic polypropylene by inducing the β-crystal formation," *Appl. Phys. Lett.*, vol. 107, no. 11, 2015, Art. no. 445502.

Part III
Doping/Blending Modification

Chapter 9
Nanoparticle Doping

9.1 Introduction

With the development of metalized film capacitors (MFCs), their applications have become more extensive, such as fields of voltage support, energy storage and power electronics. Increasingly more complex operation environments considering electrothermal stresses place more stringent requirements on the comprehensive performances of capacitors [1, 2]. MFCs are composed of overlapping polymer insulation and electrodes [3]. Based on such structures, polypropylene (PP) films as the dominant dielectric used in MFCs, play a significant role in the stable operation of capacitors in power systems [4]. It is a considerably major issue for PP films to possess superior electrical properties under long-term compound effects of high temperature and electric field.

Since Lewis [5] proposed the concept of nanodielectrics in 1994, they have made significant progress in both basic research and application fields [6]. In the field of electrical engineering, nanodielectrics usually refer to nanocomposite dielectrics obtained by introducing a certain mass fraction of nanoparticles into polymer dielectrics, whose electrical properties are generally significantly better than polymer dielectrics. Nanoparticles usually refer to 0, 1, or 2 dimensional particles or their combinations with dimensions less than 100 nm. In practical applications, there is no specific size limit, but their main characteristics are small size, large specific surface area, and physical properties different from large-sized materials. Adding them to polymers can bring significant performance enhancement, which is also the reason why nanocomposite dielectrics have received widespread attention and research [7]. Using "Dielectric Nanocomposite" as the keyword, a search was conducted in the Web of Science database to obtain the number of publications in the field of nanocomposite dielectrics in the past 20 years, as shown in Fig. 9.1. It can be seen that from 2000 to 2019, the number of publications in the field of nanocomposite dielectrics showed an explosive growth.

B. Du and M. Xiao, *Polypropylene Film for HVDC Capacitors*,
https://doi.org/10.1007/978-981-96-3029-5_9

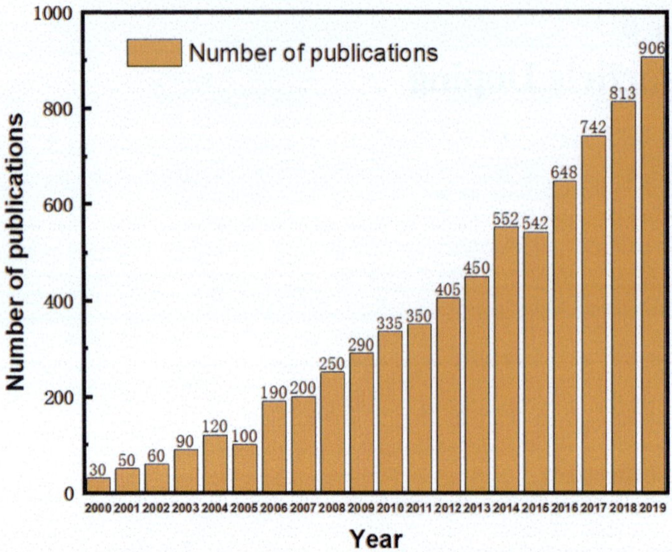

Fig. 9.1 Changes in the publication number of nanocomposite dielectrics

Studies have shown that the doping of nanoparticles can enhance the dielectric properties of polypropylene materials. The trap depth of the polypropylene sample doped with 1 wt% nano-Al_2O_3 increased by 17%, and the breakdown field strength increased by 37% [8, 9]. The DC breakdown field strength of the polypropylene sample doped with 0.5 wt% nano-ZnO increased by 37%, and the AC breakdown field strength increased by 17% [10]. After doping 5% MgO, the breakdown field strength increased by 39% [11]. It is generally believed that nanoparticles have a large specific surface area, which can introduce deep traps at the interface with the matrix to capture free charges, reduce the transmission speed of carriers, and protect the molecular chain from the impact of high-energy carriers, thereby improving the insulation performance of the material. However, the compatibility of the organic–inorganic interface is low, and the introduced nanoparticles are easily agglomerated in the dielectric. The micron-sized particles after agglomeration not only no longer have the advantage of large specific surface area, but also cause the electric field distortion near the particles, which results in the decrease of dielectric insulation performance. This is an urgent problem to be solved in future research.

In order to alleviate the agglomeration of nanoparticles, we can modify the nanoparticles, that is, introduce groups with high compatibility with polypropylene on the surface, thereby alleviating the agglomeration of nanoparticles. The results show that 3-aminopropyltriethoxysilane grafted on the surface of MgO can achieve good dispersion of nanoparticles and increase the resistivity of polypropylene film by more than 100% [12]. The polypropylene composite film prepared by surface modification of MgO nanoparticles with polypropylene grafted maleic anhydride increased the breakdown field strength of polypropylene by 50% and reduced the leakage current to 12% [13]. However, although the surface modification of nanoparticles can alleviate the agglomeration problem to a certain extent, the thickness of the capacitor film is generally less than 15 μm, and it works under a strong electric field higher than 200 kV/mm for a long time. Even small-scale agglomeration will still cause more serious electric field distortion, and the improvement of the insulation performance of the capacitor film is not stable. In addition, surface treatment does not change the significant difference in mechanical properties between the nanoparticle body and the matrix, which will lead to the formation of nanoparticles to cut the film molecules during the biaxial stretching process of polypropylene, resulting in the problem that the film cannot be formed [14].

9.2 Modification Theory

In general, the breakdown of nanocomposite dielectrics is essentially the breakdown of the polymer matrix. Therefore, the breakdown mechanism of polymer dielectrics is also applicable to nanocomposite dielectrics. The difference is that the electrical weaknesses that cause breakdown may be different. Because the dielectric constants of nanoparticles and polymers are often very different, the electric field distribution inside nanocomposite dielectrics is very uneven. Generally speaking, the breakdown strength of nanocomposite dielectrics is affected by many factors.

(1) **Trap Theory**

In a dielectric, defects or chemical centers that can capture electrons or holes can be called traps. The energy level E_t of the trap can be considered as the difference in electron affinity between the material itself and the trap [15, 16], which can be calculated as follows:

$$E_t = EA_{reff} - EA_{def} \tag{9.1}$$

In the formula: EA_{reff} is the electron affinity of the polymer; EA_{def} is the electron affinity for introducing defects or chemical centers. The essence of trap is to introduce local state energy levels in the forbidden band gap, capture carriers that transition from the conduction band or valence band, and play a role in promoting or inhibiting carrier movement. According to the energy band theory, traps in polymers can be divided into hole traps and electron traps. The hole trap is located near the valence band, and the electron trap is located near the conduction band. The deeper the trap is, the closer the energy level is to the Fermi level. Shallow traps can promote carrier migration and increase conductivity, while deep traps inhibit carrier migration and reduce conductivity. The energy level and density of the two affect the charge transport and breakdown strength in the composite dielectric [17–19].

Artbauer's free volume breakdown theory quantitatively gives the relationship between the trap energy level and the dielectric breakdown strength Eb. The smaller the maximum free path of the electron, the higher the trap energy level, and the higher the breakdown strength of the composite dielectric. In nanocomposite dielectrics, there is an interface region of tens of nanometers around the nanoparticles. The trap density and energy level in this region may be quite different from the polymer body, which affects the breakdown strength of nanocomposite dielectrics [20].

In addition, it is undeniable that the depth of the trap is related to the relative dielectric constant of the nanoparticles. The larger the relative dielectric constant of the nanoparticles, the deeper the potential well depth. In general, the deeper the trap of nanocomposite dielectrics, the higher the breakdown strength. This is because the existence of deep traps increases the activation energy of conductance, resulting in uneven distribution of electric field under DC electric field.

(2) **Blocking Theory**

Inorganic materials have stronger resistance to electrical erosion than polymers, and can more effectively inhibit or slow down the development of electrical trees, which is conducive to improving the breakdown strength of composite media, which is more obvious under low electric field. As shown in Fig. 9.2, when the electric field strength increases to a certain value, the breakdown phase begins to appear inside the nanocomposite medium. When the applied electric field increases to 185 kV/mm, the breakdown phase in the 0-dimensional nanocomposite almost penetrates the entire material, while the breakdown phase fraction in the 1-dimensional nanocomposite is still very small. That is to say, the hindering effect of 1D nanomaterials on the development of breakdown stage is obviously better than that of 0D nanoparticles.

In practice, there are usually five different structures of composite media. Compared with the electric field direction, they are: vertical arrangement of nanowires (S1), vertical arrangement of nanosheets (S2), random distribution of nanoparticles (S3), parallel arrangement of nanowires (S4) and parallel arrangement of nanosheets (S5). The results show that the arrangement of nanoparticles has an important influence on the breakdown strength of the composite dielectric. In several nanocomposite dielectrics, the breakdown strength of the composite dielectric is higher than that of the polymer only when the nanosheets are perpendicular to the electric field distribution. When the nanofibers or nanosheets are parallel to the direction of the electric field, the breakdown strength of the composite medium is lower than that of the composite medium with random dispersion of nanoparticles, and the breakdown strength of the nanofiber composite medium is the lowest. When the

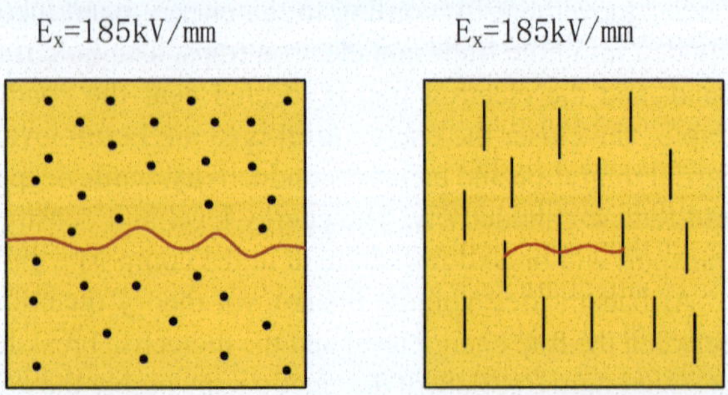

Fig. 9.2 Effect of nanoparticles and nanofibers on the breakdown performance of composite materials

nanoparticles are perpendicular to the direction of the electric field, the nanosheets are beneficial to improve the breakdown strength of the composite medium, while the nanowires cannot improve the breakdown strength of the composite medium.

(3) **Interface Effect**

The voltage applied to the multilayer dielectric is distributed according to the dielectric constant of each layer, and the high dielectric constant layer is subjected to low voltage. Taking the double-layer composite dielectric as an example, the electric field formed by the applied voltage U_0 in the A layer and the B layer has the following expression:

$$E_A = \frac{U_0 * \varepsilon_{r,B}}{\varepsilon_{r,A} d_B + \varepsilon_{r,B} d_A} \tag{9.2}$$

$$E_B = \frac{U_0 * \varepsilon_{r,A}}{\varepsilon_{r,B} d_B + \varepsilon_{r,B} d_A} \tag{9.3}$$

In the formula: d_A and d_B are the thickness of A layer and B layer respectively; $\varepsilon_{r,A}$ and $\varepsilon_{r,B}$ are the dielectric constants of A and B layers respectively.

By adjusting the thickness and dielectric constant of each layer, the electric field strength of each layer can be adjusted, so as to regulate the electric field distribution in the multilayer composite dielectric. This phenomenon can be used to design multilayer nanostructured composite dielectrics with high breakdown strength. The finite element simulation of the electric field distribution shows that the breakdown strength of the multilayer nanostructured composite dielectric is highly correlated with the difference in the dielectric constant of each layer [21, 22]. In the multilayer composite media, it should be pointed out that the electric field strength of the intermediate layer has an important influence on the breakdown strength of the multilayer composite dielectric. No matter the electric field strength of the middle layer is close to the upper layer or the lower layer, there is no significant difference in the breakdown strength of the multi-layer composite dielectric. Only when the electric field strength of the middle layer is close to the average value of the electric field strength of the upper and lower layers, the electric field strength distribution inside the multi-layer composite dielectric is the most uniform and the breakdown strength is the highest. The interlayer interface of multilayer dielectric films has a limiting effect on the movement of carriers and also plays an important role in improving the breakdown strength of films [23, 24].

(4) **Electric Field Distribution Theory**

High performance power electronic equipment insulation typically has high electrical resistivity, low dielectric constant, and high breakdown strength to reduce the impact of capacitance or leakage current on power electronic systems. However, some devices and devices, such as capacitors, cable terminals, and connectors, require high breakdown strength and high dielectric constant. In engineering, the dielectric constant of polymer media is usually improved by adding high dielectric materials such as barium titanate. However, the high dielectric constant and high conductivity of fillers form a clear contrast with the low dielectric constant and low conductivity of polymers, which will lead to local electric field distortion. According to the Poole Frenkel model [25, 26], the electron concentration ne excited by the electric field satisfies the following equation:

$$n_e \approx N_e \exp[-(E_i - \beta_{PF} E^{\frac{1}{2}})/(2kT)] \tag{9.4}$$

In the formula: N_e is the concentration of ionization center; k is the Boltzmann constant; E_i is the ionization energy or barrier height; PF is the Poole–Frenkel constant, and E is the electric field strength. The Poole–Frenkel model shows that the higher the local electric field strength at the defect or chemical center, the more the number of excited electrons, and the greater the probability of breakdown.

The maximum electric field at the interface in nanocomposite dielectrics is related to the dielectric constant of the nanoparticles [27]. The larger the dielectric constant of nanoparticles, the higher the maximum electric field at the particle/matrix interface, the more severe the electric field distortion, and the easier it is for shocks to pass through.

On this basis, the introduction of core–shell nanoparticles can significantly reduce the change in electric field concentration caused by a single high dielectric constant particle [28]. This is because the presence of the shell suppresses the electric field distortion at the interface and makes the changes in the electric field smoother. In addition, when the dielectric constant ratio between the core layer and the polymer matrix is constant, the dielectric coefficient of the shell layer has a significant impact on the electric field distribution at the interface. Generally speaking, an increase in the dielectric constant of the shell will lead to more severe electric field differentiation at the interface.

9.3 Nanoparticle Types Selecting

In domestic and foreign high voltage direct current (HVDC) projects, the rated temperature of MFCs is 85 °C currently, enduring operating temperature of PP is generally lower than 80 °C [28, 29]. There may be a large number of harmonics in the actual environment of power capacitors, which leads to overcurrent phenomenon and a gradual increase in temperature, even exceeding the allowable operating conditions. However, the insulation performances unavoidably suffer severe limitations when the environmental temperature increases.

Generally, an elevation in temperature will promote the increase of high-energy electrons and carrier migration in polymers, resulting in dielectric deterioration and accelerated insulation aging [30]. Research have shown that the breakdown strength of PP films at 85 °C was reduced to 60–80% of its breakdown strength at room temperature [31]. In addition, the loss caused by defects of the dielectric itself may lead to heating problems, further aggravating the temperature rise of the capacitor. The rupture of the molecular chains is induced, and the low-density area expands, which ultimately causes capacitor failures [32]. Such performance degradation seriously threatens social and economic benefits, thence, how to coordinate the regulation of breakdown strength and dielectric loss at high temperatures has become a major topic in recent years, showing a theoretical and practical significance in the field of MFCs.

9.3.1 Carboxylated Cellulose Nanoparticles

(1) **Effect on the Microstructure**

Studies have shown that there exists a close relationship between crystal morphology and macro properties. The introduction of β crystals results in a low degree of order and thermodynamic stability while β-PP has significant advantages in terms of mechanical properties. But β crystals must be obtained under special conditions and have a trend to transform into more stable α type [33, 34]. Therefore, more attention should be paid to the structural optimization of α crystals and the improvement of crystallinity. Based on our previous research, the breakdown characteristics can be improved without sacrificing the dielectric loss by increasing α crystal region. Results show that the average breakdown strength of films with 0.01 wt% nucleating agent increases by approximately 25.7 and 23.9% at 25 and 85 °C, respectively, compared with that of the pure PP [35]. Thus, such a method of micro-addition has certain feasibility in film modification.

Carboxylated cellulose naonocrystals (C-CNCs) are needlelike nanomaterial with high crystallinity, a proper doping of which promoted the crystallization process of film samples in this paper. The microscopic morphology became denser and more complete, resulting in a significantly increased melting enthalpy and crystallinity. The DC conductivity and breakdown of the films before and after the modification were tested at different temperatures, aiming to study the effects of C-CNCs on conduction loss and electrical strength properties. To study polarization characteristics, the permittivity and dielectric loss were compared at frequencies. With C-CNCs, the entanglement of molecular chains was promoted, and the trap sites were optimized, thereby improving the comprehensive dielectric properties. The obvious improvement in temperature resistance is expected to realize the long-term application of PP in higher temperature environments in power systems. Due to the entanglement of PP molecular chains by the needle-like superstructure of C-CNCs, the crystallization process of the film was promoted. The appropriate doping amount (0.01 wt%) improves the thermal properties of the material, and reduces the DC conductivity of the material by increasing the crystallinity and limiting the movement of carriers. Due to the regular structure of the modified film, the conduction loss is reduced, and the dielectric loss is limited at low frequencies. Through DC experiments at different temperatures, the breakdown strength of C-CNCs/PP films increased by about 51%

at 85 °C, which was basically equivalent to that of pure PP at 25 °C. This mechanism is attributed to the deep well position, which is introduced by the large crystal region, resulting in a short free path of carrier transport. This new PP film has great application potential in the field of thermoelectric composite in power system.

As shown in Table 9.1, this book will show the modification results of five C-CNCs samples with different doping contents.

Figure 9.3 shows the micro morphology of PP samples with different contents of C-CNCs. As the content increases from 0 to 0.01 wt%, the gap between the spherulites is reduced and the micro structures are significantly improved. Compared with PC00, which contains obvious structural voids and defects as is shown in Fig. 9.3a, PC10 possesses a far denser and larger crystal region. It can be illustrated that micro filling helps to form compact crystalline structures. When the doping content of C-CNCs reaches 0.03 wt% in Fig. 9.3d, the morphology has not been further ameliorated. Possible agglomeration due to the excessive addition will invalidate the crystallization control effect causing PC50 to be similar to PC00.

(2) **Effect on the Dielectric Properties**

The DC conductivity of the testing samples at different temperatures from 25 to 85 °C are shown in Fig. 9.4. When the content of C-CNCs is below 0.01 wt%, the conductivity slightly decreases compared with that of PC00, due to the increase of crystal regions and can be fundamentally attributed to the restriction of the carrier mobility. There is a close relationship between the carrier transport process and the free volume of the polymer. The free volume of the amorphous region is relatively large, thus, the binding force of molecular chains on carriers is weak. The movement of carriers in the amorphous region is freer than that in the crystalline region, resulting in greater conduction loss. The doping of C-CNCs promotes the crystallization and leads to high crystallinity, affecting the movement path of carries under the DC electric field. As for PC30 and PC50, the test data has a certain degree of uncertainty

Table 9.1 Test samples with different doping contents of C-CNCs

C–CNCs	0%	0.005%	0.01%	0.03%	0.05%
Sample	PC00	PC05	PC10	PC30	PC50

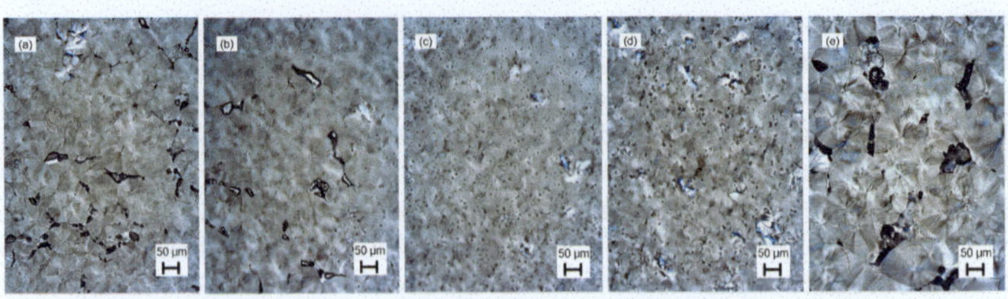

Fig. 9.3 Micro morphology of **a** PC00, **b** PC05, **c** PC10, **d** PC30 and **e** PC50 [36]. Copyright Clearance Center's RightsLink© License Number: 5811370682461

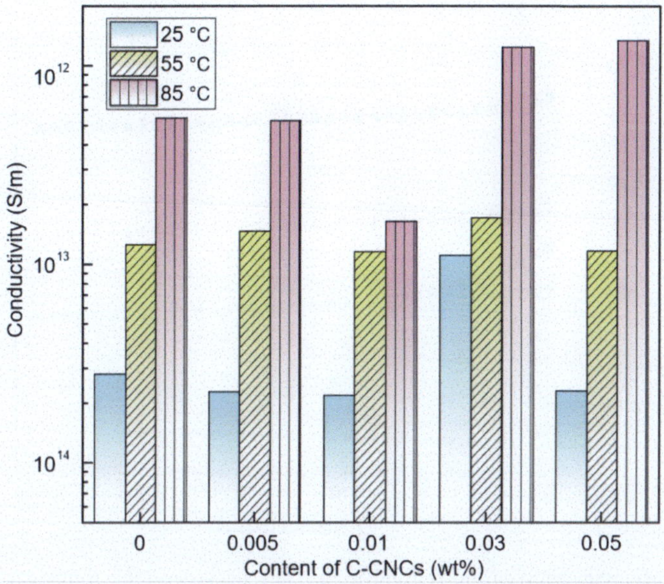

Fig. 9.4 Relationship between the conductivity and the temperature [36]. Copyright Clearance Center's RightsLink© License Number: 5811370682461

due to aggregation. Most data show that excessive addition will play a negative role, increasing the conductivity of PP films.

In addition, the environmental temperature has a great influence on the conductivity. The conductivity of the sample at 85 °C is nearly two orders of magnitude higher than that at 25 °C, because high temperature promotes the appearance of more high-energy electrons and longer free path. Compared with the effects of the temperature, the influence of the content is not obvious. Proper doping of C-CNCs can reduce the conduction loss of samples at high temperature, owing to relatively stable and dense microstructures. It is worth noting that the conductivity of PC50 is very close to that of PC00 at 25 or 55 °C, since the agglomeration causes the extreme unevenness of the sample, and the test site may not contain C-CNCs.

The permittivity at different frequencies is shown in Fig. 9.5. The permittivity of PC05 and PC10 is very close with a value of about 2.28 (50 Hz), higher than that of PC00. This is because C-CNCs have higher polarity compared with PP, and their introduction will strengthen the polarization in the sample. In addition, under the effect of applied electric field, electrons or ions in the dielectric may gather at the boundary of the matrix and filler particles, causing interface polarization. This can be attributed to different polarities and conductivities of the matrix and the filler. The dielectric stability of all modified samples has been improved, and the permittivity value is basically constant in the wide frequency range. When the content of C-CNCs is excessive, the permittivity of PC30 and PC50 does not change much compared to PC00, with an only 2% variation.

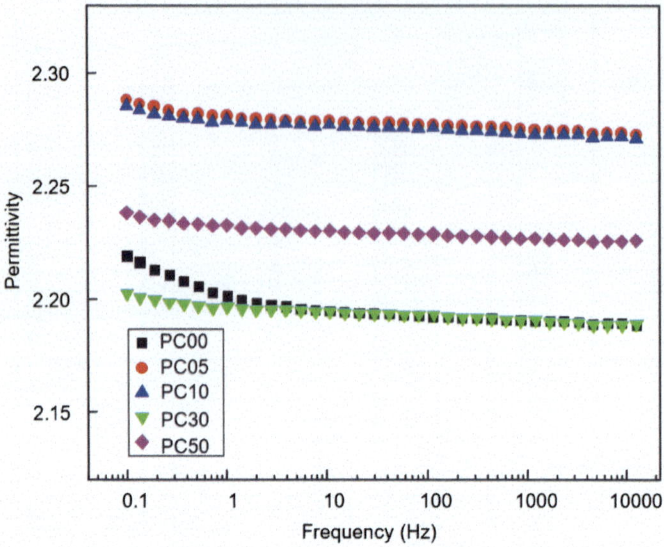

Fig. 9.5 Relationship between the permittivity and the frequency [36]. Copyright Clearance Center's RightsLink© License Number: 5811370682461

Table 9.2 shows the Weibull distribution of the breakdown strength at 25, 55, 85 °C. The scale parameter α is defined as the electric field strength corresponding to the breakdown probability of 63.2%. The electrical resistance characteristics are elevated with the content of C-CNCs below 0.01 wt%. The breakdown strength of PC10 exhibits a notable enhancement, reaching 563.1 kV/mm at 25 °C, which represents an impressive 16.25% increase compared to PC00. This substantial improvement underscores the efficacy of proper doping in augmenting the electrical properties of the material, facilitated by the optimized crystallization morphology achieved through controlled nucleating agent incorporation. The experimental data further corroborates the stability of this enhancement, indicating a consistent and reliable performance enhancement through judicious doping strategies. However, it's worth noting that excessive concentrations of C-CNCs can detrimentally impact the electrical performance of the samples. The aggregation of particles can create localized electrical weaknesses, thereby compromising the overall integrity of the material and leading to property deterioration. This underscores the importance of maintaining an optimal balance in the doping concentration to avoid adverse effects on electrical properties. At elevated temperatures, the breakdown strength trends observed remain consistent with those at room temperature. However, the influence of high experimental temperatures exacerbates the degradation of microstructures, as more electrons acquire sufficient energy to attack the polypropylene molecular chains. This accelerated degradation precipitates a significant reduction in the breakdown strength across all films. Moreover, the magnitude of breakdown reduction amplifies with increasing temperature, highlighting the criticality of ensuring electrical stability, particularly at elevated operating temperatures. Specifically, the breakdown value of PC00 decreases by 13.8% from 25 to 55 °C, and by 23.0% from 55 to 85 °C, underscoring the pronounced impact of temperature on electrical performance. This underscores the imperative of maintaining electrical stability under varying temperature conditions to ensure the reliability and longevity of polypropylene-based materials in practical applications. The dispersion of breakdown strength across samples can be attributed to the continuous evolution of microstructures under the combined influence of electrical and thermal fields. Notably, high concentrations of C-CNCs exacerbate the agglomeration phenomenon, resulting in poor film uniformity and contributing to the observed instability in breakdown strength. Thus, achieving optimal dispersion and uniformity of dopants is crucial in enhancing the electrical performance and reliability of polypropylene-based materials across a range of operating conditions.

The trap site distributions of PC00 and PC10 at different temperatures are shown in Fig. 9.6. After modification, there is a slight increase in the density of deep traps and a decrease in the density of shallow ones. The shallow trap level keeps at the basically same level while the deep trap level differences. At 85°C, the parameters of deep traps in PC10 are 0.905 eV and 0.59×10^{22} eV^{-1} m^{-3}, compared with those in PC00 (0.88 eV, 0.32×10^{22} eV^{-1} m^{-3}). When the temperature is raised, the free volume in the testing sample increases, resulting in shallow trap sites and high carrier mobility. The trap distribution is closely related to the micromorphology of the film and the overall trends of the two is consistent.

The carrier transport processes of PC00 and PC10 are shown in Fig. 9.7. First, the heterogeneous nucleation effect of C-CNCs on the crystallization process is analyzed. There are needle-like ultra-structures in C-CNCs with spherical shape. The chemical or physical effects of these structures can block the movement of molecular chains in PP, leading to the enhancement of folding and entanglement. Therefore, the formation of crystals is promoted. With the increase of crystal cores, spherulites become more compact and the size of spherulites obtain an obvious reduction.

Carriers are injected into the film under the effect of DC electric field. In PC00, there are large-size amorphous regions with shallow trap sites. Only a few carriers are captured by deep traps in the internal and boundary of the crystal region. The rest of the carriers can de-trap in an extremely short time, and then easily migrate in the direction of the electric field, which can be attributed to the weak binding force of dispersed molecular structures. For PC10, because of its compact structures, carriers can only choose to move in the minor area of amorphous regions in spherulites or gaps between spherulites. Large energy is needed for chain to break. Thus, the

Table 9.2 The scale and shape parameters at 25, 55, 85 °C

Sample (°C(kV/mm))	PC00	PC05	PC10	PC30	PC50
α-25	484.4	547.2	563.1	438.7	428.0
α-55	417.5	469.4	486.4	414.6	347.4
α-85	321.5	455.7	485.4	379.3	337.2

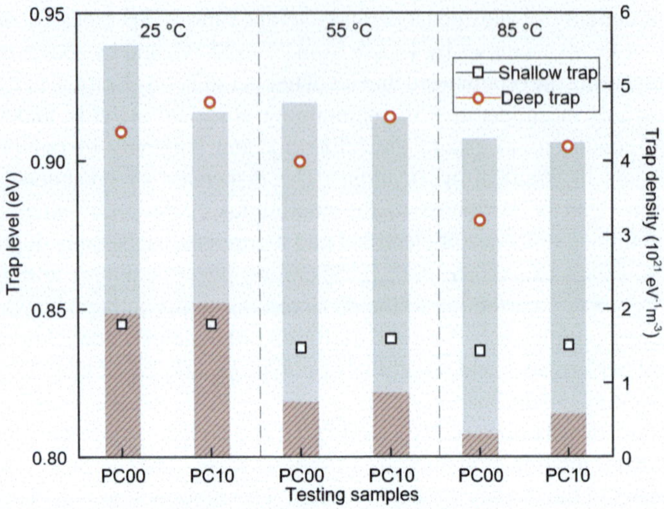

Fig. 9.6 Relationship between the trap sites and temperature [36]. Copyright Clearance Center's RightsLink© License Number: 5811370682461

breakdown strength is elevated. Furthermore, high temperature will amplify such gap of performances between PC00 and PC10 due to extremely different trap site distribution.

9.3.2 *Metal Organic Framework*

(1) **Effect on the Microstructure**

Metal organic framework is made up of inorganic metal ions and organic functional groups. It has been widely applied in lithium-ion battery, gas adsorption, solar energy conversion due to its high specific surface area and flexible design structure [37]. In this paper, the effect of MOF nanoparticles on dielectric properties of PP films was observed. The spherulite characteristics were measured by X-ray diffraction (XRD). The DC conductivity and breakdown strength were analyzed. The permittivity and dielectric loss of films were studied at the frequencies aiming to evaluate the application prospects of this modification method in the practical application. The change of energy storage density with the filler content was calculated. In addition, Quantum chemical theory was performed to analyze the effect of the nanoparticles on the energy level characteristics in the films. MOF nanoparticles reveal the great potential in reinforcing the dielectric performances of PP film for metallized film capacitors.

As shown in Table 9.3, this book will show the modification results of five samples with different doping contents.

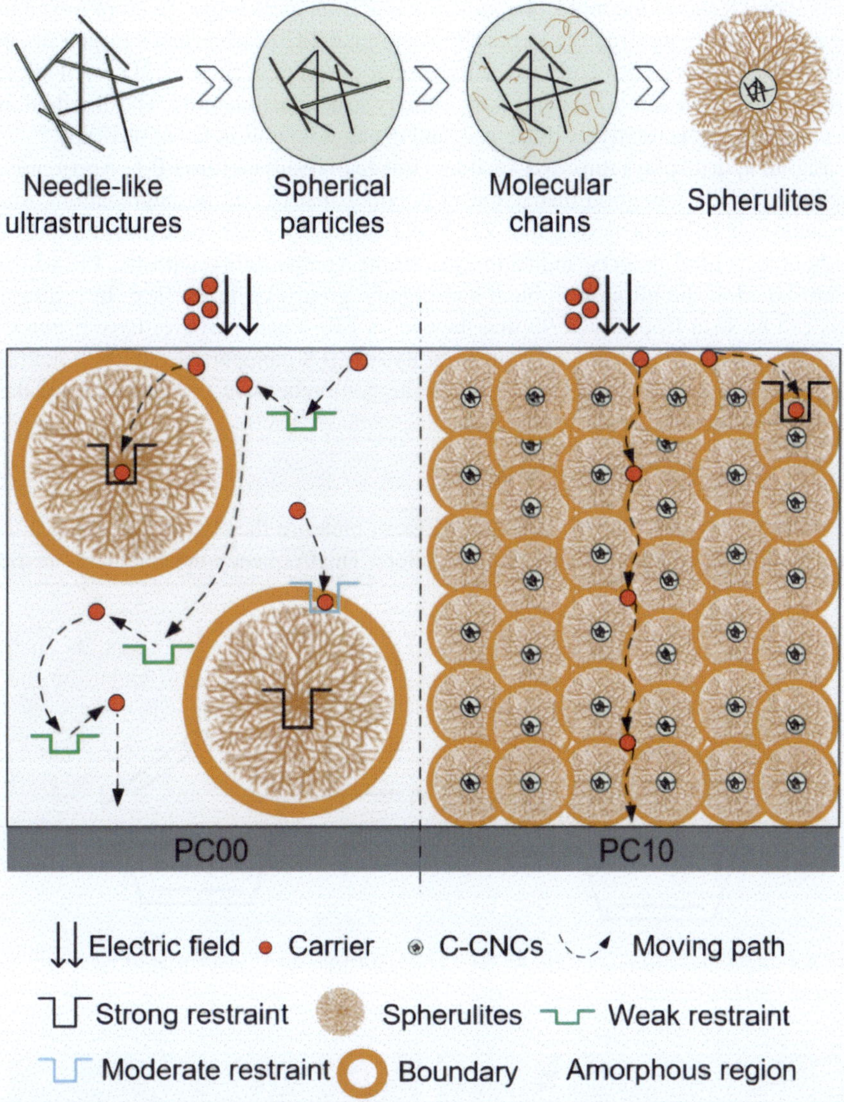

Fig. 9.7 Carrier transport process for PC00 and PC10 [36]. Copyright Clearance Center's RightsLink© License Number: 5811370682461

Table 9.3 Test samples with different doping content

ZIF-8 content	0%	0.02%	0.04%	0.06%	0.08%
Sample	PP-0	PP-2	PP-4	PP-6	PP-8

Figure 9.8 shows the molecular structure of ZIF-8, including a Zn atom stayed in the middle and a ring structure formed by a combination of carbon and nitrogen atoms around. This special structure gives its special properties, such as 3D topological structure, multiple active sites and extraordinary structural tunability. The distribution of nanoparticles is relatively dispersed, and the agglomeration is not obvious.

Figure 9.9a displays the cross-sectional image of resultant film PP-6. Apparently, the nanoparticle has good dispersion in polymer matrix and no aggregations was presented. It is worth noting that ZIF-8 is completely embedded into the matrix without structural defects, indicating the strong compatibility with the PP, which is attributed to the ultralow content and organic groups on its surface. In contrast, it is can be seen from Fig. 9.9b that the excess ZIF-8 nanoparticles have begun to agglomerate for PP-8 film, the compatibility of ZIF8 nanoparticles and PP matrix decreases. This phenomenon may affect the comprehensive performances of the composite films.

(2) **Effect on the Dielectric Properties**

Dielectric property is an important index to measure the energy storage performance of PP film for metallized film capacitors. The frequency-dependent dielectric

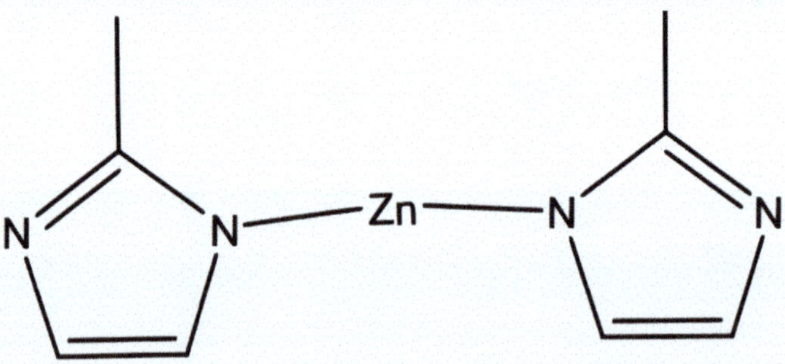

Fig. 9.8 Molecular structure of ZIF-8 [38]. Copyright Clearance Center's RightsLink© License Number: 5811410334682

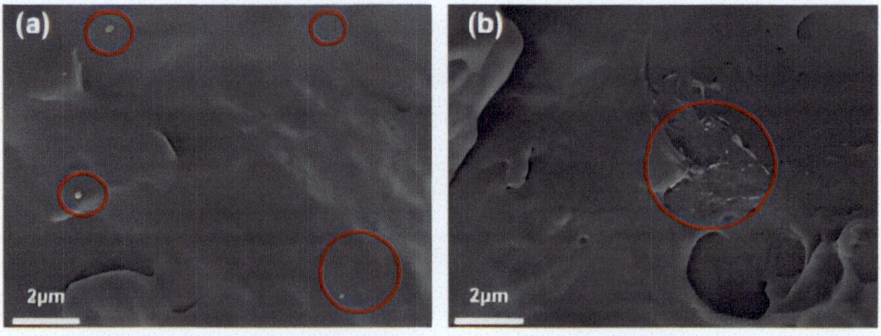

Fig. 9.9 SEM images of **a** PP-6; **b** PP-8 [38]. Copyright Clearance Center's RightsLink© License Number: 5811410334682

properties of different samples are shown in Fig. 9.10. The permittivity of PP is 2.42 at 1 kHz. After adding ZIF-8 nanoparticles, the permittivity of the composite film changed, reaching a maximum of 2.56 for PP-4. Although the permittivity of the ZIF-8 nanoparticle is low, less than 3, it can bring an increase at a loading of 0.04 wt%. That should be attributed to the polar groups in nanoparticles and the interfacial polarization brought about by its high specific surface area. That provides ideas for improving the permittivity of polymers utilizing ultralow content of nano-fillers. Meanwhile, the ZIF-8/PP films also represent excellent stability of permittivity with increasing frequency from 10^2 to 10^6 Hz. Interestingly, the ZIF-8/PP films exhibit low loss tangent. For example, the loss tangent is reduced from 0.0049 of pure PP to 0.0017 of PP-4 and 0.0021 of PP-6 at 1 k Hz. The experimental results show that ZIF-8 nanoparticles can effectively inhibit the conduction loss and joule heat of PP film during operation.

The specific data is shown in Table 9.4. When the loading of ZIF-8 nanoparticles is 0.06 wt%, the E_b of ZIF-8/PP film reaches the maximum of 589.6 kV/mm, which is 16.4% higher than that of PP (506.4 kV/mm). It can be attributed to the effect of ZIF-8 nanoparticles on the provision of deep traps to capture carriers in PP film. β denotes the shape parameter relate·d to data distribution. The β of PP-6 is 9.672, higher than that of PP-0 (8.923), which represent a stable improvement of the good insulation stability and evenly distribution.

In order to better understand the enhancement mechanism of the breakdown strength of films, the electrical conductivity of samples was measured. The conductivity of ZIF-8 films decreases first and then increases with the filler loading increasing, which is contrary to the variation of breakdown strength. The special

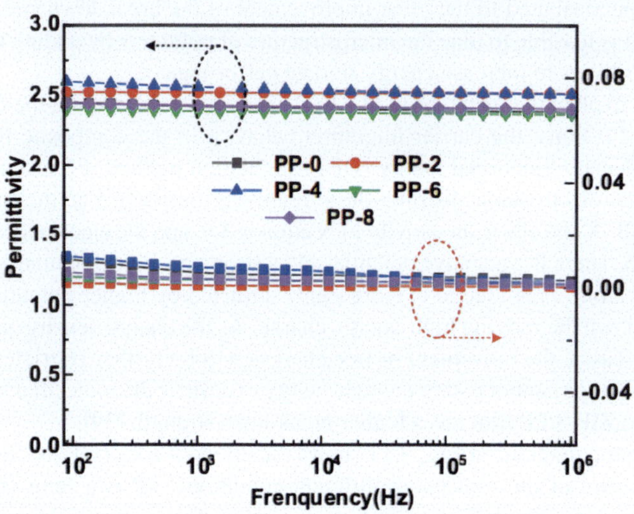

Fig. 9.10 Relationship between the permittivity, dielectric loss and frequency [38]. Copyright Clearance Center's RightsLink© License Number: 5811410334682

Table 9.4 The breakdown parameters of the samples

Sample	PP-0	PP-2	PP-4	PP-6	PP-8
Eb (kV/mm)	506.4	528.0	562.1	589.6	541.2
B	8.923	8.997	11.62	9.672	11.50

relationship between them also suggests that the increase in breakdown strength is also partly attributed to the decrease in conductivity due to the proper addition of ZIF-8 nanoparticles. The loss of capacitor films mainly comes from the leakage conductance under high field. The PP-6 has the lowest electrical conductivity, again illustrating the good effect of ZIF-8 on improving the insulating properties of the PP films.

The low energy storage density is a major factor limiting the further application of MFCs. The energy density of polymer films determines the energy density of MFCs. Based on the properties of linear polymers, the formula for the maximum energy density of PP films is shown in Equation:

$$U_e = \frac{1}{2}\varepsilon_0\varepsilon_r E^2 \tag{9.5}$$

where U_e is the energy density, ε_r is the vacuum dielectric constant, ε_0 is the dielectric constant of the sample, E represents the field strength at the cumulative breakdown probability of 63.2%. At 25 °C, the energy density of the ZIF-8/PP films is higher than that of pure PP, and the film with 0.06 wt% nanoparticle has maximum value of 3.67 J/cm^3, which is 34% higher than that of pure PP (2.74 J/cm^3). The reason for the increase can be attributed to the great improvement of the breakdown strength and permittivity. It is feasible to tune the microstructure of polymers by adding ultra-low levels of nano-fillers to increase energy storage performances.

In order to explain the mechanism of ZIF-9 nanoparticles changing the dielectric properties of PP films, the carrier migration behavior in the composite films was analyzed by density functional theory (DFT) simulation method. PP is a non-polar polymer, so its electrostatic distribution is relatively uniform. For the molecular structure of ZIF-8, the electronegativity of N atom is 3.0, and the electronegativity of Zn atom is 1.6. There is a negative potential and a strong positive potential zinc atom around the N atom. Free charge is more easily attracted by higher potential, while ZIF-8 slows down the moving high-energy charge. As the charge density around the nanofiller increases, the movement of free charges will be further restricted. In this case, the free charge cannot gather enough energy to destroy the molecular structure, and finally the ZIF-8/PP film has a higher breakdown strength [39].

The schematic diagram of Fig. 9.11 shows the effect of the crystallization of PP on the breakdown of the film under different conditions. PP is a semi-crystalline polymer with a large number of crystalline and amorphous regions, the breakdown strength of these two regions are quite different. Under the high electric fields, the charge comes from electrode injection and internal excitation, carriers are more inclined to move along the interfaces between crystalline and amorphous regions

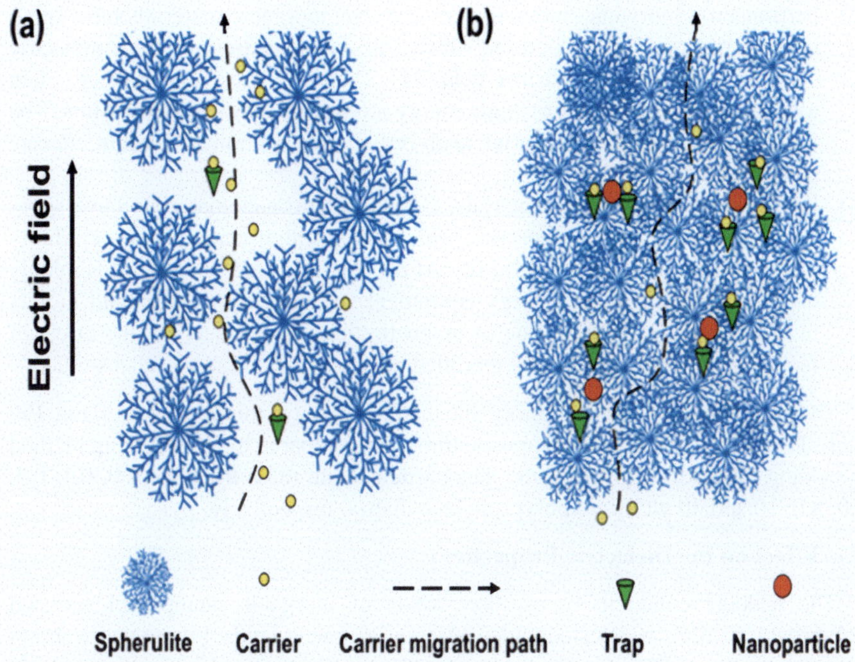

Fig. 9.11 Schematic diagram of the breakdown in **a** PP and **b** ZIF-8/PP [38]. Copyright Clearance Center's RightsLink© License Number: 5811410334682

[39]. Meanwhile, charges are more likely to accumulate around interfaces, which leads to partial discharges. In conclusion, the addition of ZIF-8 nanoparticle can improve the crystallization behavior of PP to reduce the interfaces by reducing the crystal size and the gap between the crystals, instead of increasing its crystallinity.

9.3.3 Inorganic Nanoparticles

(1) **Effect on the Microstructure**

There are many methods to classify the types of inorganic materials. In this section, according to its role in the structural design of electrical equipment, the following three inorganic materials are mainly studied:

(1) Insulating inorganic materials. Insulating inorganic materials generally have a wide band gap to ensure that they still have a low charge excitation intensity under a strong electric field [40]. Excellent mechanical properties and good insulation properties are the main reasons why these materials are widely used in the production of electrical equipment. The insulating inorganic material analyzed in this section is silicon dioxide (SiO_2).

(2) Ferroelectric inorganic materials. Ferroelectric inorganic materials have spontaneous polarization properties, which can exhibit extremely high dielectric constant under external electric field [41]. Therefore, these materials are often used in the development of high energy storage density capacitor films. The ferroelectric inorganic material analyzed in this section is barium titanate ($BaTiO_3$).
(3) Ferromagnetic inorganic materials. Ferromagnetic nanomaterials have spontaneous magnetization properties, which exhibit high magnetic susceptibility under external magnetic fields [42]. At present, this kind of material is mainly used in electromagnetic energy equipment such as transformers and motors to improve the utilization rate of magnetic field. In this section, the effect of ferromagnetic material Fe_3O_4 modified polypropylene film was also studied.

The particle size of the three inorganic materials used in this section is 20 nm. PP/SiO_2, PP/$BaTiO_3$, PP/Fe_3O_4 composite films were prepared by hot-pressing method by adding three kinds of inorganic nanoparticles with mass fraction of 0, 0.1, 0.5, 1.0, 5.0, 10 and 40 wt% into polypropylene matrix, respectively.

(2) **Effect on the Dielectric Properties**

The leakage conductivity of three kinds of inorganic nanoparticles doped polypropylene films was tested under different magnetic fields. Figure 9.12a shows the conductivity of PP/SiO_2 composite film. The conductivity of the modified film is optimal at 1 wt%, which is about 25% lower than that of pure PP film under different magnetic fields. When the SiO_2 content is less than 1 wt%, the conductivity of the composite film gradually decreases with the increase of the doping amount. When the doping amount exceeds 5 wt%, the conductivity of the film begins to rise sharply. The conductivity of the composite film with 40 wt% doping is 350–480% of that of pure PP.

Figure 9.12b shows the conductivity of PP/$BaTiO_3$ composite film. Different from doped SiO_2, doped $BaTiO_3$ cannot reduce the conductivity of PP film under different doping amount and magnetic induction intensity. The conductivity of PP/SiO_2 composite film is 64 times higher than that of pure PP film and 16 times higher than that of PP/SiO_2 composite film with the same content under 12T magnetic field and 40 wt% doping amount.

Figure 9.12c shows the conductivity of PP/Fe_3O_4 composite film. With the increase of Fe_3O_4 content from 0 to 0.5 wt%, the conductivity of the composite films decreased gradually. Compared with the pure PP film, the conductivity of the modified PP film doped with 0.5 wt% Fe_3O_4 is reduced to about 30%, and its optimization effect is better than that of SiO_2 doping. If the Fe_3O_4 content continues to increase, the conductivity of the composite film begins to increase, and its growth rate is between SiO_2 and $BaTiO_3$ doping. The conductivity of PP/Fe_3O_4 film is 8.4 times that of PP/SiO_2 and 50% that of PP/$BaTiO_3$ when the doping amount is 40 wt% in 12T magnetic field.

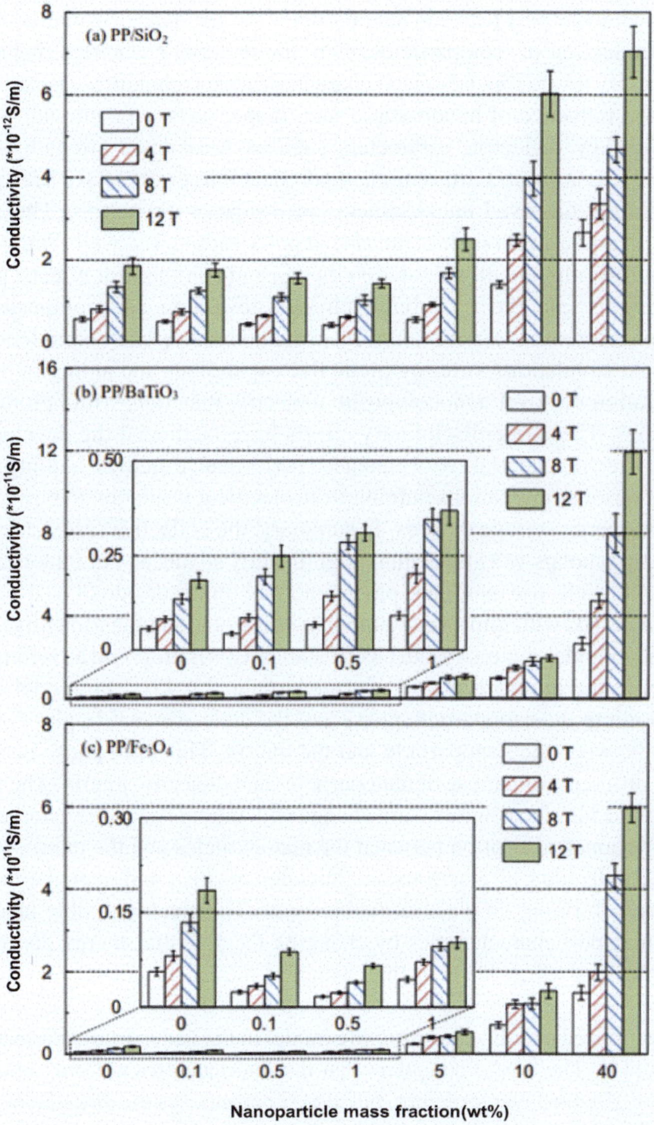

Fig. 9.12 Relationship between the conductivity and the magnetic field intensity

Research aimed at enhancing the insulation properties of dielectric materials through nanoparticle doping has garnered significant attention in both domestic and international academic communities. Scholars worldwide have conducted extensive investigations into the physical and chemical structural characteristics of nanocomposite dielectrics, offering various theoretical models to elucidate the mechanisms underlying the observed improvements in insulation properties. These studies delve into the intricate interplay between nanoparticles and the dielectric matrix, exploring how the introduction of nanoparticles alters the material's electrical behavior and performance. By leveraging advanced characterization techniques, researchers have elucidated the influence of nanoparticle size, shape, surface functionalization, and dispersion on key dielectric parameters such as breakdown strength, dielectric constant, and loss tangent. Furthermore, theoretical frameworks have been proposed to rationalize the observed enhancements in insulation properties. These models encompass diverse aspects such as interfacial polarization, Maxwell–Wagner-Sillars effects, nanofiller alignment, percolation theory, and confinement effects, providing valuable insights into the underlying physics governing nanocomposite dielectric behavior. Moreover, experimental investigations have been complemented by computational simulations, offering predictive capabilities and aiding in the design and optimization of novel nanocomposite dielectric materials with tailored insulation properties. These interdisciplinary efforts have facilitated the development of innovative dielectric materials with enhanced performance metrics, holding immense promise for various applications ranging from electrical insulation in power systems to capacitive energy storage devices. In summary, the collective research endeavors undertaken by scholars worldwide have significantly advanced our understanding of nanocomposite dielectric materials, paving the way for the design of next-generation insulation materials with improved reliability, efficiency, and functionality. Based on the theory of colloidal chemistry and electrostatics, Lewis proposed an electric double layer model in the interface region of nanoparticles [5, 43]. The model considers that in addition to the nanoparticle phase and the dielectric matrix phase, there is a third phase between the nanoparticle and the matrix. This third phase has physical parameters different from those of nanoparticles and dielectric matrix. The dielectric properties of the interface phase do not change with the change of the external electric field. The frictional adsorption between the nanoparticles and the matrix molecular chain or the ionization of electrons accumulates positive and negative ions in the interface layer, forming an electric double layer. The electric double layer affects the carrier transport characteristics by changing the potential energy distribution in the nanodielectric, which leads to the change of the macroscopic properties of the composites.

Tanaka proposed a multi-core structure model in the nanoparticle/dielectric interface region [44]. The model considers that the interface area not only contains the above electric double layer structure, but also superimposes the three-layer structure of the bonding layer, the binding layer and the loose layer from the inside to the outside. The bonding layer is a part of the nano-particles and the dielectric matrix, which is closely connected, and the thickness is about 1 nm. The bound layer is composed of polymer molecular chain segments tightly bound by the bonding layer and nanoparticles, and the thickness is about 2–10 nm. There is a weak coupling between the relaxation layer and the binding layer. The molecular chain and aggregation structure are different from the polymer matrix, and the thickness of the layer is generally tens of nanometers.

If the size of the nanoparticle is small enough, the nanodielectric composed of the nanoparticle can exhibit obvious quantum properties. At this time, the exciton will be tightly bound inside the nanoparticle, thus forming a limiting effect on charge transport [45]. In this case, the doped nanoparticles are called quantum dots (QDs), whose size is determined by the exciton Bohr radius of the matrix material. The quantum dot size of metal materials is calculated to be less than a few nanometers, and the quantum dot size of semiconductor and insulator materials is in the order of tens of nanometers [46].

The results show that the conductivity of PP film is reduced to a certain extent after proper doping of SiO2 nanoparticles, which confirms that the interface between nanoparticles and matrix can form charge trapping traps and reduce the carrier mobility of the film. When the doping amount gradually increases, the average distance between the particles is shortened, and the agglomeration problem of the particles is aggravated. The rapid increase of the effective radius of the nanoparticles makes the small size effect disappear, and the interface area overlaps a lot, and the efficiency of trapping charge in the interface area is significantly reduced. Therefore, after adding a higher content of SiO_2 nanoparticles, the conductivity of the composite film increases exponentially. When the doping particles are replaced by $BaTiO_3$ or Fe_3O_4, the conductivity of the composite films is greatly different, which indicates that in addition to the influence of the interface between the nanoparticles and the polymer matrix on the carrier transport behavior, the dielectric properties of the composite films are also related to the structure of the dopant itself.

The depth of charge traps introduced by the nanoparticles is determined by the band structure of the nanoparticles. When the carrier is transported to the vicinity of the nanoparticles in the conduction band of the polymer, because the band gap of the nanoparticles is narrower than that of the polymer, the electrons enter the conduction band of the nanoparticles after releasing energy to the outside world. Because the size of the quantum dots is small, the charge cannot tunnel or jump between the quantum dots when a small amount of doping, and the charge is thus captured. The depth of the charge trap introduced by the nanoparticle body is determined by the band gap difference between the polymer matrix and the nanoparticle, that is:

$$E_{trap} = \frac{(E_{g1} - E_{g2})}{2} \tag{9.6}$$

E_{trap} is the trap depth, E_{g1} and E_{g2} are the band gap of polymer matrix and nanoparticles, respectively. Different energy level structures will cause different capture and excitation. When the band gap of nanoparticles is wide, the depth of charge trap is shallow. At this time, the ability of nanoparticles to capture charge is weak, and the trapped charge may frequently fall into the trap. However, due to its wider band gap, the charge in the valence band of nanoparticles is not easy to excite into the conduction band. As the band gap of the nanoparticles gradually narrows, the charge traps gradually deepen, and the probability of valence band excitation gradually increases. When the nanoparticle is a conductor, the depth of charge trap reaches the maximum, but at the same time, the charge can migrate freely in the nanoparticle without excitation. The band gap of PP is 9.69 eV. The band gap and the corresponding trap depth

Table 9.5 Band gap of SiO_2, $BaTiO_3$ and Fe_3O_4

Nanoparticles	Band gap (eV)	Trap depth (eV)
SiO_2	8.9	0.395
$BaTiO_3$	3.3	3.195
Fe_3O_4	0.1	4.795

of SiO_2, $BaTiO_3$ and Fe_3O_4 nanoparticles are calculated by Formula (8–9), as shown in Table 9.5.

When SiO_2 is doped with shallow traps, the charge trapping effect of the nanoparticle body is weak, and the limitation of carrier transport mainly depends on the charge traps introduced by the interface effect. When Fe_3O_4 is doped with deep traps, not only the interface traps of nanoparticles can capture charge, but also the capture effect of nanoparticles on charge is greatly enhanced. Therefore, the optimization effect of Fe_3O_4 is better than that of SiO_2 when a small amount of doping. However, due to the large amount of free charge in Fe_3O_4, once the doping amount is increased to the percolation threshold, the conductivity of the composite film will increase significantly. Therefore, the conductivity of PP/Fe_3O_4 film is several times higher than that of PP/SiO_2 film at higher doping content.

The properties of PP/$BaTiO_3$ thin films are between PP/Fe_3O_4 and PP/SiO_2. The shallow traps make the binding effect of charge weak, and the narrow band gap increases the excitation probability of charge. In general, the conductivity of PP/$BaTiO_3$ films with a small amount of doping can be maintained at the same order of magnitude before modification, but when the doping amount increases, the conductivity increases most significantly in the three nano-dielectrics.

With the increase of the doping amounts of nanoparticles, the dielectric constant of the composite films increased to varying degrees. Among them, $BaTiO_3$ thin film has the most significant effect on the improvement of dielectric constant. Under 40 wt% doping, the dielectric constant of the composite film increases from 2.3 to about 30. This is attributed to the extremely high dielectric constant of $BaTiO_3$ ferroelectric nanomaterials. Because the dielectric constant of Fe_3O_4 is about 20, the dielectric constant of PP/Fe_3O_4 composite film reaches about 4.5 at 40 wt% doping amount, which is significantly higher than that of pure PP. In contrast, SiO_2 has the weakest effect on the dielectric constant of the composite film, and the dielectric constant is increased to about 2.6 at the maximum doping amount.

In a non-magnetic environment, different contents of SiO_2 will lead to a decrease in the breakdown field strength of the composite film. This shows that although the introduction of SiO_2 can inhibit charge transport and reduce conductivity to a certain extent, SiO_2 particles will produce strong enough electric field distortion for micron-sized polypropylene films, which will cause partial discharge at the internal defects of the film until the breakdown through the film is formed. When the magnetic field is gradually enhanced, a small amount of SiO_2 doping can improve the performance by about 7%. This is because a large amount of charge in the pure PP film is excited under a strong magnetic field, which accelerates the degradation process of the material. The charge introduced by SiO_2 can inhibit the damage of high-energy carriers to the film insulation to a certain extent. However, on the whole, the comprehensive effect of interface traps on improving the performance of the film is not obvious. Especially when the doping amount is increased, the distortion effect of the electric field still dominates, which further reduces the breakdown field strength of the composite film.

The breakdown field strength of PP/$BaTiO_3$ film fluctuates slightly at low doping amount, but when the doping amount increases, the breakdown field strength decreases sharply. The extremely high dielectric constant of $BaTiO_3$ makes the problem of electric field distortion inside the composite film more serious. Although the traps introduced by $BaTiO_3$ are deeper than SiO_2, it is not enough to make up for the insulation shortcomings caused by significant differences in dielectric parameters.

Fe_3O_4 has the best breakdown field strength optimization effect among the three types of nanoparticles. At the content of 0.1 and 0.5 wt%, Fe_3O_4 can increase the breakdown field strength of the composite film by 7–20%. Compared with SiO_2, Fe_3O_4 has a deeper trap depth, which makes Fe_3O_4 have a stronger charge trapping effect. Compared with $BaTiO_3$, the dielectric constant of Fe_3O_4 is lower, and the electric field distortion effect caused by Fe_3O_4 in the composite film is weaker. The two reasons together lead to the use of an appropriate amount of Fe_3O_4 to optimize the insulation of PP films can improve their performance.

After mastering the dielectric constant and breakdown field strength characteristics of three kinds of nanoparticles doped polypropylene films, the theoretical energy storage density of the composite films can be estimated by using the formula. It must be pointed out that the dielectric loss of $BaTiO_3$ and Fe_3O_4 doped polypropylene films is much higher than that of PP bulk, and the actual energy storage density of the three groups of films, especially PP/$BaTiO_3$ and PP/Fe_3O_4 composite films, is lower than the calculated value. However, the existing test methods cannot realize the in-situ characterization of the charge and discharge characteristics of the capacitor film in a strong magnetic field environment. Therefore, this section mainly analyzes the energy storage characteristics of the three nanocomposite films. The theoretical energy storage density of each film under different magnetic induction intensities was calculated by using the dielectric constant and DC breakdown field strength of each composite film at 1 kHz frequency, as shown in Fig. 9.13. With the increase of magnetic field, the energy storage density of PP film decreases slightly, which is mainly attributed to the decrease of breakdown field strength caused by magnetic field. After doping SiO_2, the energy storage density of the film is slightly improved, because the interfacial polarization increases the dielectric constant of the film, and the charge trap enhances the insulation performance of the film. When the doping amount increases, the breakdown field strength of the PP/SiO_2 film begins to decrease, resulting in a corresponding decrease in the energy storage density of the film. The energy storage density of the composite film increased significantly after doping $BaTiO_3$, which was mainly attributed to the high dielectric constant of $BaTiO_3$. Because $BaTiO_3$ also reduces the breakdown field strength of the film, the energy storage density of the film fluctuates slightly under low doping. The energy storage density of PP/Fe_3O_4 film increased first and then decreased. When a small amount of doping, the dielectric constant and breakdown field strength of

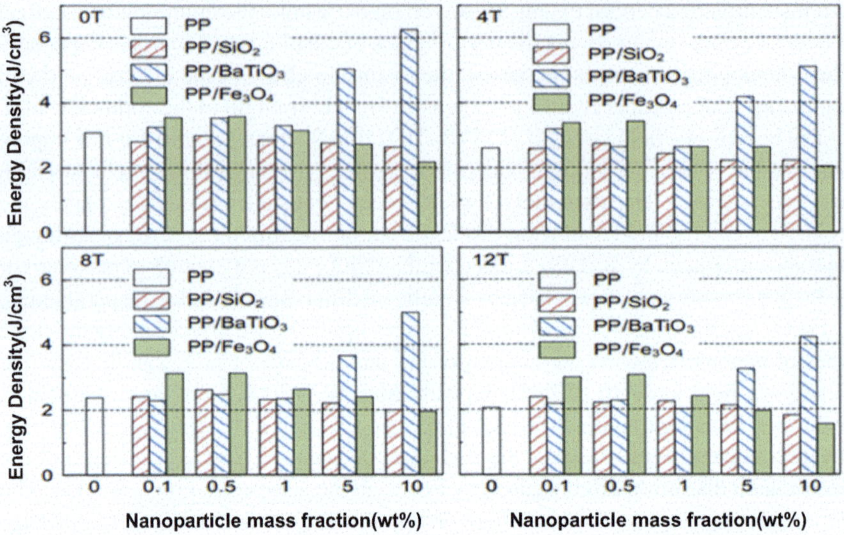

Fig. 9.13 Relationship between the energy storage density and the magnetic field intensity

the composite film are optimized at the same time, while when a large amount of doping, the decrease of breakdown field strength dominates.

9.4 Nanoparticles Surface Modifying

9.4.1 Surface Morphology Regulating

Due to the small particle size and extremely high surface energy, nanoparticles are prone to agglomeration and introduce micropores in the polymer. These two factors will cause the uneven distribution of the electric field inside the composite medium, resulting in a weakness that causes breakdown, resulting in a breakdown strength lower than the polymer matrix. Using small organic molecules or polymers to modify the surface of nanoparticles [47], defects can be reduced, trap energy levels and densities can be regulated, thereby improving the breakdown strength of nanocomposite dielectrics. Among them, coupling agent [48] and phosphoric acid molecules [49] are the most commonly used nanoparticle modifiers. For example, the modification of silica nanoparticles by silane coupling agent not only improves the dispersion of nanoparticles in polyethylene, but also changes the supramolecular structure of polyethylene [50]. The modified silica nanoparticles can inhibit the space charge injection of polyethylene composite dielectric, so that the composite dielectric has higher DC breakdown strength [51].

When the side chain of the grafted polymer is longer, the breakdown strength of the composite dielectric is lower. This is because the side chains of the grafted polymer are arranged more closely, and the interaction force with the polymer matrix is weak. When the side chain of the grafted polymer is short, the breakdown strength of the composite dielectric is high. This is due to the loose arrangement of the side chains of the grafted polymer, which can form a strong interfacial interaction with the polymer matrix, which is conducive to improving the breakdown strength. When the macromolecules grafted on the surface of the nanoparticles are the same as the matrix polymer, it is more conducive to enhancing the interface compatibility, and the breakdown strength of the nanocomposite medium may be more obvious.

A typical example is that Zhou Yao et al. [52] modified MgO nanoparticles with maleic anhydride grafted polypropylene to enhance the breakdown and energy storage performance of PP. The breakdown strength of the obtained product at 120°C reaches 501 MV/m, which is 122% and 110% of that of pure PP and silane coupling agent modified composite materials, respectively. The results of the thermally stimulated current test indicate that the trap energy levels and density formed on the material surface are high, which can effectively suppress the migration of charge carriers and improve the breakdown strength of the composite dielectric [53].

Surface modification of nanoparticles holds promise not only for enhancing their dispersion within the polymer matrix but also for finely tuning the trap energy levels and densities within the composite dielectric. These adjustments play a pivotal role in augmenting the breakdown strength of the composite dielectric, thereby rendering surface modification a widely adopted strategy in the quest to enhance dielectric performance. By tailoring the surface chemistry and functionality of nanoparticles, researchers can effectively mitigate issues related to agglomeration and promote uniform dispersion within the polymer matrix. This enhanced dispersion facilitates more uniform electrical properties throughout the material, contributing to improved breakdown strength and overall dielectric performance [54]. Furthermore, surface modification enables precise control over the trap energy levels and densities within the composite dielectric. These traps can influence charge carrier mobility and distribution, ultimately impacting the dielectric breakdown behavior. Through strategic surface engineering of nanoparticles, researchers can modulate these trap characteristics to optimize breakdown strength and mitigate potential failure mechanisms. Despite the significant strides made in leveraging surface-modified nanoparticles to enhance breakdown strength, it is essential to acknowledge the ongoing debate surrounding the underlying mechanisms at play. While surface modification has demonstrated promising results in many cases, there remains a lack of consensus regarding the precise mechanisms through which nanoparticles influence breakdown strength. Indeed, it is not uncommon for modified nanoparticles to inadvertently reduce the breakdown strength of polymers, underscoring the complexity of the underlying processes involved. As such, elucidating the fundamental mechanisms governing the interaction between nanoparticles and polymer matrices during the breakdown process remains a critical objective for future research endeavors. By unraveling these intricate mechanisms, researchers can not only refine existing methodologies for enhancing breakdown strength but also unlock new avenues for innovation in the design and development of advanced composite dielectric materials. Thus, while surface modification represents a valuable tool in the quest for improved dielectric performance, continued exploration and investigation are necessary to fully realize its potential and address remaining challenges in this field.

9.4.2 Dimension and Arrangement Regulating

See Fig. 9.14.

Nanoparticles can be divided into zero-dimensional nanoparticles, one-dimensional nanowires and two-dimensional nanosheets according to their dimensions. As the dimension increases, the probability of nanoparticles blocking carrier migration in the direction perpendicular to the electric field increases. According to the blocking effect of nanoparticles on the electrical breakdown of composite dielectrics, high-dimensional nanoparticles can be selected to regulate their distribution in the polymer matrix, and nanocomposite dielectrics with high breakdown strength can be designed and prepared [55, 56].

In addition, Li et al. [57] prepared three dimensions of nano-alumina, and found that the breakdown strength of alumina nanosheet composite medium was 489 MV/m, which was 146% and 127% of the breakdown strength of nanoparticle composite medium and nanowire composite medium, respectively. Compared with nanoparticles and nanowires, alumina nanosheets are more effective in inhibiting the development of electrical trees, thereby effectively improving the breakdown strength of composite dielectrics [58, 59]. In addition, it is also found that the nanosheets play

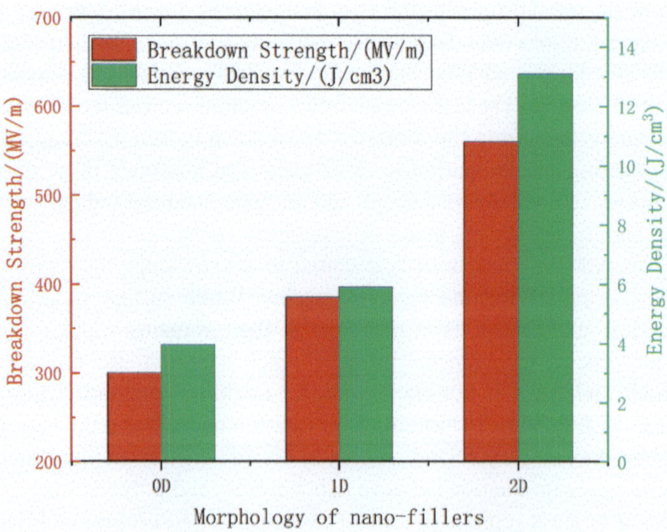

Fig. 9.14 Comparison of the dielectric performance with different composite dielectrics

the role of uniform electric field, avoiding the decrease of breakdown strength caused by local electric field concentration. The above results show that two-dimensional nanosheets are more effective than one-dimensional nanowires and zero-dimensional nanoparticles in improving the breakdown strength of composite dielectrics.

The above results show that regulating the dimension and arrangement of nanoparticles can regulate the internal electric field distribution of nanocomposite dielectrics, suppress the leakage current, and improve the breakdown strength of nanocomposite dielectrics. Compared with nanowires and nanoparticles, nanosheets can more effectively improve the breakdown strength of composite dielectrics. When the nanosheets are arranged along the plane parallel to the film in the composite dielectric, the breakdown strength of the composite dielectric is the best.

9.5 Summary

In this chapter, polypropylene composite films were obtained by adding a small amount of nanoparticles to PP or LCBPP materials with the aim of regulating the crystal morphology and microforms in the sample films. Through nanoparticle doping, the microforms of the polypropylene films were improved, which effectively optimized the macroscopic electrical properties of the films, and are expected to be applied in the field of thin-film capacitors. Based on the contents of this chapter, the following conclusions can be drawn:

The insulating property of dielectric is positively correlated with the trap energy level. Doping a certain amount of nanoparticles will increase the trap depth in the composite dielectric, and the presence of deep traps will increase the activation energy of the material and enhance its ability to capture free charge, thus improving its dielectric properties such as breakdown field strength.

The insulating properties of the dielectric are related to the free movement distance of free charges in the dielectric. Doping nanoparticles in the dielectric can hinder the movement of excited charges to a certain extent, thus reducing the average free range of free charges and improving the insulating properties of the material.

The effect of nanomaterials on the polymer matrix includes both physical (morphology, supramolecular structure, mechanical properties, etc.) and chemical (cross-linking, etc.) properties. There are interfacial zones in the nanocomposite medium that are physically and chemically different from the polymer matrix, and the high specific surface area of the nanoparticles gives the composite medium an interfacial zone with high volume fraction. By regulating the thickness and dielectric constant of each layer it is possible to regulate the electric field strength that each layer is subjected to, thus regulating the electric field distribution in multilayer composite dielectrics.

References

1. P. Guido and R. Maurizio, "Pulse handling capability of energy storage metallized film capacitors," *IEEE Trans. Plasma*, vol. 28, no, 5, pp. 1603–1606, 2000.
2. A. Kartal, "Review of film capacitor trends and design changes as a result of improved technologies in power electronics," PCIM Europe 2017; International Exhibition and Conference for Power Electronics, Intelligent Motion, Renewable Energy and Energy Management, Nuremberg, Germany, 2017, pp. 1–3.
3. R. W. Brown, "Linking corrosion and catastrophic failure in low-power metallized polypropylene capacitors," *IEEE Trans. Device Mat Reliab.*, vol. 6, no. 2, pp. 326–333, 2006.
4. E. J. Barshaw, "High Energy Density (HED) Biaxially-Oriented Poly-Propylene (BOPP) capacitors for pulse power applications," *IEEE T. Magn.*, vol. 43, no, 1, pp. 223–225, 2007.
5. T. J. Lewis, "Nanometric dielectrics," *IEEE Trans. Dielect Elect. Insul.*, vol. 1, no. 5, pp. 812–825, 1994.
6. X. Huang, J. Zhang and P. Jiang, "Thermoplastic insulation materials for power cables: history and progress," *High Volt. Engineering*, vol. 44, no. 5, pp. 1377–1398, 2018.
7. Z. M. Dang, J. K. Yuan, S. H. Yao and R. J. Liao, "Flexible nanodielectric materials with high permittivity for power energy storage," *Adv. Mater.*, vol. 25, no. 44, pp. 6334–6365, 2013.
8. Y. Zhu, P. Jiang and X. Huang, "Theory and method for improving breakdown strength of nanocomposite dielectrics," Proceedings of the Chinese Society of Electrical Engineering, vol. 41, no. 1, pp. 183–199+409, 2021.
9. D. R. Xie, "Classified effects of nanofillers on DC breakdown and partial discharge resistance of polypropylene/alumina nanocomposites," *IEEE Trans. Dielect Elect. Insul.*, vol. 26, no. 3, pp. 698–705, 2019.
10. J. G. Gao, "Effect of zinc oxide nanoparticle size on the dielectric properties of polypropylene-based nanocomposites," *Polym. Eng. Sci.*, vol. 63, no. 7, pp. 2204–2217, 2023.
11. Y. Zhou, J. Hu, B. Dang and J. L. He, "Effect of different nanoparticles on tuning electrical properties of polypropylene nanocomposites," *IEEE Trans. Dielect Elect. Insul.*, vol. 24, no. 3, pp. 1380–1389, 2017.
12. Y. Zhou, J. L. He, J. Hu and B. Dang, "Surface-modified MgO nanoparticle enhances the mechanical and direct-current electrical characteristics of polypropylene/polyolefin elastomer nanodielectrics," *J. Appl. Polym. Sci.*, vol. 133, no. 1, 2016, Art. no. 42863.
13. Y. Zhou, "Enabling high-performance polypropylene nanocomposites with interfacial deep traps," *IEEE Trans. Dielect Elect. Insul.*, vol. 30, no. 1, pp. 484–487, 2023.
14. W. F. Liu, L. Cheng and S. T. Li, "Review of electrical properties for polypropylene based nanocomposite," *Compos. Commun.*, vol. 10, pp. 221–225, 2018.
15. M. Meunier, N. Quirke and A. Aslanides, "Molecular modeling of electron traps in polymer insulators: Chemical defects and impurities." *J. Chem. Phys.*, vol. 115, no. 6, pp. 2876–2881, 2001.
16. M. Meunier, N. Quirke, "Molecular modeling of electron trapping in polymer insulators," *J. Chem. Phys.*, vol. 113, no. 1, pp. 369–376, 2000.
17. T. Takada, Y. J. Hayase, Y. Tanaka and T. Okamoto, "Space charge trapping in electrical potential well caused by permanent and induced dipoles for LDPE/MgO nanocomposite," *IEEE Trans. Dielect Elect. Insul.*, vol. 15, no. 1, pp. 152–160, 2008.
18. M. Xiao Z. Y. Zhang and B. X. Du, "Dielectric performance improvement of polypropylene film by hierarchical structure design for metallized film capacitors," *J. Phys. D: Appl. Phys.*, vol. 57, no. 38, pp. 5501–5509, 2024.
19. H. L. Liu, B. X. Du, M. Xiao, and Z. Y. Ran, "Improved conductivity and breakdown performance of polypropylene film by parylene blending for power capacitor," *IEEE Trans. Dielect Elect. Insul.*, vol. 29, no. 3, pp. 997–1004, 2022.
20. S. T. Li, D. M. Min, W. W. Wang and G. Chen, "Linking traps to dielectric breakdown through charge dynamics for polymer nanocomposites," *IEEE Trans. Dielect Elect. Insul.*, vol. 23, no. 5, pp. 2777–2785, 2016.

21. Z. H. Shen, "High-throughput phase-field design of high-energy-density polymer nanocomposites," *Adv. Mater.*, vol. 30, no. 2, 2018, Art. no. 1704380.
22. Y. F. Wang, "Ultrahigh electric displacement and energy density in gradient layer-structured BaTiO3/PVDF nanocomposites with an interfacial barrier effect," *J. Mater. Chem. A*, vol. 5, no. 22, pp. 10849–10855, 2017.
23. M. Mackey, D. E. Schuele, L. Zhu and E. Baer, "Layer confinement effect on charge migration in polycarbonate/poly (vinylidene fluorid-co-hexafluoropropylene) multilayered films," *J. Appl. Phys.*, vol. 111, no. 11, 2012, Art. no. 113702.
24. M. Mackey et al, "Enhanced breakdown strength of multilayered films fabricated by forced assembly microlayer coextrusion," *J. Phys. D: Appl. Phys.*, vol. 42, no. 17, 2009, Art. no. 175304.
25. E. Bear and L. Zhu, "50th anniversary perspective:Dielectric phenomena in polymers and multilayered dielectric films," *Macromolecules*, vol. 50, no. 6, pp. 2239–2256, 2017.
26. J. Frenkel, "On pre-breakdown phenomena in insulators and electronic semi-conductors," *Physical Review*, vol. 54, no. 8, pp. 647–648, 1938.
27. J. P. Calame, "Finite difference simulations of permittivity and electric field statistics in ceramic-polymer composites for capacitor applications," *J. Appl. Phys.*, vol. 99, no. 8, 2006, Art. no. 084101.
28. Z. P. Li et al, "High dielectric constant polycarbonate/nylon multilayer films capacitors with self-healing capability," *ACS Appl. Energy Mater.*,vol. 1, no. 4, pp. 867–875, 2019.
29. L. Cheng et al, "Online degradation of biaxial-orientated polypropylene film from HVDC filter capacitors," *IEEE Trans. Dielect Elect. Insul.*, vol. 26, no. 1, pp. 26–33, 2019.
30. J. Ho and T. R. Jow, "High field conduction in biaxially oriented polypropylene at elevated temperature," *IEEE Trans. Dielect Elect. Insul.*, vol. 19, no. 3, pp. 990–995, 2012.
31. T. X. Ju, I. Treufeld, M. Wolak, M. Ponting, E. Baer and L. Zhu, "Effect of glass transition temperature on enhanced fielectric breakdown dtrength and lifetime of multilayer polymer films," *ACS Appl. Mater. Interfaces*, vol. 16, no. 1, pp. 795–806, 2024.
32. M. Horak and P. Mach, "Study of electrical ageing of polypropylene film capacitors using third harmonics measurement," 2014 IEEE 20th International Symposium for Design and Technology in Electronic Packaging, Bucharest, Romania, 2014, pp. 55–59.
33. A. Kahouli et al, "Structure effect of thin film polypropylene view by dielectric spectroscopy and X-ray diffraction: Application to dry type power capacitors," *J. Appl. Polym. Sci.*, vol. 132, no. 39, 2015.
34. W. Zhang et al, "Effect of β-crystals on the mechanical and electrical properties of β-nucleated isotactic polypropylene," *IEEE Trans. Dielect Elect. Insul.*, vol. 26, no. 3, pp. 714–721, 2019.
35. Z. Y. Ran et al, "Effect of crystallization regulation on the breakdown strength of metallized polypropylene film capacitors," *IEEE Trans. Dielect Elect. Insul.*, vol. 28, no. 1, pp. 175–182, 2021.
36. Z. Y. Ran, B. X. Du and M. Xiao, "High-temperature breakdown performance improvement of polypropylene films based on micromorphology control," *IEEE Trans. Dielect Elect. Insul.*, vol. 28, no. 5, pp. 1547–1554, 2021.
37. Z. M. Zhao, J. W. Ding, R. M. Zhu, and H. Pang, "The synthesis and electrochemical applications of core-shell MOFs and their derivatives," *J. Mater. Chem. A*, vol. 7, no. 26, pp. 15519–15540, 2019.
38. K. Chen, B. X. Du and M. Xiao, "Improving dielectric properties of polypropylene film for metallized film capacitors by MOF nanoparticle," Tianjin University-IET Electrical & Information Engineering Doctoral Forum, Tianjin, China, 2023, pp. 62–67.
39. Z. Y. Ran, B. X. Du, M. Xiao, H. L. Liu, and J. W. Xing, "Effect of crystallization regulation on the breakdown strength of metallized polypropylene film capacitors," *IEEE Trans. Dielect Elect. Insul.*, vol. 28, no. 1, pp. 175–182, 2021.
40. C. Yang et al, "Improving the performance of dielectric nanocomposites by utilizing highly conductive rigid core and extremely low loss shell," *J. Phys. Chem. C*, vol. 124, no. 24, pp. 12883–12869, 2020.

41. X. Y. Huang, P. K. Jiang and L. Y. Xie, "Ferroelectric polymer/silver nanocomposites with high dielectric constant and high thermal conductivity," *Appl. Phys. Lett.*, vol. 95, no. 24, 2009, Art. no. 242901.
42. V. Segal, A. Rabinovich, D. Nattrass, K. Raj and A. Nuneset, "Experimental study of magnetic colloidal fluids behavior in power transformers," *J. Magn. Magn. Mater.*, vol. 215, pp. 513–515, 2000.
43. 44. T. J. Lewis, "Interfaces are the dominant feature of dielectrics at the nanometriclevel," *IEEE Trans. Dielect Elect. Insul.*, vol. 11, no. 5, pp. 739–753, 2004.
44. 45. T. Tanaka, M. Kozako, N. Fuse and Y. Ohki, "Proposal of a multi-core model for polymer nanocomposite dielectrics," *IEEE Trans. Dielect Elect. Insul.*, vol. 12, no. 4, pp. 669–681, 2005.
45. 46. R. Rossetti, S. Nakahara and L. E. Brus, "Quantum size effects in the redox potentials,resonance Raman spectra, and electronic spectra of CdS crystallites in aqueoussolution," *J. Chem. Phys.*, vol. 79, no. 2, pp. 1086–1088, 1983.
46. 47. K. Kalyanasundaram, E. Borgarello, D. Duonghong and M. Grätzel, "Wasserspaltung durch Bestrahlung kolloidaler CdS-Lösungen mit sichtbarem Licht; Unterdrückung der Photokorrosion durch RuO2†," *Angew. Chem. Int. Ed.*, vol.93, no.11, pp.1012–1013, 1981.
47. K. K. Nanda, A. Maisels, F. E. Kruos, H. Fissan and S. Stappert, "Higher surface energy of free nanoparticles," *Phys. Rev. Lett.*, vol. 91, no. 10, 2003, Art. no. 106102.
48. 49. X. Y. Huang and P. K. Jiang, "Core-shell structured high-k polymer nanocomposites for energy storage and dielectric applications," *Adv. Mater.*, vol. 27, no. 3, pp. 546–554, 2015.
49. 50. Q. Q. Zhang et al, "Enhanced dielectric tunability of Ba0.6Sr0.4TiO3/poly (vinylidene fluoride) composites via interface modification by silane coupling agent," *Compos. Sci. Technol.*, vol. 129, pp. 93–100, 2016.
50. 51. P. Kim et al, "High energy density nanocomposites based on surface-modified BaTiO3 and a ferroelectric polymer," *ACS Nano*, vol. 3, no. 9, pp. 2581–2592, 2009.
51. X. Y. Huang, P. K. Jiang and Y. Yin, "Nanoparticle surface modification induced space charge suppression in linear low density polyethylene," *Appl. Phys. Lett.*, vol. 95, no. 24, 2009, Art. no. 242905.
52. 53. L. Zhang et al, "Effect of nanoparticle surface modification on breakdown and space charge behavior of XLPE/SiO2 nanocomposites," *IEEE Trans. Dielect Elect. Insul.*, vol. 21, no. 4, pp. 1554–1564, 2014.
53. 54. K. Yang, X. Y. Huang, Y. H. Huang, L. Y. Xie and P. K. Jiang, "Fluoro-polymer@BaTiO3 hybrid nanoparticles prepared via raft polymerization:Toward ferroelectric polymer nanocomposites with high dielectric constant and low dielectric loss for energy storage application," *Chem. Mater.*, vol. 25, no. 11, pp. 2327–2338, 2013.
54. 55. Y. Zhou et al, "Interface-modulated nanocomposites based on polypropylene for high-temperature energy storage," *Energy Storage Mater.*, vol. 28, pp. 255–263, 2020.
55. A. C. Ashcraft, R. M. Eichhorn, and R. G. Shaw, "Laboratory studies of treeing in solid dielectrics and voltage stabilization of polyethylene," 1976 IEEE International Conference on Electrical Insulation, Montreal, QC, Canada, 1976, pp. 213–218.
56. 57. Y. H. Gao, X. Y. Huang, D. M. Min, S. T. Li and P. K. Jiang, "Recyclable dielectric polymer nanocomposites with voltage stabilizer interface: Toward new generation of high voltage direct current cable insulation," *ACS Sustain. Chem. Eng.*, vol. 7, no. 1, pp. 513–525, 2019.
57. 58. X. Y. Huang, Q. Zhang and P. K. Jiang, "Influence of nanoparticle shape on electrical properties of linear low density polyethylene nanocomposites," *Proceedings of the CSEE*, vol. 36, no. 24, pp. 6606–6612, 2016.
58. 59. Y. K. Zhu, H. Yao, P. K. Jiang, J. D. Wu, X. Zhu and X. Y. Huang, "Two-dimensional high-k nanosheets for dielectric polymer nanocomposites with ultrahigh discharged energy density," *J. Phys. Chem. C*, vol. 122, no. 32, pp. 18282–18293, 2018.
59. X. Zhang, "Polymer nanocomposites with ultrahigh energy density and high discharge efficiency by modulating their nanostructures in three dimensions," *Adv. Mater.*, vol. 30, no. 16, 2018, Art. no. 1707269.

Chapter 10
Ring Compounds Doping

10.1 Introduction

The loss of polymer materials is closely related to dielectric properties such as breakdown and the internal charge transport behavior of polymers, and adding chemical additives containing specific functional groups to introduce chemical trap energy levels into polymers is an effective way to improve the dielectric properties of polymers. Since the 60s of the twentieth century, chemical additives have been used to inhibit the long-term electrical aging of polymer insulating materials and improve the electrical resistance of insulating materials. The modifier that can capture high-energy electrons is called thermos-electron trap, which has the ability to regulate charge transport and improve the dielectric properties of insulating materials, and is one of the most widely studied organic modifiers.

At present, there are few studies on the application of PC to improve the loss and breakdown characteristics of polypropylene films, the mechanism of the influence of PC on the charge transport behavior of polypropylene films to improve the macroscopic electrical properties is not clear, the feasibility of PC applied to capacitor insulating polypropylene film lacks in-depth and systematic analysis, and whether it can use its ability to regulate charge transport to improve the loss and breakdown performance of capacitor insulating polypropylene film is not yet known.

10.2 Modification Theory

Most Polyphenyl ring compounds (PC) have low ionization energy, which makes them easier to be excited or ionized than polymer molecules, and have a high electron affinity that allows them to trap high-energy electrons. Under the action of strong electric field, the ionization energy and electron affinity energy of PC can be excited or ionized before the polymer molecules are hit by high-energy electrons,

B. Du and M. Xiao, *Polypropylene Film for HVDC Capacitors*,
https://doi.org/10.1007/978-981-96-3029-5_10

weakening the destruction of polymer molecular chains by high-energy electrons [1], and then improving the electrical resistance of polymer insulating materials. From the mechanism of action of the thermos-electron trap, the trap of polyphenyl ring chemical type can be restored to the original state without being consumable, so it has extremely high research value. Studies have shown that polyphenyl ring chemicals (naphthalene, anthracene, etc.) [2] can trap electrons, which can significantly improve the DC breakdown field strength of polymers, and the attachment of specific molecular groups to the anthracene molecular chain can introduce deep traps, enhance its electron capture ability, and further improve the breakdown strength of polymer materials [3]. Benzene ring-containing compounds such as acetophenone derivatives, chalcone derivatives and benzophenone derivatives can improve the DC breakdown strength of polyethylene and cross-linked polyethylene [4]. the PC have a significant effect on the breakdown lifting of polymer insulating materials and the suppression of electric branches due to their ability to capture thermal electrons and regulate charge transport.

In this chapter, the different PC are added to polypropylene to study the regulatory effect of polypropylene film by adding PC, the influence of the regulatory trap energy level on the loss and breakdown characteristics of polypropylene film under the operating environment of DC support capacitors is analyzed, and the mechanism of the action of PC on the loss and breakdown characteristics of polypropylene film is discussed. At present, Fig. 10.1a–c are several commonly used PC, which are expressed as PP-PC1, PP-PC2 and PP-PC3, respectively [5]. The effects of adding different amounts of PC (Fig. 10.1d) on the dielectric properties of PP are additionally elaborated, recording PP-1*, PP-2*, PP-3*, PP-4 and PP-5*.

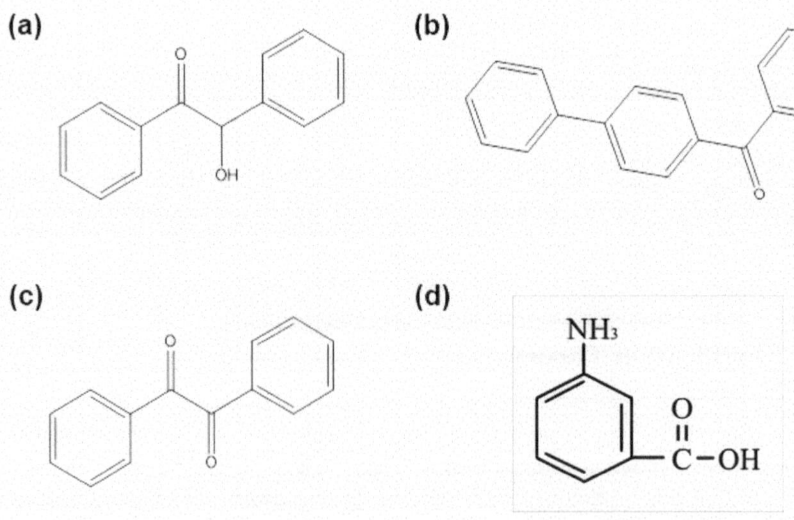

Fig. 10.1 Molecular structure of PC

10.3 Ring Compounds Doped Polypropylene

Due to the similarity of the structure of PCs, only the effects of different contents on the permittivity and dielectric dissipation factor of polypropylene films were investigated, and the results are shown in Fig. 10.2. Observing Fig. 10.2a, at the same frequency, with the increase of PC content, the dielectric constant of the polypropylene sample generally shows an increasing trend, but the increase is very small, the permittivity of the pure polypropylene sample is about 2.2, the permittivity of PP2* and PP3* is about 2.23, even for the highest content of PP5*, the dielectric constant at the same frequency is only 2.28. The dipole turning polar of the polar group and the interfacial polarization of the PC and PP matrix together increase the permittivity of the polypropylene sample due to the extremely low addition content under the electric field. In addition, it can be seen that the permittivity of the film remains basically stable with frequency, showing the typical non-polar characteristics of polypropylene, and the polarization form in the matrix is mainly instantaneous. Observing Fig. 10.2b, the variation law of the dielectric loss factor of the five samples is basically the same, and the difference is not large. In the test frequency band, with the increase of frequency, the loss factor shows a trend of first decreasing and then slightly increasing. As the frequency increases, some polarization forms gradually fail to keep up with the frequency changes, and it is too late to establish, and the loss decreases. In addition, although there is little difference in the dielectric loss of each sample, it can be found that the dielectric loss of PP5* is basically the largest at the same frequency. Therefore, it can be found that trace amounts of aromatic compounds can cause a small increase in the dielectric constant and dielectric loss of the polypropylene sample, but the effect is not significant.

The conduction mechanism can be explained by hopping conduction. The process that carriers transport from one localized state to another localized state is referred to as "hopping transport". The conduction current is related to the temperature, electric field, carrier density, and carrier mobility. The carrier mobility is related to the hopping distance and energy of carriers. The current density J is given as follow [6, 7]:

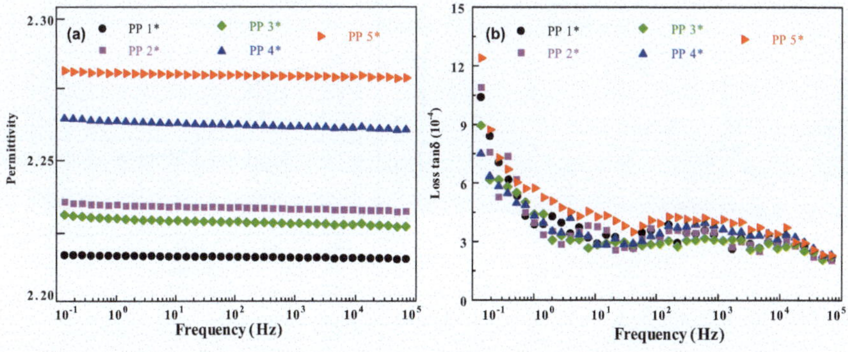

Fig. 10.2 Relationship between the permittivity, dielectric loss factor and the frequency

$$J = 2nedv\exp\left(-\frac{eW}{k_B T}\right)\sinh\left(\frac{edE}{2k_B T}\right) \tag{10.1}$$

where n is the carrier density, e the electric charge of the carriers, d the hopping distance, n the frequency of thermal vibration, W is the activation energy, k_B is the Boltzmann's constant, T the temperature, and E is the electric field. DC conduction current was systematically recorded across a temperature range spanning 25–85 °C, enabling the calculation of DC conductivity derived from these measurements. This methodical approach provided insights into the temperature-dependent behavior of DC conduction current, thereby facilitating the accurate determination of DC conductivity values. Figure 10.3 illustrates the interplay among DC conductivity, electric field strength, and temperature. Notably, DC conductivity exhibits a pronounced sensitivity to temperature, demonstrating a consistent rise as temperature increases. This relationship underscores the significant impact of temperature on DC conductivity, highlighting a direct correlation where conductivity escalates with higher temperatures. The DC conductivity is relatively lower at temperatures from 25 to 45 °C. When the temperature increases from 45 to 85 °C, the DC conductivity increases significantly. The charge carriers can obtain more kinetic energy at a higher temperature, and the transport process of carriers can be promoted. In Fig. 10.3b, the conductivity decreases first and then decreases with the increase of PC content. The conductivity current of PP is mainly contributed by impurities, and when a very small amount of PC is added, local trap points are introduced in PP (which can be confirmed in subsequent studies) to trap carriers. The doping of PC hinders carrier transport and slows down the electron conduction process, resulting in a decrease in DC conductivity. However, at the same time, for PP, PC is also the introduction of a class of impurities, and when the addition content increases, it will also ionize under the action of electric field/thermal field to produce carriers, which aggravates the increase of conductivity current. It can be seen that the appropriate amount of addition can ensure the improvement of the performance of the sample.

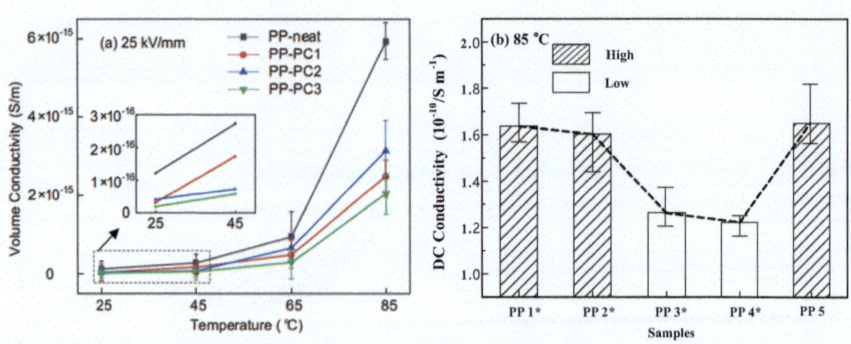

Fig. 10.3 Relationship between the conductivity and the temperature [5]

The DC conductivity increases at a high temperature and high electric field, which means more conduction loss. The increased conduction loss will lead to a higher temperature rise in the power capacitor. The vicious cycle between conduction loss and temperature rise speeds up the breakdown of PP film. Therefore, decreasing the DC conductivity is beneficial for the operation of power capacitors. The DC breakdown strength of the samples was tested at 85 °C, and the data were analyzed using the Weibull distribution, as shown in Fig. 10.4. The characteristic breakdown strength is defined as the breakdown field strength when the breakdown probability is 63.2%. The characteristic breakdown strength of films under both positive and negative voltages is improved by the addition of polycyclic compounds. It can be seen that the breakdown strength of PP-PC3 is highest. Based on the breakdown theory of high-energy electrons, the molecules of polymers attacked by high energy electrons will break and form low-density domains. The breakdown of polymers will then initiate from these low-density domains. Therefore, the protection of PP molecular chains from the attack of high-energy electrons is of great importance for improving breakdown strength. Polycyclic compounds can protect PP molecular chains from the attack of high-energy electrons, which improves the breakdown strength.

In addition, in Fig. 10.4b, it can be seen that the breakdown field strength of the sample increases first and then decreases with the increase of PC content. The breakdown strength of pure polypropylene film PP1* was 539 kV/mm at 25 °C, and the breakdown field strength of PP4* and PP5* showed a downward trend after adding 0.005 and 0.01% mass fraction PC, respectively. This phenomenon can be explained by the precipitation of small molecules at high temperatures. In order to prolong the life of polymer materials in a changeable and complex working environment, various additives are usually added to the preparation of materials, and it is found that during the use of materials, additives will migrate and precipitate inside the materials, thereby affecting the service life of materials, especially in high-temperature working environments. Therefore, when the test temperature increases from 55 to 85 °C, some

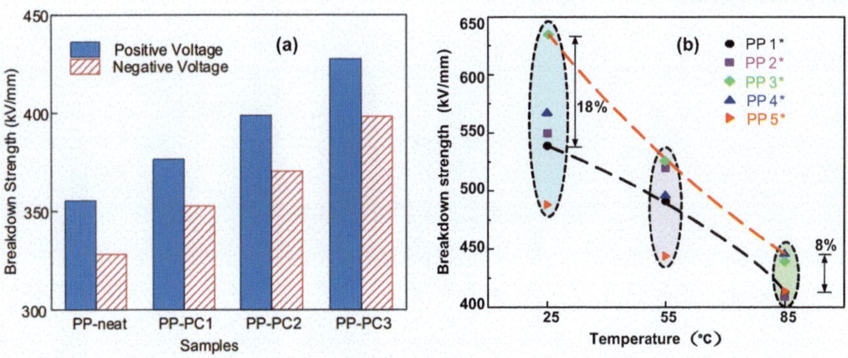

Fig. 10.4 Characteristic breakdown strength of the samples [6]. Copyright Clearance Center's RightsLink© License Number: 5812490187530

PC precipitates from the possible substrate and migrates to the surface of the film, resulting in a decrease in the breakdown performance modification effect.

The results show that for linear polymer materials (such as PP), the energy storage density can be calculated by Eq. 10.2, which obtains a theoretical value, but the calculated value based on this formula is not much different from the measured value, which has great reference significance. Therefore, considering the actual cost performance, the theoretical value calculation method is used for the subsequent study of energy storage characteristics.

$$U_e = \frac{1}{2}\varepsilon_0\varepsilon_r E^2 \tag{10.2}$$

ε_0 and ε_r are the vacuum and relative permittivity of the material, and E is the electrical strength. Based on the above formula, the measured dielectric constant and breakdown field strength value of each sample are brought into the formula to calculate the theoretical maximum energy storage density of the sample, in the range of 25–55 °C, the dielectric constant of polypropylene is almost unchanged, so the influence of temperature can be ignored, and the dielectric constant at room temperature can be used as the calculated value when calculating the energy storage density at different temperatures. Since the electric field strength substituted into the formula is the breakdown field strength, and the actual working field strength of the thin film cannot reach the breakdown field strength, the calculated result is the theoretical maximum energy storage density of the thin film, and the actual energy storage density cannot reach this calculated value, which only has certain reference significance. The calculation results are shown in Fig. 10.5.

Figure 10.5 contain the theoretical maximum energy storage densities of the samples at 25, 55, and 85 °C. It can be found that the trend of energy density of energy storage is consistent with the trend of breakdown strength in Fig. 10.4. This

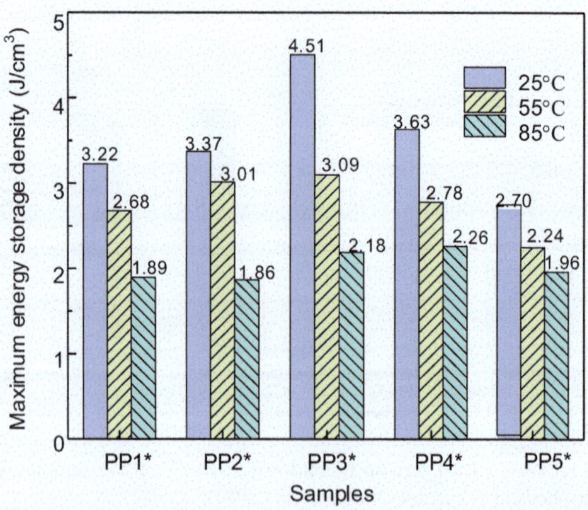

Fig. 10.5 Relationship between the maximum energy storage density and the temperature

is because PC has little effect on the dielectric constant and more on the breakdown strength, so the calculation result is determined to a certain extent by the breakdown properties of the film. At 25 °C, PP3* had the highest energy storage density (4.19 J/cm^3), which was 43% higher than that of pure PP, and the energy density of other samples was similar to or slightly higher than that of pure PP. As the test temperature increases, the theoretical maximum energy storage density decreases. As the temperature rises from 25 to 85 °C, the energy storage density (PP1*) of pure polypropylene decreases by 40.8% and PP3* by 51.7%. Therefore, the influence of temperature on the energy storage characteristics is huge, mainly because of the large influence of temperature on the breakdown performance of polypropylene films.

The trap level distribution directly affects the carrier transport process, which is closely related to the conductivity and breakdown strength [8]. In order to understand the effects of polycyclic compounds, the trap energy distribution needs to be analyzed. The trap level distribution of the samples was obtained by the method of surface potential attenuation. Since the trap level distribution at different temperatures has a similar trend, Fig. 10.6 shows the trap level distributions at 25 and 85 °C. It can be found that there are two types of traps (deep trap and shallow trap), the deep trap energy level is roughly between 0.80–0.95 eV, and the shallow trap energy level is between 0.70–0.80 eV. The trap energy density presents two peaks. The lower energy level corresponding to the energy density peak is defined as the shallow trap level (STL), and the higher energy level corresponding to the energy density peak is defined as the deep trap level (DTL). Charge carriers can easily trap and de-trap from the STLs; therefore, the STLs will promote the conduction process. It is hard for charge carriers to de-trap from the DTLs, which will suppress the transport of carriers. The trapped carriers in DTLs can form an electric field opposite to the applied electric field, which can inhibit the injection of carriers from the electrode.

In Fig. 10.6a, the trap energy density (TED) of STL in PP film is highest at 25 °C. The STL of PP-PC are deeper than the STL of PP-neat, and the TEDs of STL of PP-PC are lower than that of PP-neat. It also can be obtained that the DTL of PP-PC are deeper than the DTL of PP-neat. PP-PC3 presents the deepest DTL, and the

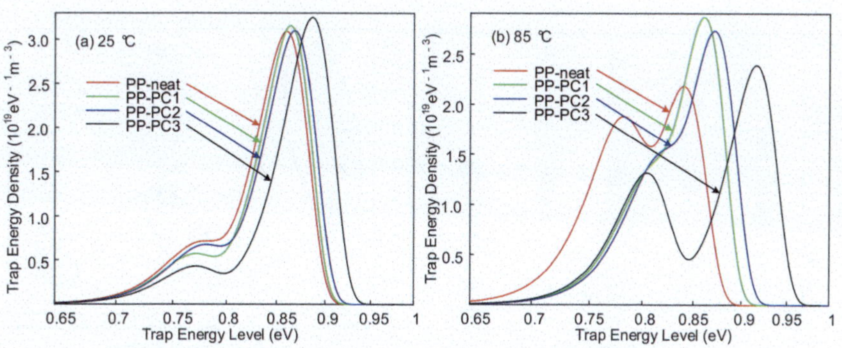

Fig. 10.6 Relationship between the trap level distribution and the temperature [5]

TED of DTL of PP-PC3 is the highest as well. The trap energy distribution at 85 °C presents a similar trend with the trap energy distribution at 25 °C. Both the STL and DTL of PP films modified by polycyclic compounds are deeper than that of PP-neat at 85 °C. Furthermore, the PP-PC3 presents the deepest DTL at 85 °C. Based on the results of the trap level distribution shown in Fig. 10.6, it can be inferred that the polycyclic compounds can introduce DTLs. The introduced DTLs by polycyclic compounds can suppress the transport of carriers, which will result in a decreased DC conductivity.

In addition, with the increase of PC content (Fig. 10.7), the deep trap characteristics showed a certain change, for the PP2* and PP3* samples, the deep trap density showed an increasing trend (8.12 to 10.88, 13.98 $\times$ 10^{22} eV^{-1} m^{-3}), the number of deep traps located in the energy level center increased significantly, and the trap energy level center increased slightly (0.929 eV to 0.935, 0.947 eV), but the increase was small, among which PP3* For PP4* and PP5*, the density of deep traps began to show a downward trend, and the energy level center did not change much. In addition, the shallow trap density of the film did not show significant change, but the shallow trap energy level center increased, from 0.829 eV to 0.852, 0.849, 0.843 and 0.855 eV. This suggests that the right amount of PC addition introduces the highest number of deep traps inside the polypropylene material. The deep trap site can trap the carriers and inhibit the migration of the carriers, thereby maximizing the breakdown strength of PP films.

The hopping conduction theory holds that the carrier hopping transport in a localized state is the main process of the conduction current. The expression of the current density also indicates that the conduction current is closely related to the charge carrier mobility. The carrier mobility reflects the average transport rate of charge carriers. The carrier mobility at different temperatures was obtained by the method of surface potential attenuation, as shown in Fig. 10.8. It can be obtained that the temperature greatly affects the carrier mobility, and the carrier mobility increases at high temperatures. More energy can be obtained by charge carriers at higher temperatures, which makes charge carriers detrap much easier. The increased trapping probability leads to increased transport of the charge carrier from one trap to

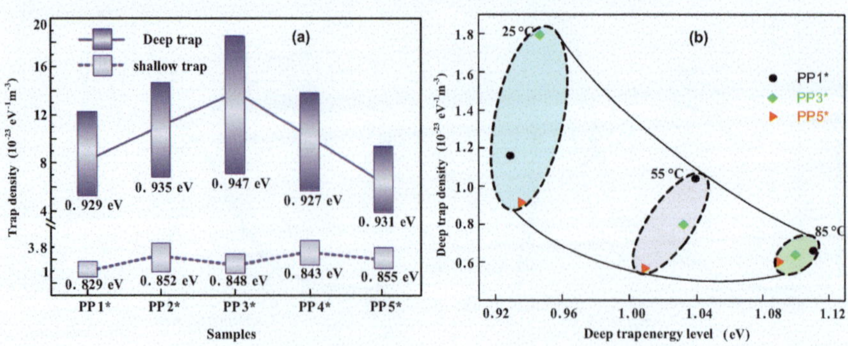

Fig. 10.7 Relationship between the trap distribution characteristics and the temperature

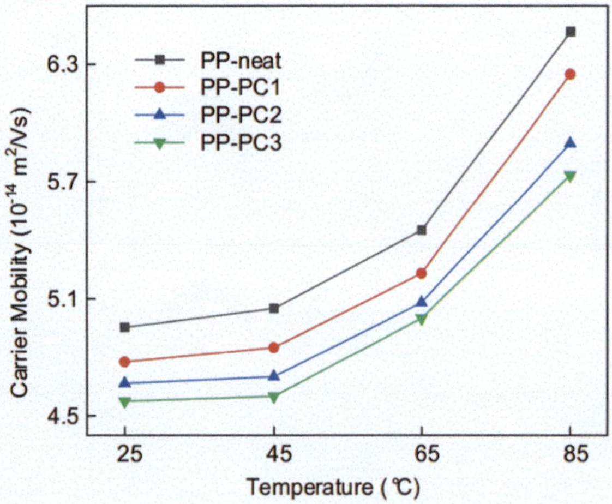

Fig. 10.8 Relationship between the carrier mobility and the temperature [5]

another trap, which increases the carrier mobility. Since there is a positive correlation between conductivity and carrier mobility, the trend of carrier mobility with temperature explains the rise of conductivity with temperature.

It also can be obtained that the carrier mobility of PP-PC is lower than that of PP-neat. The decreased carrier mobility can be explained by the introduced DTLs. It is hard for trapped charges to detrap from DTLs because the trapped charges need more energy to detrap from the DTLs. The PP-PC3 presents the lowest carrier mobility, which is related to the deepest DTLs of PP-PC3. The decreased carrier mobility leads to the lower DC conductivity of PP-PC. The lowest carrier mobility explains the mechanism of the lowest DC conductivity of PP-PC3 at different temperatures.

A quantum chemistry calculation was employed to explore the effect mechanism of polycyclic compounds [9]. The DOS, energy level distribution, and MOs of the PP-neat and polycyclic compounds are shown in Fig. 10.8. Comparing the DOS of PP-neat and PC, the polycyclic compounds have some electron traps and hole traps in the range of band gaps of PP-neat, which means polycyclic compounds can introduce electron traps and hole traps. Figure 10.9(a_2–d_2) presents the energy level distribution. The highest occupied molecular orbital (HOMO) and lowest unoccupied molecular orbital (LUMO) levels are marked with red and blue lines, respectively. The LUMOs of PC are negative and much lower than the LUMO of PP-neat, and the HOMOs of all polycyclic compounds are higher than the HOMO of PP-neat. The energy level distribution also indicates that polycyclic compounds can introduce energy levels in PP. The DOS and energy level distribution verify the experimental results of SPD in Fig. 10.6.

Figure 10.9(a_3–d_3) presents the MOs of LUMO and HOMO. The MOs present the wave function characteristics of an electron in the molecules, which is related to the probability of electron occurrence. For PP-neat in Fig. 10.9(a_3), the MO at LUMO locates at the surface of the molecular chain, indicating that an injected electron will travel outside the backbone of the molecular chain. The injected electrons prefer

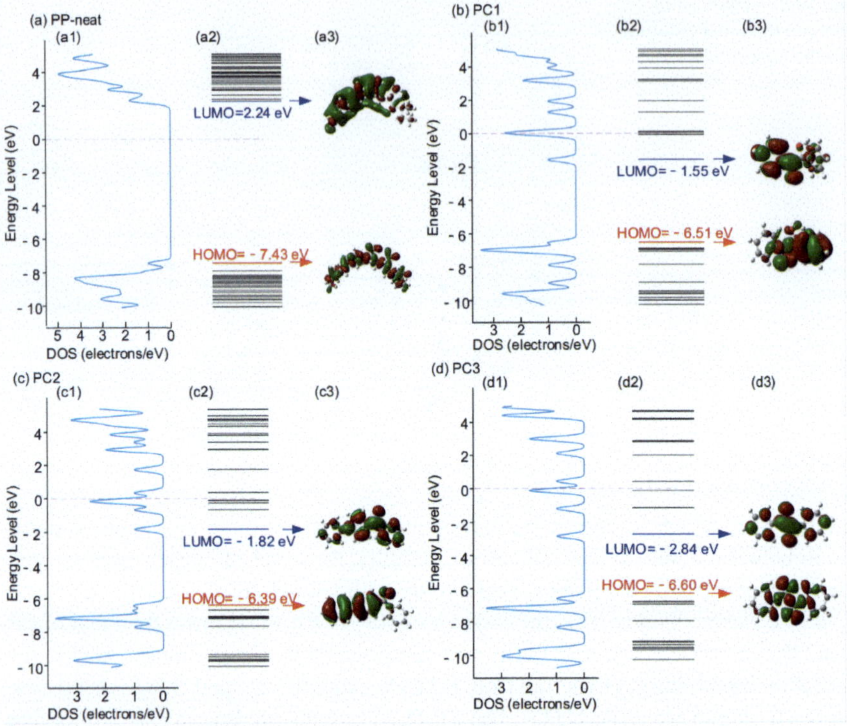

Fig. 10.9 (a_1–d_1) DOS, (a_2–d_2) energy level distribution, and (a_3–d_3) MOs [5]

to travel among the backbones, such as the free volumes among chains or voids, where they can easily accelerate and become high-energy electrons [10]. The PP molecular chain will break when attacked by high-energy electrons. The MO at the HOMO level distributes inside the backbone, indicating that the orbital electrons of PP-neat at the HOMO level travel inside the molecular chains, making it easy for PP-neat to break. The MOs at the LUMO and HOMO levels of polycyclic compounds distribute around the molecular chains. Benzene rings and carbonyl groups are the centers that attract additional electrons. The benzene ring structure forms a large π bond through SP2 hybridization, forming a chemical trap with a large charge capture cross-section and strong capture ability. The carbonyls in polycyclic compounds also act as a chemical trap. The electron affinity χ is the energy difference between the vacuum level and the LUMO level. As a result, the polycyclic compounds with positive χ are easier to attract electrons than PP-neat. Polycyclic compounds can absorb and attenuate the energy of high-energy electrons, and the absorbed energy will be released by unharmful methods, such as lattice vibration. The high-energy electrons then become low-energy electrons, which cause little destruction to PP molecular chains.

This explains that the addition of polycyclic compounds can capture high-energy electrons and protect the polymer matrix under high electric fields. The protection of molecular chains is beneficial for the improvement of the electrical breakdown strength of PP film. The carbonyl group and benzene ring increase the electron affinity χ of polycyclic compounds, as discussed in the preceding paragraph. The contribution of the carbonyl group is greater than that of the benzene ring in increasing the capturing ability of electrons; that is to say, the electron affinity. PC3 has the highest electron affinity. The higher the electron affinity, the more protection for PP molecular chains. As a result, the polycyclic compounds are easier to attract hole charges than PP-neat.

The 3D electrostatic potential distribution of PP-neat and polycyclic compounds was calculated. In Fig. 10.10 the warm colors represent negatively charged, and the cold colors represent positively charged, respectively. The orbital electron deviates in molecular chains for the difference in electronegativity between the carbon and hydrogen in PP-neat. For the polycyclic compounds, both the warm colors and cold colors were deepened, especially at the regions nearing benzene rings and carbonyl groups, meaning that both the electrons and holes can be captured much more easily.

The comprehensive analysis of density of states (DOS), energy levels, molecular orbitals (MOs), and 3D potential distribution provides compelling evidence supporting the capability of polycyclic compounds to capture high-energy electrons, thereby elucidating the mechanism underlying the alteration of energy level distribution. Leveraging insights from thermal electron theory further enriches our understanding of the intricate processes governing the behavior of electrons within

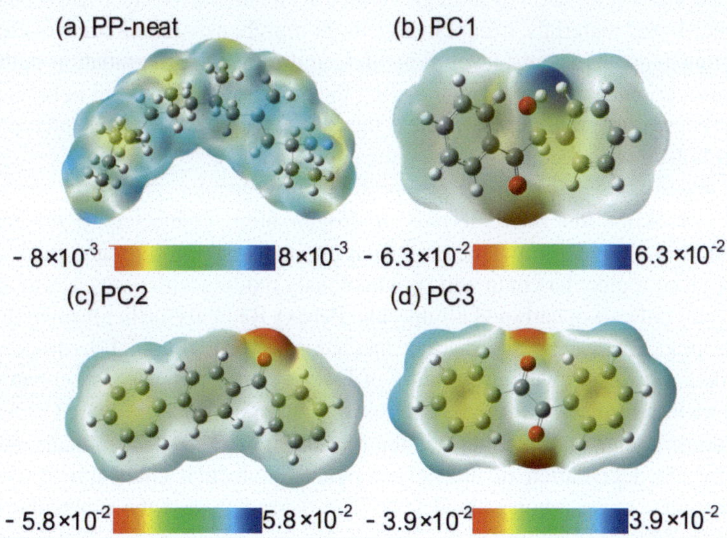

Fig. 10.10 Three-dimensional (3D) electrostatic potential of **a** PP-neat, **b** PC1, **c** PC2, and **d** PC3 [5]

polymers subjected to electrical stress [11]. According to thermal electron theory, electrons injected into the polymer from the electrode experience rapid trapping by localized traps following one or several scattering events. Subsequent to trapping, these electrons undergo a transition from a high-energy state to a lower-energy state, with the released energy imparted to another electron, thereby elevating it to a high-energy thermal state. This cascade of high-energy electron interactions serves as a precursor to molecular bond disruption, precipitating the generation of free radicals within the polymer matrix. The relentless bombardment of molecular chains by high-energy thermal electrons instigates the formation of free radicals, thereby catalyzing the degradation and expansion of molecular chains, ultimately culminating in the formation of low-density regions within the polymer matrix. This progressive degradation process inexorably paves the path towards breakdown initiation. Of particular significance is the role played by tertiary carbon atoms and tertiary hydrogen atoms within polypropylene molecular chains. Tertiary carbon atoms, characterized by their lone attachment to hydrogen atoms, exhibit heightened reactivity, rendering them susceptible to bond cleavage when subjected to bombardment by high-energy thermal electrons. Consequently, the rupture of bonds between tertiary carbon and hydrogen atoms facilitates the generation of carbon radicals, thereby expediting the degradation process of polypropylene molecular chains.

The accelerated degradation ensuing from the formation of free radicals underscores the critical role played by high-energy thermal electrons in instigating breakdown phenomena within polymer matrices. By elucidating the intricate interplay between electron bombardment, bond cleavage, and free radical generation, these findings not only deepen our understanding of breakdown mechanisms but also pave the way for the development of strategies aimed at enhancing the electrical robustness of polymer-based materials. In conclusion, the comprehensive elucidation of electron capture mechanisms, energy level dynamics, and molecular degradation pathways represents a crucial step towards unraveling the complexities inherent in breakdown phenomena within polymer dielectrics, thereby facilitating targeted interventions aimed at bolstering dielectric performance and reliability.

The degradation process of polypropylene molecular chains is shown in Fig. 10.11a. After the active tertiary hydrogen atom connected to the tertiary carbon atom in the polypropylene molecular chain is lost, the carbon radicals in the polypropylene molecular chain will spontaneously oxidize with oxygen, which may form peroxy radicals or carbonyl groups, and Peroxy Radicals further form hydroperoxide groups. There exist two distinct mechanisms through which hydroperoxide groups decompose, one is the formation of H–O–O* free radicals, accompanied by the degradation of the polypropylene molecular chain into an unsaturated group and a new carbon radical, and the other is the formation of H–O* free radicals, accompanied by the degradation of the polypropylene molecular chain into a terminal aldehyde and a new carbon radical [12]. The degradation of the molecular chain entails its expansion, resulting in the creation of a low-density region that eventually culminates in structural breakdown.

PC have low ionization energy, easy to be excited or ionized, their electron affinity energy is higher, and has a large capture area, so PC are easier to capture hot electrons,

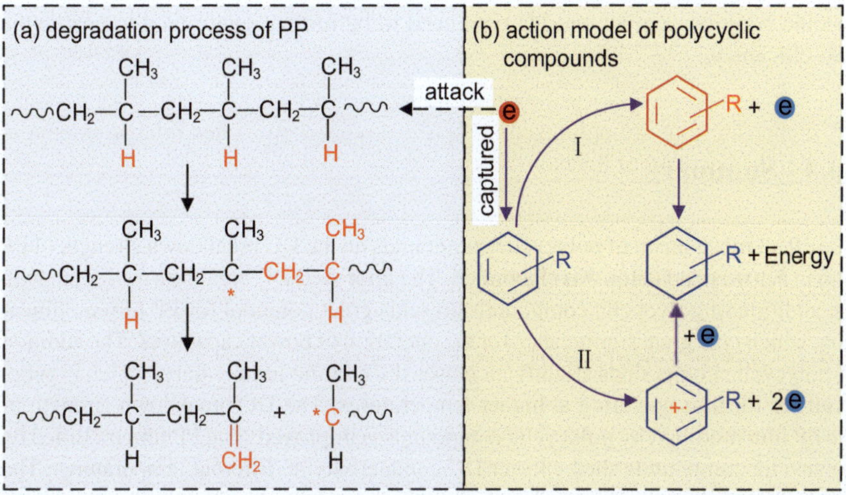

Fig. 10.11 Process of action of polyphenyl ring compounds

and many factors are affected together, under the strong electric field, PC will be bombarded by hot electrons before the polypropylene molecular chain, weakening the destruction of the polypropylene molecular chain by high-energy electrons, thereby improving the breakdown field strength of PP film [13]. The interaction process between the PC and the thermal electrons is shown in Fig. 10.11b. In addition, the excited state of the PC after excitation can return to the original state and release the energy in a harmless form such as vibration, while the positive free radicals formed by the ionization of the PC can be recombined with electrons to the original state. The PC is excited or ionized and then returns to its original state after releasing energy, so that the PC is not consumable and can have long-term effect. At the same time, the PC introduce deep trap energy levels in the polypropylene film, which reduces the mobility of the carriers, resulting in a decrease in the conductivity of the polypropylene film.

In general, PC has the following effects: (I) the deep trap site introduced by PC can trap high-energy carriers and convert them into free carriers by absorbing energy, which will greatly reduce the disruption of molecular chains by high-energy carriers [14]. (II) PC can eliminate the terminal free radicals of the broken pp molecular chain to repair the molecular chain, alleviate the formation of low-density regions, and improve the breakdown field strength. It is worth noting that some PCs may be ionized under the electrothermal composite field, and small molecules with free radicals will reduce molecular stability, promote the formation of low-density regions, and lead to distortion of the local electric field [15]. In conclusion, the precipitation and ionization properties of PC can lead to a decrease in the efficiency of PC at high temperatures. Based on the above analysis, micro-filling of aromatic compounds with suitable conjugated structures is beneficial to improve the insulation performance, but the specific addition content and PC type need to be further determined to ensure the best efficiency.

10.4 Summary

The effect mechanism of polycyclic compounds on the DC breakdown strength of PP film in a power capacitor was elaborated. The improved DC breakdown strength with the addition of polycyclic compounds presents great potential for PP film modification, which is of great significance for the operation of power capacitors. The addition of polycyclic compounds slightly increases the melting temperature, which is beneficial for PP film operating at higher temperatures. The DC breakdown strength of the PP film modified by polycyclic compounds is improved. The PP film modified by polycyclic compounds shows lower DC conductivity at different temperatures. The trap level distribution is affected by the addition of polycyclic compounds through the measurement, which is closely related to the breakdown strength and conductivity. The DOS, energy level, MOs, and 3D potential distribution obtained by the quantum chemistry calculation analyze the effect mechanism of polycyclic compounds. The benzene rings and carbonyls of polycyclic compounds act as charge capture centers, which limits the carrier mobility and results in a decreased conductivity. The higher capability of capturing high-energy electrons of polycyclic compounds protects the molecular chains, which increases the DC breakdown strength of PP film.

The research results reveal that the selected polycyclic compounds have great potential for use in PP film modification for power capacitor applications. The PP films in this paper were made by hot pressing. Furthermore, PP film used in power capacitors is biaxially oriented polypropylene (BOPP) film. In future research, the effect of polycyclic compounds on BOPP film will be studied to analyze the influence of the bidirectional drawing process on polycyclic compounds.

References

1. Y. Yamano, "Roles of polycyclic compounds in increasing breakdown strength of LDPE film," *IEEE Trans. Dielect Elect. Insul.*, vol. 13, no. 4, pp. 773–781, 2006.
2. S. L. Nunes and M. T. Shaw, "Mechanism of protection of polyethylene by anti-treeing additives," Conference on Electrical Insulation & Dielectric Phenomena - Annual Report 1981, Whitehaven, PA, USA, 1981, pp. 238–243.
3. M. Jarvid et al, "Tailored side-chain architecture of benzil voltage stabilizers for enhanced dielectric strength of cross-linked polyethylene," *J. Polym. Sci.*, vol. 52, no, 16, pp. 1047–1054, 2014.
4. J. Ho, R. Ramprasad and S. Boggs, "Effect of alteration of antioxidant by UV treatment on the dielectric strength of BOPP capacitor film," *IIEEE Trans. Dielect Elect. Insul.*, vol. 14, no. 5, pp. 1295–1301, 2007.

5. R. R. Xu, J. W. Xing, B. X. Du, M. Xiao and J. Li, "Improved breakdown strength of polypropylene film by polycyclic compounds addition for power capacitors," *Materials*, vol. 14, no. 5, 2021, Art. no. 1185.
6. X. B. Du, J. W. Xing, M. Xiao, Z. Y. Ran, H. L. Liu and J. N. Dong, "Improved dielectric properties of polypropylene films based on aromatic compounds," 2021 International Conference on Electrical Materials and Power Equipment, Chongqing, China, 2021, pp. 1–4.
7. B. X. Du, C. L. Han, Z. L. Li and J. Li. "Improved DC conductivity and space charge characteristics of XLPE for HVDC cable application: effect of voltage stabilizers," *IEEE Access*, vol. 7, pp. 66576–66583, 2019.
8. B. X. Du, Z. H. Hou, J. Li and Z. L. Li, "Effect of graphene nanoplatelets on space charge and breakdown strength of PP / ULDPE blends for HVDC cable insulation," *IEEE Trans. Dielect Elect. Insul.*, vol. 25, no.6, pp. 2405–2412, 2018.
9. J. Li, H. C. Liang and B. X. Du, "Mechanism of deep trap sites in epoxy/graphene nanocomposite using quantum chemical calculation," *IEEE Trans. Dielect Elect. Insul.*, vol. 26, no. 5, pp. 1577–1580, 2019.
10. W. Wang, T. Takada, Y. Tanaka and S. T. Li, "Space charge mechanism of polyethylene and polytetrafluoroethylene by electrode/dielectrics interface study using quantum chemical method," *IEEE Trans. Dielect Elect. Insul.*, vol. 24, no. 4, pp. 2599–2606, 2017.
11. K. C. Kao, "New theory of electrical discharge and breakdown in low-mobility condensed insulators", *J. Appl. Phys.*, vol. 55, no. 3, pp. 752–755, 1984.
12. G. Zhang, C. Nam, L. Petersson, J. Jämbeck, H. Hillborg, and T. C. Mike Chunget, "Increasing polypropylene high temperature stability by blending polypropylene-bonded hindered phenol antioxidant," Macromolecules, vol. 51, no. 5, pp. 1927–1936, 2018.
13. D. Tan, L. Zhang, Q. Chen and P. Irwin, "High-temperature capacitor polymer films," *J. Mater. Sci.*, vol. 43, no. 12, pp. 4569–4575, 2014.
14. A. C. Ashcraft, R. M. Eichhorn and R. G. Shaw, "Laboratory studies of treeing in solid dielectrics and voltage stabilization of polyethylene," 1976 IEEE International Conference on Electrical Insulation, Montreal, QC, Canada, 1976, pp. 213–218.
15. Y. Hayase, T. Osada and T. Takada, "Numerical estimation of trap depth in polymeric materials using molecular orbital method," 2009 IEEE Conference on Electrical Insulation and Dielectric Phenomena, Virginia Beach, VA, USA, 2009, pp. 161–164.

Chapter 11
Cyclic Olefin Copolymer Blending

11.1 Introduction

As shown in the previous analysis, the problem of decreasing breakdown field strength and increasing dielectric loss of polypropylene films at high temperatures is becoming increasingly prominent [1], so it is significant to improve the high-temperature dielectric properties of polypropylene films. Currently, polypropylene modification methods can be divided into two categories: physical modification and chemical modification [2]. Physical modification refers to optimizing the dielectric properties of polypropylene through doping or blending methods, including nanoparticle doping, nucleating agents, and benzene ring compound filling [3, 4]. Chemical modification refers to changing the molecular chain structure of polypropylene through chemical reactions to improve its dielectric properties, including graft modification and copolymerization [5, 6]. Among them, physical modification is still the most commonly used modification method for polypropylene films for capacitors [7, 8]. Still, it is necessary to solve the compatibility problem between fillers and polypropylene [9, 10]. Copolymers of cycloolefin (COC), as highly heat-resistant polymers with high glass transition temperatures and good compatibility with polypropylene, provide a feasible way to improve the dielectric properties of polypropylene [11].

11.2 Modification Theory

COC is an amorphous polymer consisting of only the elements C and H [12]. The molecular structure formula is shown in Fig. 11.1. COCs have excellent dielectric properties, stable chemical properties, and good heat resistance and are widely used in packaging, medical devices, and electronic equipment [13, 14]. COCs were first documented in the 1950s and commercialized in the late 1980s by the Ziegler–Natta

B. Du and M. Xiao, *Polypropylene Film for HVDC Capacitors*,
https://doi.org/10.1007/978-981-96-3029-5_11

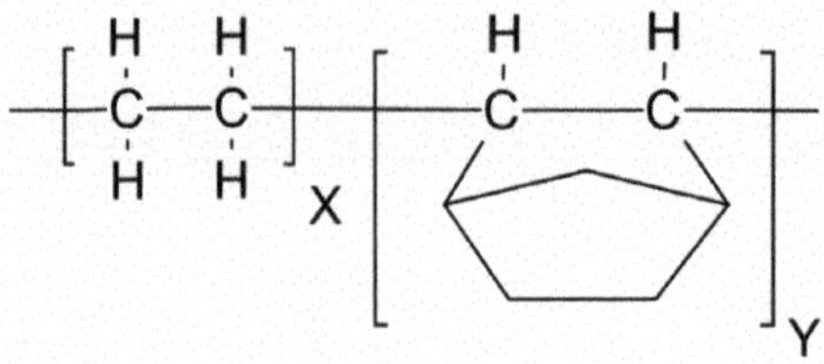

Fig. 11.1 Molecular structure formula of COC [11]. Copyright Clearance Center's RightsLink© License Number: 5811390213898

catalytic reaction [15]. COCs are produced by the metallocene-catalyzed reaction of 2-norbornene, a bicyclic olefin formed by ethylene and cyclopentadiene, with ethylene. 2-norbornene is larger than ethylene and has a rigid bridging ring structure that prevents crystallization, which is randomly incorporated into the main chain of the polymer molecule, resulting in an entirely amorphous structure of the COC [16].

Its glass transition temperature determines the thermodynamic properties of polypropylene (PP) [17]. When the temperature is higher than the glass transition temperature, the overall performance of the polymer appears to decline. The glass transition temperature of polypropylene film is about – 20 °C, and the flexibility of the molecular chain is very high and easy to fold or move at high temperatures, which is one of the essential reasons for the failure of polypropylene film insulation at high temperatures [18]. Cycloolefin copolymers have tall glass transition temperatures (Tg), up to 214 °C, when the content of ethylene/tricyclopentadiene copolymer is 41 mol% [19]. Therefore, COCs can potentially improve polypropylene's high-temperature dielectric properties [20].

The rigidity of the molecular chain mainly determines the glass transition temperature (Tg) of COC, and the greater the rigidity of the chain segment, the higher its glass transition temperature [21]. The bridging ring structure in norbornene can increase the rigidity of the molecular chain, so the glass transition temperature of COC depends on the content of norbornene in the main chain; when the COC mainly consists of alternating ethylene and norbornene units, the glass transition temperature is lower; and when the COC contains long norbornene sequences, the glass transition temperature is higher [22, 23].

11.3 Content Adjusting

Poor compatibility of polymer dielectric materials mixed will form an obvious layered interface; under the action of the applied electric field, the charged particles in the dielectric gathered at the interface, and the internal local electric field distortion is serious, resulting in composite materials being prone to insulation breakdown [24, 25]. At the same time, the high content of cyclic olefin will appear as an agglomeration phenomenon affecting the composite film performance [26]. Therefore, by characterising the fundamental physicochemical properties of the composite films, such as microstructure, crystalline properties and thermomechanical properties, as well as the evolution of dielectric constant and dielectric loss factor at different frequencies, the breakdown and conductive loss characteristics of the composite films at different temperatures were analyzed [27, 28]. The role of the content of COCs on the dielectric properties of polypropylene films was revealed [29].

11.3.1 Effect on the Microstructure

The glass transition temperature of the COC was 110 °C. The effect of different contents of COC on the properties of the composite films is investigated below. The contents in the composite film were set at 10, 20, 30, 40, and 50 wt% in the composite film. The film samples with PP/COC content ratios of 100/0, 90/10, 80/20 70/30, 60/40, 50/50 were referred to as PP, PP90, PP80, PP70, PP60, PP50, and PP60, respectively.

The microscopic morphology of PP/COC composite film sections was observed by SEM, and the effect of COC content on the microstructure was investigated [30]. Figure 11.2 illustrates the cross-section topography of the test film after quenching in a liquid nitrogen environment. As can be seen from the figure, the cross-section of the pure PP film is smooth and gentle, indicating that the polypropylene material has very few internal impurities, and the quenching process causes a small amount of surface bumps. The cross-section morphology of the composite film was related to the content of COC. When the content of COC was low (10 and 20 wt%), the cross sections of PP90 and PP80 were relatively smooth, with no obvious delamination and boundary structure, indicating that the polypropylene and the COC were embedded in each other, and the compatibility between the two was good; When the content of COC increased to 30 wt%, the composite film cross-section of the obvious layered structure defects; when the mass fraction continues to grow to 50 wt%, the content of COC is too high in the composite film is not evenly distributed, PP50 cross-section of the apparent agglomeration phenomenon. The compatibility of polypropylene and filler directly determines the insulating properties of the composite film; poor compatibility will introduce electrical weaknesses, leading to a reduction in dielectric properties, while a small amount ($\leq$ 20 wt%) of COC is compatible with polypropylene and overcomes the critical challenges of physical modification.

(1) **Crystalline Properties**

As a quintessential semi-crystalline polymer, polypropylene (PP) within the context of PP/COC composite films exhibits crystalline characteristics that profoundly impact the dielectric properties of the resulting composite. Thus, an in-depth exploration of the crystalline attributes of PP/COC composite films assumes paramount importance owing to its implications for material performance and application viability. The crystallization behavior of isotactic polypropylene, a primary constituent of PP/COC composites, is governed by a myriad of intrinsic properties intrinsic to the polymer chains themselves. These properties encompass the relative molecular mass of the polypropylene chains, their degree of regularity, symmetry,

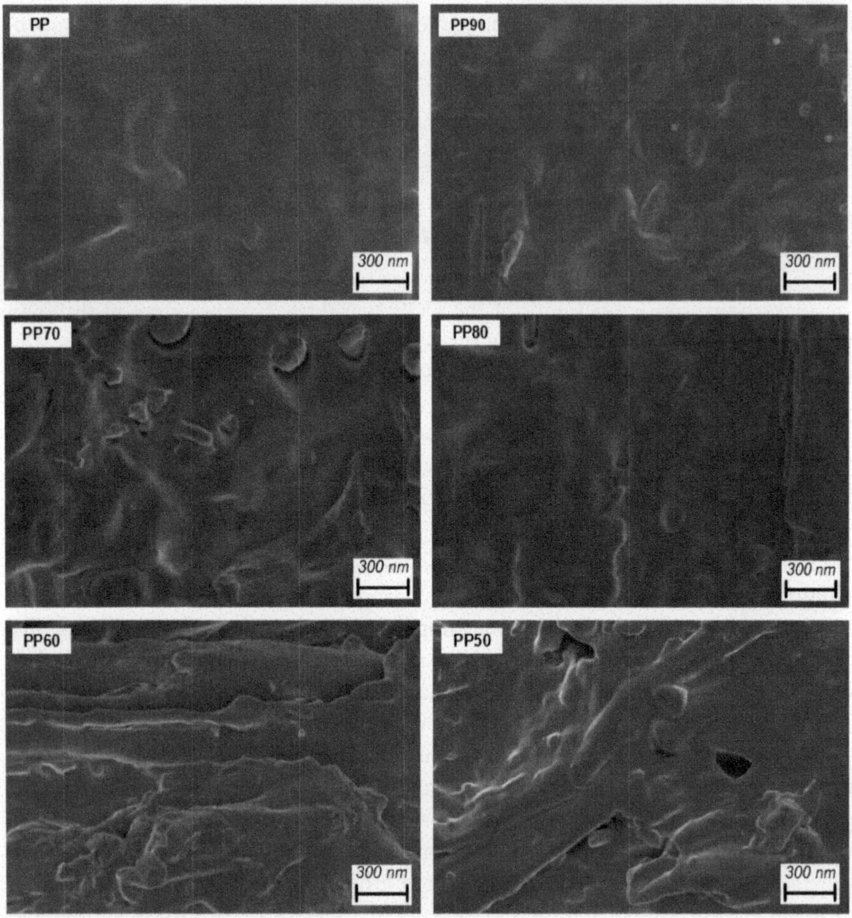

Fig. 11.2 Micro-morphological characteristics of composite films [11]. Copyright Clearance Center's RightsLink© License Number: 5811390213898

flexibility, and the distribution of branched chains. The intricate interplay of these factors dictates the kinetics and thermodynamics of the crystallization process, thereby shaping the resultant crystalline morphology and characteristics. Furthermore, the crystallization behavior of isotactic polypropylene is intricately intertwined with external environmental factors, which exert a pronounced influence on the crystallization process. Experimental parameters such as temperature, applied stress, and the presence of impurities exert a discernible impact on the nucleation and growth kinetics of crystalline domains within the polymer matrix. Variations in these external conditions can precipitate significant alterations in crystalline morphology, size, and distribution, thereby modulating the dielectric properties of the composite film. The crystalline characteristics of PP/COC composite films encompass a diverse array of attributes, ranging from crystal shape and size to grain morphology and overall crystallinity. These parameters collectively contribute to the macroscopic dielectric behavior of the composite film, influencing factors such as electrical conductivity, breakdown strength, and insulation performance. Understanding and elucidating the crystalline characteristics of PP/COC composite films represent a pivotal step towards optimizing their dielectric properties and harnessing their full potential in diverse applications. By delineating the intricate interplay between polymer structure, processing conditions, and crystallization kinetics, researchers can devise tailored strategies for enhancing the electrical robustness and performance of PP/COC composite films, thereby unlocking new frontiers in dielectric engineering and applications. Observation of the crystalline morphology of thin film specimens at the initial moment, at the beginning, and after crystallization using hot-stage polarizing microscopy. As shown in Fig. 11.3, the lamellar crystal bundles of the spherical crystals of the test films are all arranged in a radial pattern, typical of polypropylene α crystals. The initial state of the polypropylene film has a smooth crystalline morphology because of the low content of impurities in the polypropylene film; with the increase of the content of COC, the initial state of the composite film gradually increases the roughness of the crystalline morphology.

PP specimens crystallized with the most significant number of nuclei at the beginning of crystallization, and the spherical crystals were smaller in size and concentrated in distribution. In contrast, the composite film' initial number of spherical crystals decreased with increased COC content. When the content of COC is 10 and 20 wt%, the number of nuclei at the beginning of crystallization of the composite film is less than that of PP, the distribution of spherical crystals is more dispersed, and the boundary of spherical crystals is gradually fuzzy; when the content of COC is increased to more than 30 wt%, the number of nuclei of the composite film initial nuclei is further reduced, the dispersion of spherical crystals is increased, and the boundary of spherical crystals is wholly disappeared.

As can be seen from the figure, the morphology of PP samples after crystallization is visible, and the boundary between spherical crystals is clear. Still, there is an amorphous defect between spherical and spherical crystals (i.e., the black area in the red circle in the figure). The crystalline morphology of the composite film, when the crystallization is completed, is related to the content of COC; when the content of COC is 10 and 20 wt%, the number of spherical crystals of the composite film is less than that of the PP sample, which reduces the stacking of spherical crystals, and the crystal structure is denser, and the defects in the amorphous zone are decreased significantly; however, when the content of COC copolymer is increased to more than 30 wt%, the unstructured nature of COC copolymer will be reduced, and the defects in the amorphous zone will be reduced. Due to copolymers' amorphous nature, excessive doping drastically reduces the number of spherical crystals in the composite film, reducing the thickness of the crystal region.

In summary, an appropriate amount of COC can reduce the defects in the amorphous region of the composites and improve the crystalline morphology of polypropylene films. Still, excessive doping will significantly reduce the crystallinity of the composite film.

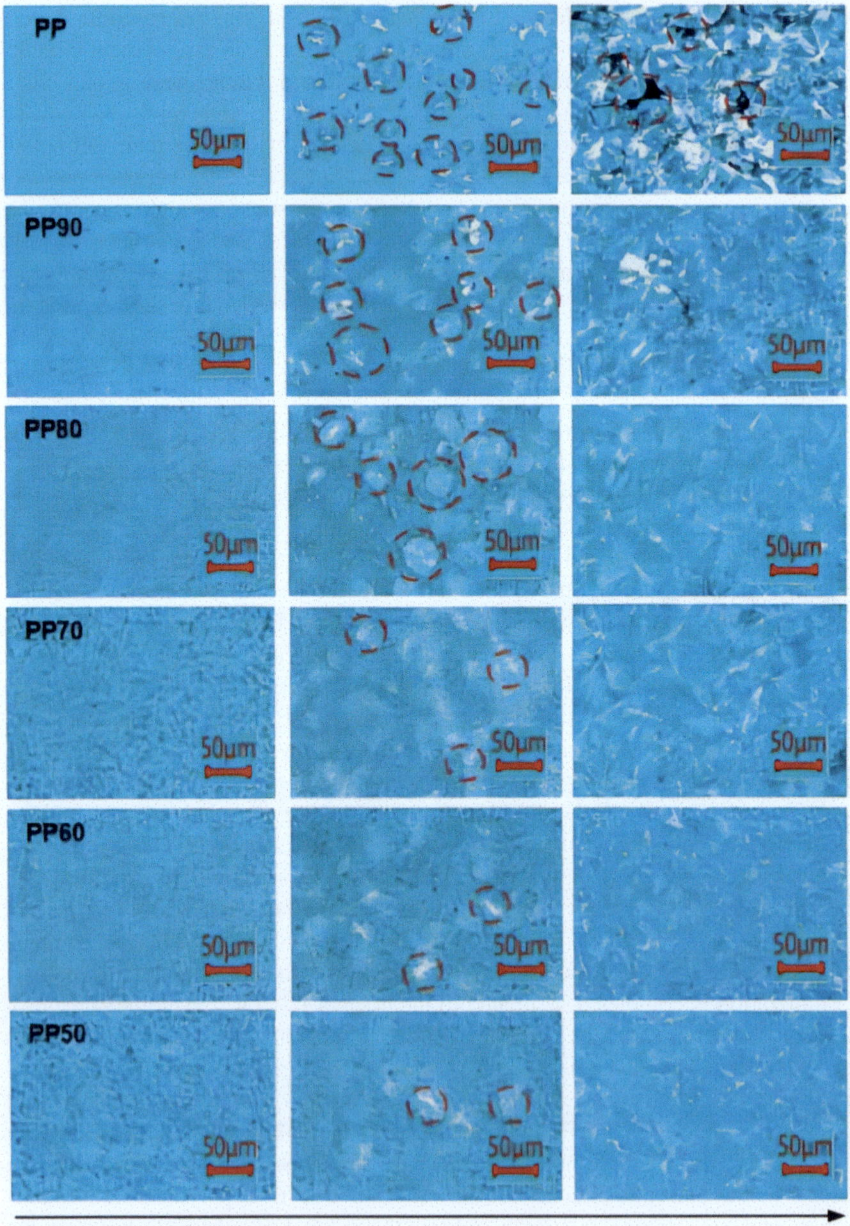

Fig. 11.3 Crystalline morphology of composite films [11]. Copyright Clearance Center's RightsLink© License Number: 5811390213898

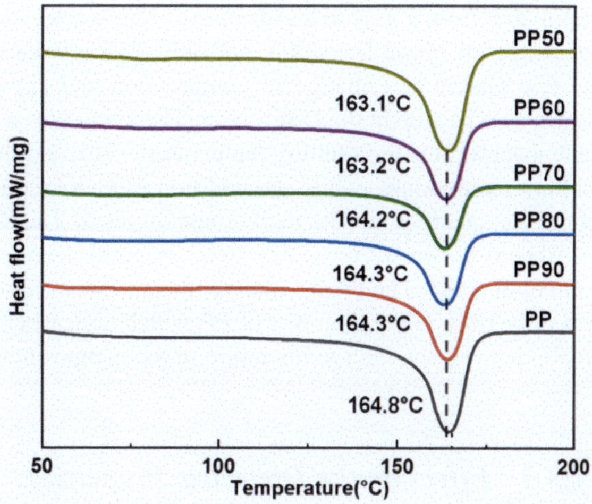

Fig. 11.4 Crystallization and melting characteristic curves of composite films [11]. Copyright Clearance Center's RightsLink© License Number: 5811390213898

(2) **Thermodynamic Properties**

The DSC experimental results can be obtained by repeatedly raising and lowering the temperature of the film specimens in the temperature range of 25–250 °C. Figure 11.4 show the crystallization and melting curves of the composite film, respectively. The DSC curves of the film specimens have obvious heat absorption or exothermic peaks (depending on whether the melting or crystallization curve is observed), which is due to the fact that polypropylene is semi-crystalline in the composite film.

When a polymer changes from the crystalline state to the molten state, a first-order phase transition occurs, resulting in discontinuous changes in entropy and volume, i.e., there is a latent heat of phase transition during the phase transition. The DSC experimental principle is based on the change of heat capacity as a function of temperature, and the first-order phase change is accompanied by the absorption or release of latent heat, while the melting and crystallization of polymers belong to the first-order phase change, so the DSC curves of the tested films show absorption peaks.

From the crystallization characteristic curve, it can be seen that the trend of the exothermic peaks of the film specimens is similar. The temperature corresponding to the exothermic peak is the crystallization temperature (Tc), and the crystallization temperature of PP is the highest; the crystallization temperature of the PP/COC composite film is lower than that of PP, and decreasing with the increase of the content of added COC. The addition of COC reduces the number of initial nuclei of the composite film, resulting in lower energy required for crystallization and hence lower crystallization temperature of the composite film. However, the effect of COC doping is limited, and the crystallization temperature of the composite film is as low as 111.4 °C, which is only 2.3% lower than that of PP.

The melt curves of the film samples all contain a heat absorption peak and a Tg step, which are typical DSC melt curves of semi-crystalline polymers. The transition of polymer from glassy to highly elastic state is a secondary phase transition without latent heat of phase transition, but only discontinuous change of heat capacity of the two phases and change of compression and expansion coefficients, which is shown as a Tg step in the DSC curve. The temperature corresponding to the peak of heat absorption is the melting temperature (Tc), and the PP sample has the highest melting temperature, while the addition of COC reduces the melting temperature of the composite film. The melt temperatures of The PP/COC composite film were distributed between 163.1 and 164.3 °C, and decreased with the increase of the content of COC. The lowest melt temperature was found in the PP50 sample, but it was only 1.0% lower than that of PP, which indicates that the content of COC has a small effect on the melt temperature of the composite film.

11.3.2 Effect on the Dielectric Properties

(1) **Permittivity**

The dielectric constant is an essential physical quantity that characterizes the polarization of a dielectric under the action of an applied electric field and is expressed by the symbol ε, which represents the magnitude of the ability of the dielectric material to hold an electric charge. The dielectric constant of the thin film specimen was tested using a broadband dielectric spectrometer with a frequency range of 10^{-1} Hz ~ 104 Hz and a testing temperature of 25 °C, and the experimental results are shown in Fig. 11.5. To study the effect of the content of cyclic olefin copolymer on the dielectric constant of polypropylene films, pure cyclic olefin copolymer (COC) films were also tested in this section. As can be seen from the figure, the dielectric constants of the film specimens all decreased slightly with increasing test frequency. The polarization of microscopic particles causes the generation of dielectric constant. When the frequency of the applied electric field is very high, the dipole steering inside the PP/COC composite film cannot keep up with the change in the electric field. Some of the dipoles are difficult to steer, which reduces the contribution of polarization to the permittivity of the composite film. The overall performance shows that the permittivity decreases with the increase in frequency. The composite film's dielectric constant increases with COC's content because the pure COC film has the highest dielectric constant of 2.69 at 50 Hz, which is 1.22 times higher than that of polypropylene film at the same frequency. When the content of COC is 20 wt%, the dielectric constant of PP80 is 2.42. From the above analysis, it can be seen that the addition of COC increases the dielectric constant of polypropylene films, which helps enhance the composite films' maximum theoretical energy density.

(2) **Dielectric Loss Factor**

Dielectric loss is the energy loss of an insulating material due to conductivity and polarization under the action of an applied electric field. Under the action of an alternating electric field, the residual angle between the angle between the current

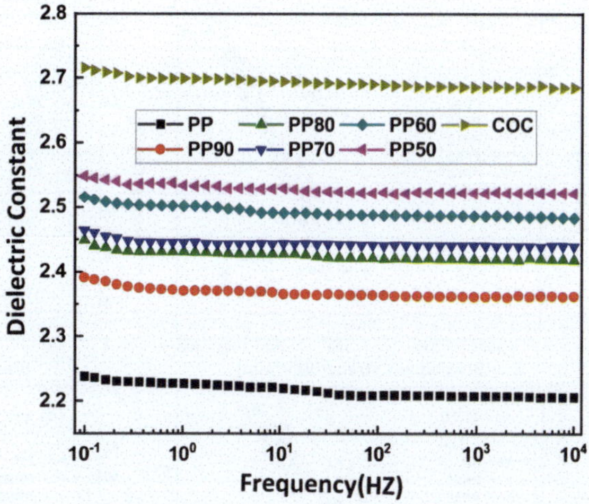

Fig. 11.5 Relationship between the dielectric constant and the frequency [11]. Copyright Clearance Center's RightsLink© License Number: 5811390213898

phase and the voltage phase flowing in a dielectric material is known as the dielectric loss angle, and the symbol is δ. The dielectric loss factor is an important performance parameter of an insulating material and indicates the size of the dielectric polarization loss.

As shown in Fig. 11.6, the dielectric loss angle factors of the tested films all show a decreasing and then increasing trend with the increase of the applied electric field frequency. This is because when the applied electric field frequency is shallow (tends to 0 when), the dielectric material internal polarization can keep up with the electric field changes, resulting in a small change in the dielectric constant, so the dielectric loss factor will gradually decrease with the increase in frequency. Then, as the electric field frequency rises, the dielectric material of the relaxation of the polarization in the frequency reaches a specific value and can't keep up with the changes in the electric field. The relaxation polarization of the permittivity contribution the contribution of relaxation polarization to the dielectric constant gradually decreases, and the dielectric constant decreases with the frequency increase. However, the increase of the steering frequency still increases the polarization loss, and the tan δ shows a trend of growing with the frequency increase in this frequency range. Polypropylene and COC are both non-polar materials with low dielectric loss factor values of 2.64×10^{-4} and 4.34×10^{-4} at 50 Hz, respectively, resulting in a slight difference in the dielectric loss angle tangent value of the composite films over the frequency range tested. When the content of COC is 10 wt% and 20 wt%, the dielectric loss factors of PP90 and PP80 at 50 Hz are only 4.4×10^{-5} and 5.8×10^{-5} higher than those of PP, which indicates that the effect of COC on the dielectric loss factor of polypropylene films is small (Fig. 11.7).

(3) **DC Breakdown Strength**

As shown in Fig. 11.6, the breakdown strength of the film specimens all show a decreasing trend with increasing temperature. This is because high temperatures

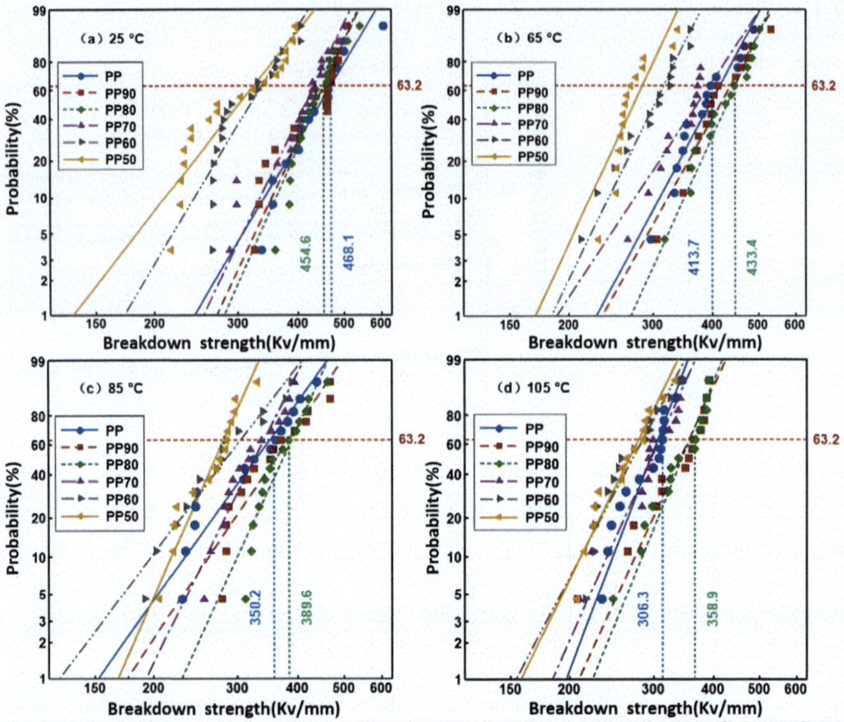

Fig. 11.6 Relationship between the breakdown strength and the temperature

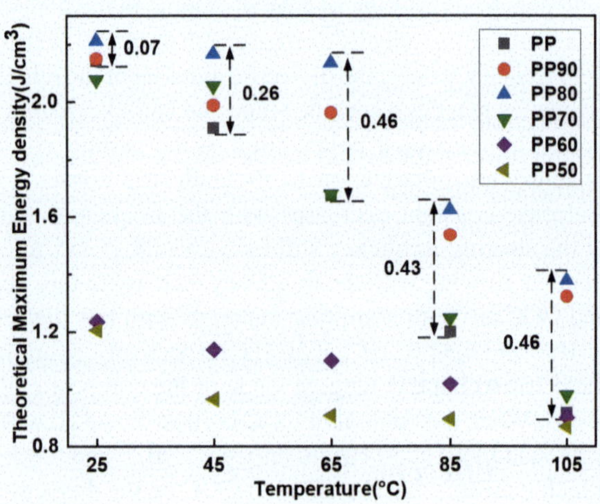

Fig. 11.7 Relationship between the theoretical maximum energy density and the temperature [11]. Copyright Clearance Center's RightsLink© License Number: 5811390213898

intensify the thermal movement of polymer molecular chains, the free volume inside the composite film increases, the average free range of free electrons increases, and the destructiveness of high-energy carriers to polymer molecular chains increases. In addition, due to problems such as the manufacturing process and improper transport, electrical weaknesses such as bubbles and impurities can exist inside the material. Under the combined effect of high temperature and high electric field, the electric field at the electrical weaknesses undergoes distortion, which accelerates the aging process of the material and leads to a decrease in the breakdown strength of the film with the increase in temperature. Based on the experimental results, a small amount of COC can reduce the degradation of the high-temperature breakdown performance of polypropylene films. In addition, the shape parameter of PP80 was the highest during the whole temperature rise process, which remained between 8.30 and 9.78, indicating the best stability of the breakdown performance of PP80. However, when the amount of COC was increased to more than 30 wt%, the high-temperature breakdown performance of the composite films was reduced due to the uneven distribution of COC.

(4) **DC Conductivity**

The dielectric loss of polypropylene film in a DC electric field is mainly the conductivity loss, which is positively correlated with the conductivity. The conductivity of solid dielectric materials can be classified into ionic and electronic conductivity according to the type of carriers. Polar dielectric has a large ionic conductivity, but polypropylene and COC are non-polar dielectric, ionic conductivity is mainly impurity ionic conductivity, material impurities are generally residual various additives and catalysts; electronic conductivity current at room temperature is relatively weak, this is because of polypropylene film in the number of free electrons is small, electrons are mainly from the outside world from the injection of metal electrodes.

Figure 11.8 illustrates the DC conductivity of the test films at different temperatures; as shown in the figure, the conductivity of the film specimens all increase with increasing temperature. On the one hand, when the external temperature increases, the kinetic energy gained by the electrons in the metal electrode increases due to the thermoelectric emission effect, and when the kinetic energy exceeds the escape work, the number of electrons injected into the interior of the film by the electrodes increases dramatically, leading to an increase in the electronic conductivity; on the other hand, under the compound effect of high temperature and applied electric field, the ion ionic ionization activity of the internal impurities is strengthened to increase the number of charged particles in the interior of the film, leading to the increase in ionic conductivity. This indicates that the heat resistance of COC is better, the molecular chain is less likely to migrate at high temperatures, and the ring structure in the molecular chain also hinders the movement of polypropylene molecules, which inhibits the increase of the free volume, leading to significant inhibition of the conductivity of the composite film.

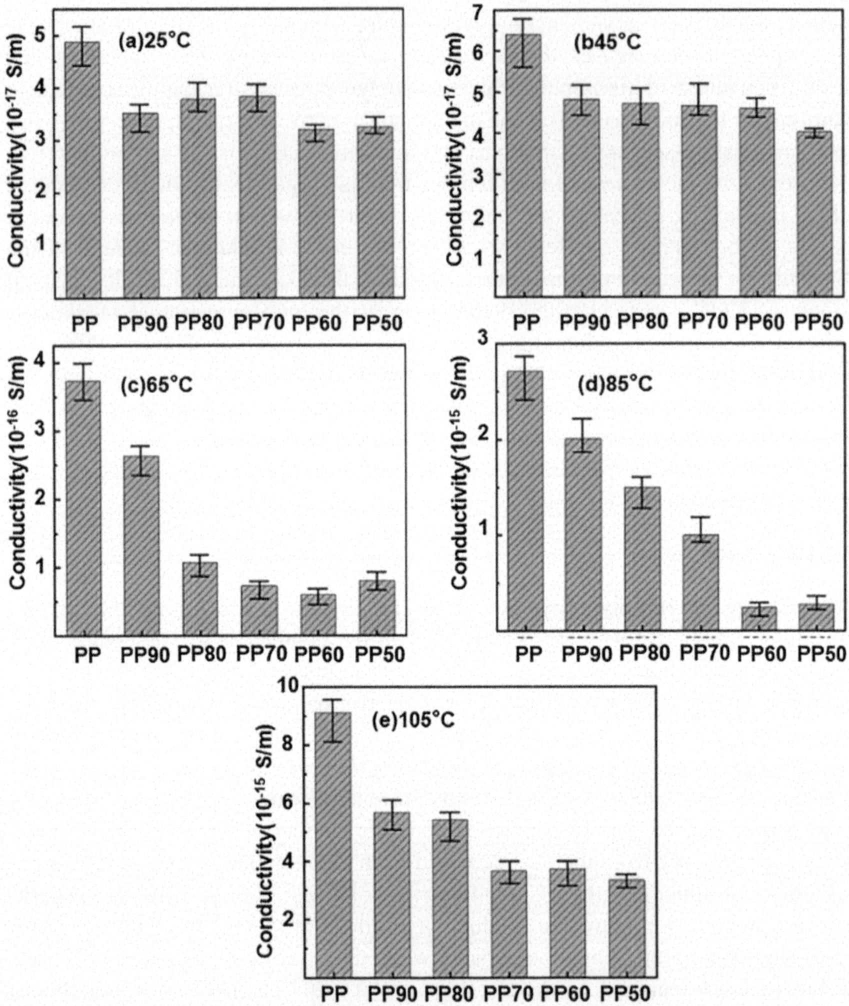

Fig. 11.8 Relationship between the conductivity and the temperature

(5) **Mechanistic Analysis**

PP/COC composite film will impact the composite film's primary physicochemical properties, such as microstructure, crystalline, and thermodynamic properties. At the same time, the content of PP/COC copolymer and glass transition temperature also have an impact on the dielectric properties of the composite film in response to the problem of the decrease in the breakdown field strength and the rise in the dielectric loss of polypropylene films at high temperatures.

The compact molecular chain arrangement in polypropylene spherical crystals decreases the mean free range, and carriers migrate mainly within the amorphous region. In addition, the movement of molecular chains is complex due to the excessive intermolecular forces within the spherical crystals, so the amorphous region's structure mainly influences the polypropylene properties at high temperatures. The polypropylene molecular chain has low internal rotational barriers. The molecular chain in the amorphous region is easy to rearrange under the action of high temperature, which increases the average free range and energy of the carriers, resulting in polypropylene molecular chain by the collision of the airlines are more likely to break, the macroscopic manifestation of polypropylene film high-temperature breakdown strength of the reduction.

The high-temperature breakdown performance of PP90 and PP80 is better than that of PP when the addition of COC is 10 and 20 wt%; on the one hand, the addition of COC makes the morphology of polypropylene crystalline region more compact, which reduces the defects within the film; on the other hand, COC contains cyclic structure branched chain, which increases the rigidity of polymer molecular chain and improves the thermal stability of composite films. In PP90 and PP80 composite films, polypropylene and COC are well compatible, the molecular segments of the two polymer phases penetrate and entangle with each other, and the introduction of the ring structure restricts the migration of the polypropylene molecular chains at high temperatures, which reduces the effect of temperature on the free volume. As a result, the carrier free range of PP90 and PP80 is shorter, suppressing carrier impact on the polypropylene molecular chains and macroscopically manifested in the optimized dielectric properties of the composite films at elevated temperatures. However, as the content of COC continues to increase, the compatibility between polypropylene and COC deteriorates, and the internal defects of the composite films increase, leading to a decrease in the breakdown strength and an increase in the conductivity loss of the composite films. Based on the above analysis, the high-temperature dielectric properties of polypropylene can be effectively improved by doping appropriate amounts of COCs.

11.4 Glass Transition Temperature Adjusting

The heat resistance of the composite films is not only related to the content of COC, but also depends on the glass transition temperature (T_g) of COC. Therefore, by characterizing the basic physicochemical properties of the composite films, such as the microstructure, crystalline properties and thermodynamic properties, as well as the evolution of dielectric constants and dielectric loss factors at different frequencies, the effect of the T_g of COC on the breakdown field and DC conductivity of the polypropylene films was analyzed. The effect of T_g of COC on the breakdown field strength and DC conductivity of polypropylene is analyzed, and the mechanism of T_g of COC on the dielectric properties of composite films is revealed by combining the crystalline properties and dielectric properties.

11.4.1 *Effect on the Microstructure*

As mentioned earlier, it was noted that when filler levels were low (≤ 20 wt%), COCs exhibited excellent compatibility with polypropylene, thereby enhancing the dielectric properties of the composite films. Accordingly, COCs featuring different glass transition temperatures (Tg) were incorporated into formulations with 10 and 20 wt% additives (Table 11.1).

(1) **Crystalline Morphology**

When the glass transition temperature (Tg) of COC is low (≤ 134 °C), a minor quantity of COC is evenly dispersed within the polypropylene matrix, thereby enhancing the crystalline morphology of PP/COC composite films. The spherical crystal size of PP6 samples is smaller, the crystal structure is more intensive, and the defects between spherical crystals are reduced. In addition, the Tg of COC directly determines the content of cyclic structure inside the composite film; the higher the Tg, the more cyclic structure inside the film, and the cyclic structure can effectively impede the movement of polypropylene molecular chains, resulting in a more orderly molecular chain arrangement, and amorphous defects are reduced. However, when the Tg of COC was increased to 178 °C, the COC dispersion in polypropylene films deteriorated due to COC's rigidity, so the amorphous defects in PP7 and PP8 samples increased significantly (Fig. 11.9).

(2) **Thermodynamic Properties**

The crystallization and melt characteristics of different PP/COC composite films are shown in Fig. 11.10. The pure polypropylene film (PP0) has the highest crystallization temperature (Tc of 164.8 °C). In contrast, the crystallization temperatures of the composite films are lower than that of PP0, which are distributed in the temperature range of 112.6–113.5 °C, and show a tendency of increasing and then decreasing with the increase of T_g. When the glass transition temperature of COC is 178 °C, the crystallization temperature of PP8 is the lowest. Still, it is only 1.23% lower than

Table 11.1 Basic parameters of different PP/COC composite films

Samples	T_g (°C)	Content ratio (PP/COC)
PP0	– 20	100/0
PP1	78	90/10
PP2	78	80/20
PP3	110	90/10
PP4	110	80/20
PP5	134	90/10
PP6	134	80/20
PP7	178	90/10
PP8	178	80/20

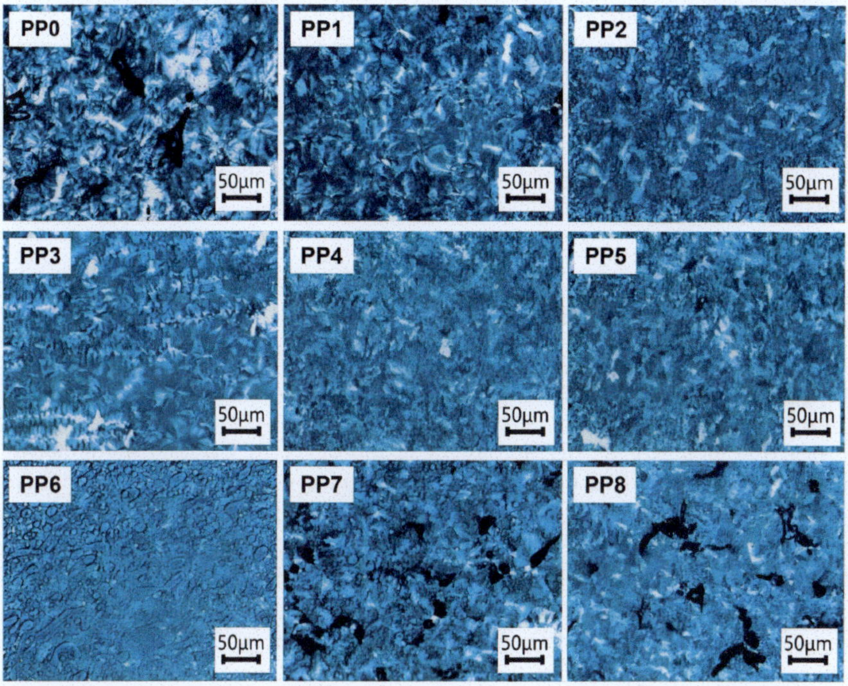

Fig. 11.9 Crystalline morphology of composite films

PP0, indicating that the effect of T_g of COC on the crystallization temperature of composite films is small.

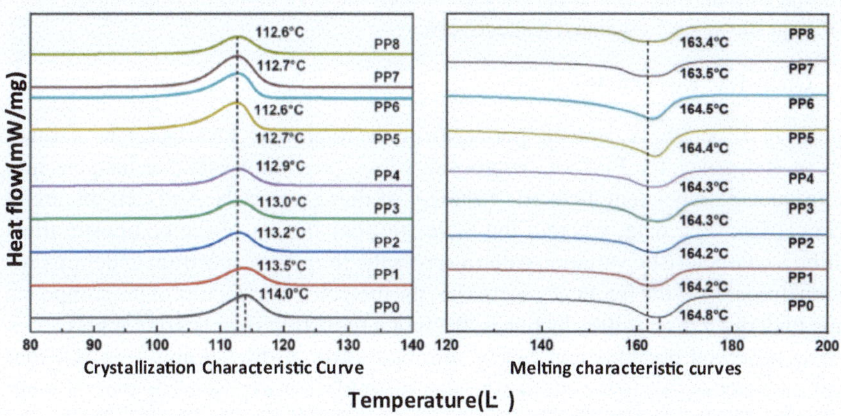

Fig. 11.10 Crystallization and melting characteristic curves of composite films

As shown in the figure, PP0 has the highest melting temperature (Tm of 164.8 °C), while adding COC reduces the melting temperature of the composite film. The melt temperatures of the composite films were distributed between 163.4 and 164.5 °C and showed a tendency to increase and decrease with the increase of the COC Tg. When Tg is 178 °C, PP8 has the lowest melting temperature, but it is only 0.80% lower than that of PP0, which indicates that the influence of COC Tg on the melting temperature of the composite film is small.

11.4.2 Effect on the Dielectric Properties

(1) **Permittivity**

To describe more graphically the law of action of Tg of COC on the dielectric constant of polypropylene films. Pure cyclic olefin copolymer films with different Tg, COC-8007, COC-7010, COC-5013, and COC-6017 were used as a control group. Figure 11.11 demonstrates the dielectric constants of the composite films at different frequencies. As the figure shows, the dielectric constants of the tested films all decrease with increasing frequency of the electric field. In the frequency range of 10^{-1} to 10^4 Hz, the dielectric constants of the composite films are located between the pure polypropylene films and the pure COC films and increase with the increase of the content of the COC, which is because the dielectric constants of the pure COC are all higher than those of the pure polypropylene films. When the olefin copolymer's Tg increases, the composite film's dielectric constant decreases. The number of cyclic side-chain structures within the COC increases with the increase in Tg, increasing the rigidity of the molecular chain and a decrease in the polarity of the composite film. When the Tg of the COC is 134 °C, the dielectric constants of PP5 and PP6 at 50 Hz are 2.34 and 2.42, which are 0.13 and 0.21 higher than those of PP0 at the same frequency, respectively.

(2) **Dielectric Loss Factor**

Figure 11.12 offers a detailed portrayal of the dielectric loss factor as a function of frequency for PP/COC composite films, providing valuable insights into the frequency-dependent dielectric behavior of these materials. A discernible trend emerges from the data, wherein the dielectric loss factors of the composite films exhibit a characteristic variation pattern across the frequency spectrum under investigation. Initially, as the frequency increases, the dielectric loss factors of the composite films undergo a discernible decline, reflecting a reduction in polarization loss.

The observed phenomenon can be ascribed to the restricted ability of polymer chains to react swiftly to the alternating electric field when exposed to higher frequencies. This limitation results in a decrease in polarization contributions and subsequently lowers the overall dielectric losses experienced. However, beyond a certain threshold frequency, the dielectric loss factors of the composite films begin to ascend, indicating a resurgence in polarization losses attributed to interfacial polarization

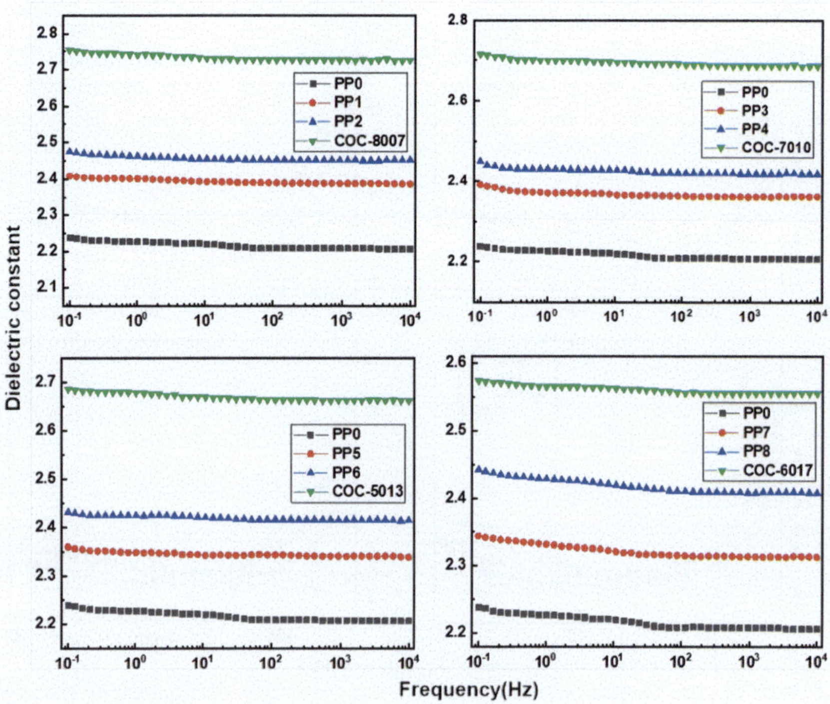

Fig. 11.11 Relationship between the dielectric constant and the frequency

and other high-frequency phenomena. This upward trend is indicative of enhanced molecular mobility and interfacial polarization effects at elevated frequencies, which contribute to heightened dielectric losses in the material.

A comparative analysis reveals that the dielectric loss factors of the composite films surpass those of pure polypropylene films, owing to the intrinsic polarization losses associated with the COC component. Notably, within the frequency range of 10^1 to 10^4 Hz, the dielectric loss factor curves of the tested films exhibit convergence, with the polarization losses of the composite films closely resembling those of pure polypropylene films. Furthermore, the influence of COC's glass transition temperature (Tg) on the dielectric loss factor of the composite films is found to be negligible, as evidenced by the marginal disparity in dielectric loss factors between PP/COC composite films with varying COC Tg values. This observation underscores the dominant role of other factors, such as polymer morphology and interfacial interactions, in dictating the dielectric performance of the composite membranes.

In essence, Fig. 11.12 provides a comprehensive exploration of the complex interactions among frequency, material composition, and molecular dynamics, which collectively influence the dielectric loss characteristics of PP/COC composite films. Through a detailed analysis of these frequency-dependent phenomena, researchers gain invaluable perspectives into the fundamental mechanisms driving dielectric

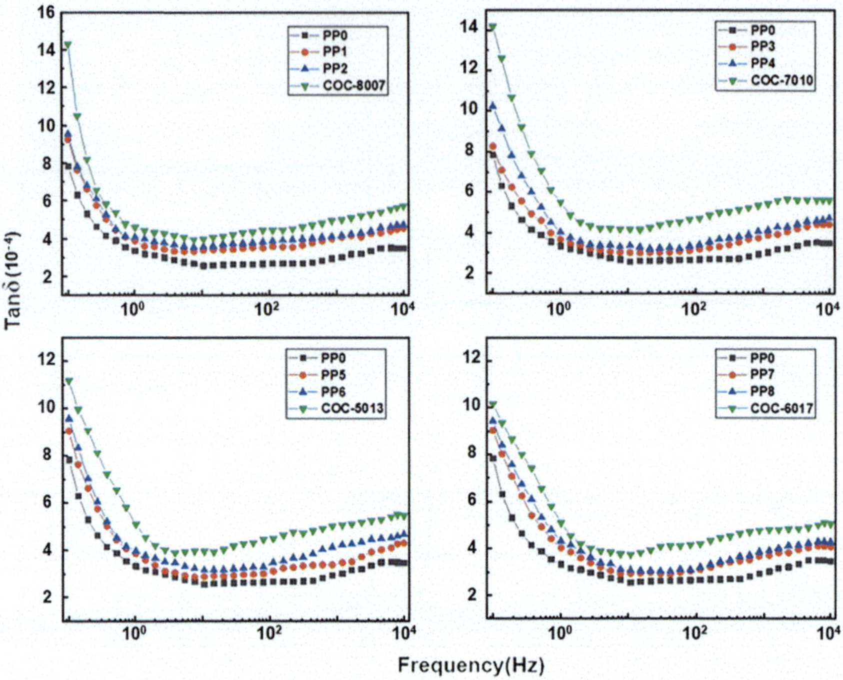

Fig. 11.12 Relationship between the dielectric loss factor and the frequency

relaxation processes, thus facilitating informed strategies for the development and enhancement of composite materials across a broad spectrum of electrical applications.

(3) DC Breakdown Strength

Figures 11.13 offer a comprehensive visualization of the Weibull breakdown probability plots for various composite films across a spectrum of temperatures, including 25, 85, 105, and 125 °C. A discernible trend emerges from the data, wherein the breakdown field strength values of the composite films exhibit a consistent pattern of behavior across the temperature gradient under investigation. At ambient conditions (25 °C), the breakdown field strength values of the composite films reach their zenith, reflecting optimal dielectric performance under standard operating conditions. However, as the temperature escalates, a progressive decline in breakdown field strength becomes evident, with higher temperatures correlating with diminished dielectric robustness.

This observed decrease in breakdown field strength can be attributed to the concurrent elevation of two key contributing factors: the energy of hot electrons and the free volume within the composite films. As the temperature rises, thermal energy imparts greater kinetic energy to electrons within the material, augmenting their propensity for high-energy interactions and facilitating molecular chain breakage.

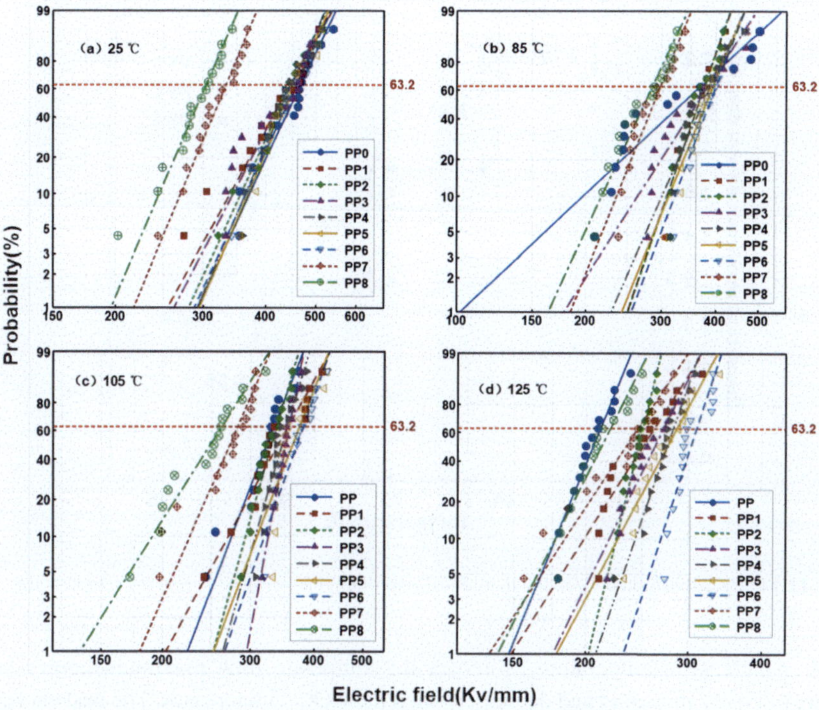

Fig. 11.13 Relation between the breakdown strength and the temperature

Simultaneously, the expansion of free volume within the composite films amplifies the mobility of molecular segments, promoting their displacement and facilitating the initiation of breakdown phenomena. This combination of heightened electron energy and increased molecular mobility exacerbates the vulnerability of the material to breakdown, manifesting macroscopically as a reduction in breakdown field strength. The interplay between temperature, electron energy, and molecular dynamics underscores the intricate relationship between thermal effects and dielectric performance in composite materials. By delineating these underlying mechanisms, researchers can gain valuable insights into the thermal stability and reliability of composite films, thereby informing the design and optimization of dielectric systems for diverse applications. In summary, the Weibull breakdown probability plots provide a quantitative framework for assessing the impact of temperature on the breakdown characteristics of composite films. Through a holistic understanding of the underlying phenomena, researchers can devise targeted strategies for mitigating thermal effects and enhancing the dielectric resilience of composite materials, thereby advancing the frontier of electrical insulation and reliability engineering.

By calculating the maximum theoretical energy density of the composite films as shown in Fig. 11.13, it is found that their maximum theoretical energy density decreases with temperature, which is attributed to the decrease in breakdown strength

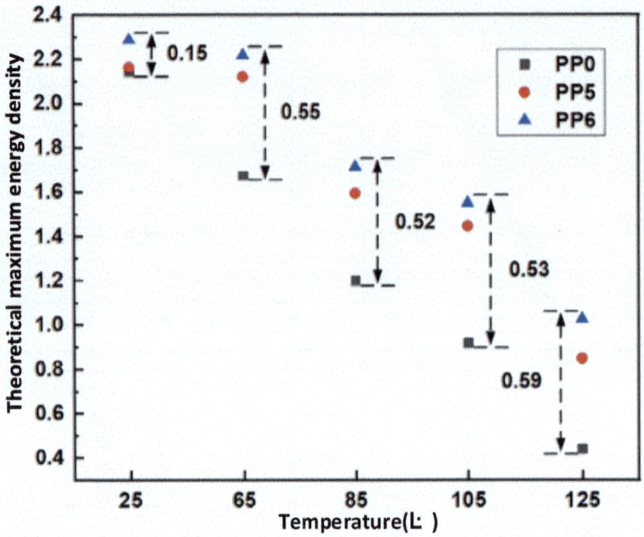

Fig. 11.14 Relationship between the theoretical maximum energy density and the temperature

with an increase in temperature. Moreover, the composite films' maximum theoretical energy density increases and decreases when the COC Tg increases. The best energy storage performance of the composite films is achieved when the COC Tg is 134 °C, so Fig. 11.14 demonstrate the maximum theoretical energy density variation with temperature only for PP5 and PP6. Among them, the maximum theoretical energy density of PP6 at 25 °C is 2.29 J/cm^3, which is 1.07 times that of PP0 at the same temperature, and when the temperature rises to 125 °C, the maximum theoretical energy density of PP6 increases to 2.25 times of that of PP0. However, when the Tg of the COC increases further, the energy storage performance of the composite film decreases.

(4) **DC Conductivity**

As shown in Fig. 11.15, the DC conductivity of polypropylene films increases exponentially with temperature, while the conductivity of composite films with added COCs is much less affected by temperature. High temperatures intensify the movement of molecular chains in the amorphous region of polypropylene, increasing the number of carriers and the mean free range, leading to a significant increase in the conductive current. In contrast, the rigid ring structure in the COC can limit the thermal movement of the molecular chains, and the composite film has lower conduction losses at high temperatures than the pure polypropylene film. When the Tg of the COC is 134 °C, the conductivity of PP6 at 125 °C is 7.43 × 10^{-15} S/m, which is 9.0% of that of PP0 at the same temperature. The conductivity of the composite films gradually decreased with the increase of COC Tg. At a lower additive amount, the COC with lower Tg has poorer heat resistance, less cyclic side chains inside

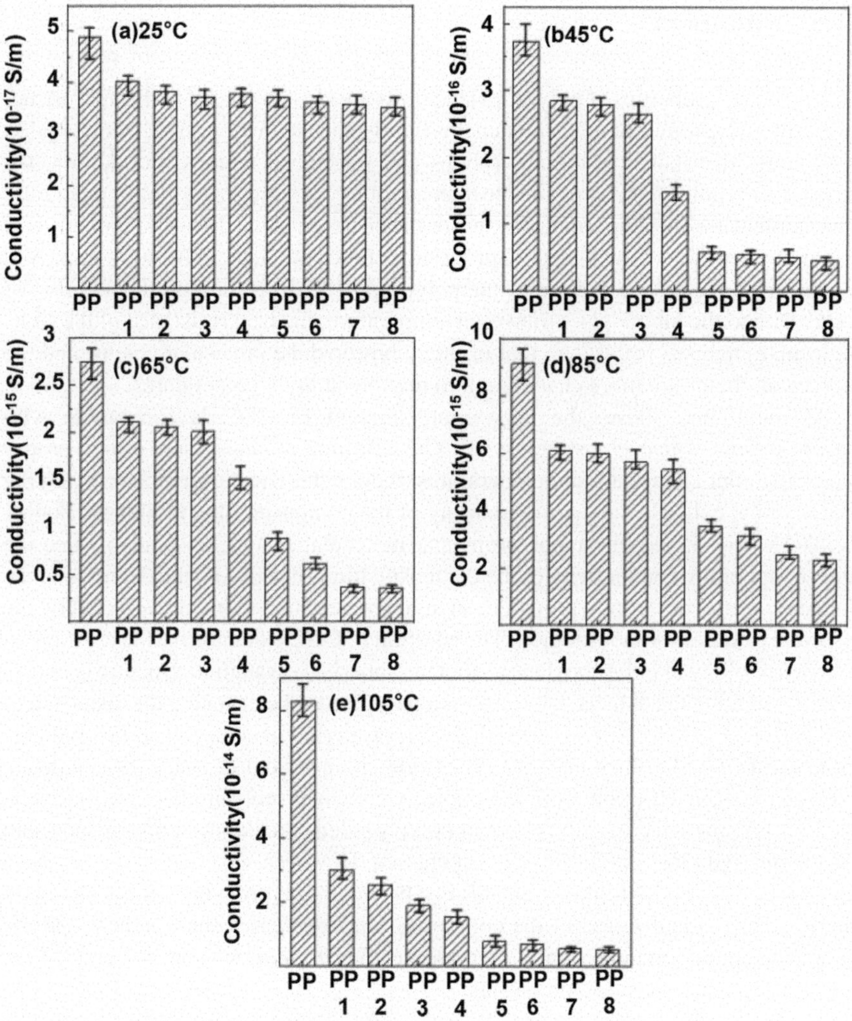

Fig. 11.15 Relationship between the conductivity and the temperature

the film, and less inhibition of carrier migration. In contrast, the COC with higher Tg has more vital intermolecular chain forces, more cyclic side chains, and more potent inhibition of carrier migration. Hence, the conductivity of the composite film decreases.

11.5 Summary

The dielectric properties of polypropylene films are closely related to the microstructure, with polypropylene molecular chains tightly packed in the crystallization region but loosely distributed in the amorphous and interfacial regions. In addition, the migration of molecular chains and the increase of free volume in the high-temperature environment are also important reasons for the decrease in the dielectric properties of polypropylene films. The dielectric properties of the composite films were improved by changing the microstructure of the polypropylene films through the addition of COC. Due to the higher glass transition temperature, the heat resistance of the COC molecular chains is better than that of the polypropylene molecular chains, and the molecular chains are more challenging to migrate at high temperatures.

As mentioned above, the appropriate amount of COC is compatible with polypropylene. The cyclic structure of COC will limit the movement of polypropylene molecular chains at high temperatures, reduce the effect of temperature on the free volume, enhance the thermal stability of the composite film, inhibit the Poole–Frenkel effect of the linear polypropylene molecular chains, and thus reduce the high-temperature conductivity of the composite film. In addition, the decrease in the dielectric properties of polypropylene at high temperatures is mainly affected by the structure of the amorphous region. The addition of COC can improve the crystallization morphology of polypropylene so that the spherical crystalline structure becomes denser, reducing the defects in the amorphous region and weakening the distortion of the local electric field to improve the dielectric properties of composite films. With the increase of COC Tg, the number of cyclic side-chain structures inside the composite film increases, and the polypropylene molecular chain is more hindered, so the cyclic olefin copolymers increase the effect of polypropylene breakdown performance, and the conductivity loss is significantly suppressed. However, when the glass transition temperature is too high, the compatibility between polypropylene and COC deteriorates, the crystallization of the composite films decreases substantially, and the defects in the amorphous region increase, leading to a significant decrease in the high-temperature dielectric properties.

References

1. M. Xiao, K. X. Wang, Y. S. Zhao and B. X. Du, "Effect of UV irradiation time on dielectric performance of PP film for HVDC capacitor," *IEEE Trans. Dielect Elect. Insul.*, vol. 31, no. 3, pp. 1268–1275, 2024.
2. R. W. Brown, "Linking corrosion and catastrophic failure in low-power metallized polypropylene capacitors," *IEEE Trans. Device Mat Reliab.*, vol. 6, no. 2, pp. 326–333, 2006.
3. Z. Y. Ran, B. X. Du and M. Xiao, "High-temperature breakdown performance improvement of polypropylene films based on micromorphology control," *IEEE Trans. Dielect Elect. Insul.*, vol. 28, no. 5, pp. 1547–1554, 2021.

4. H. L. Liu, B. X. Du, M. Xiao, Z. Y. Ran and T. Tanaka, "Breakdown strength and energy density improvement of polypropylene by parylene deposition for film capacitor," *IEEE Trans. Dielect Elect. Insul.*, vol. 29, no. 3, pp. 981–989, 2022.
5. T. Umemura, K. Abe, K. Akiyama and D. Couderc, "Thermal aging behavior of BOPP films," *IEEE Trans. Dielect Elect. Insul.*, vol. EI-22, no. 6, pp. 735–7434, 1987.
6. H. L. Liu, B. X. Du and M. Xiao, "Improved energy density and charge discharge efficiency of polypropylene capacitor film based on surface grafting," *IEEE Trans. Dielect Elect. Insul.*, vol. 28, no. 5, pp. 1539–1546, 2021.
7. F. C. Chiu, "A review on conduction mechanisms in dielectric films," *Adv. Mater. Sci. Eng.*, vol. 2014, 2014, Art. no. 578168.
8. M. Ieda, "Electrical conduction and carrier traps in polymeric materials," *IEEE Trans. Dielect Elect. Insul.*, vol. EI-19, no. 3, pp. 162–178, 1984.
9. M. L. Gasperi, "Life prediction modeling of bus capacitors in AC variable frequency drives," *IEEE Trans. Ind. Appl.*, vol. 41, no. 6, pp. 1430–1435, 2005.
10. M. Xiao, M. D. Zhang and B. X. Du, "Crystallization regulation of long-chain branched polypropylene on dielectric performance and energy density for metallized film capacitors," *High Volt.*, vol. 8, no. 5, pp. 921–930, 2022.
11. M. Xiao, K. L. Fan, M. D. Zhang, B. X. Du, Z. Y. Ran and Z. L. Xing, "Effect of cyclic olefin copolymer on dielectric performance of polypropylene films for capacitors," *IEEE Trans. Dielect Elect. Insul.*, vol. 29, no. 6, pp. 2266–2273, 2022.
12. J. Q. Wang, H. X. Guo, S. X. Zeng, J. Q. Du, Q. Zhang and K. Wang, "Thin, largescale processed, high-temperature resistant capacitor films based on polypropylene/cycloolefin copolymer blend," *Chem. Eng. J.*, vol. 492, 2024, Art. no. 152237.
13. Y. Yoshida et al, "Living copolymerization of ethylene with norbornene catalyzed by bis (pyrrolide-imine) titanium complexes with MAO," *J. Am. Chem. Soc.*, vol. 126, no. 38, pp. 12023–12032, 2004.
14. A. L. McKnight and R. M. Waymouth, "Ethylene/norbornene copolymerizations with titanium CpA catalysts," *Macromolecules*, vol. 32, no. 9, pp. 2816–2825, 1999.
15. Z. L. Li, Z. Y. Zhong and Du B X, "Dielectric relaxation and trap-modulated DC breakdown of polypropylene blend insulation," *Polymer*, vol. 185, 2019, Art. no. 121935.
16. G. H. Motlagh, A. N. Hrymak and M. R. Thompson, "Properties of a carbon filled cyclic olefin copolymer," *J. Polym. Sci.*, vol. 45, no. 14, pp. 1808–1820, 2007.
17. M. Hong, L. Cui, S. R. Liu and Y. S Li, "Synthesis of novel cyclic olefin copolymer (COC) with high performance via effective copolymerization of ethylene with bulky cyclic olefin," *Macromolecules*, vol. 45, no. 13, pp. 5397–5402, 2012.
18. T. Scrivani, R. Benavente, E. Pérez and J. M. Pereña, "Stress-Strain behaviour, microhardness, and dynamic mechanical properties of a Series of ethylen-norbornene copolymers," *Macromol. Chem. Phys.*, vol. 202, no. 12, pp. 2547–2553, 2001.
19. J. Forsyth et al, "Influence of the polymer microstructure on the thermal properties of cycloolefin copolymers with high norbornene contents," *Macromol. Chem. Phys.*, vol. 202, no. 5, pp. 614–620, 2001.
20. A. Gopanna et al, "Investigation of mechanical, dynamic mechanical, rheological and morphological properties of blends based on polypropylene (PP) and cyclic olefin copolymer (COC)," *Eur. Polym. J.*, vol. 108, pp. 439–451, 2018.
21. M. Šlouf, J. Kolařík and L. Fambri, "Phase morphology of PP/COC blends," *J. Appl. Polym. Sci.*, vol. 91, no. 1, pp. 253–259, 2004.
22. A. Pegoretti, J. Kolarık, L. Fambri and P. Amabile, "Polypropylene/cycloolefin copolymer blends: effects of fibrous phase structure on tensile mechanical properties," *Polymer*, vol. 44, no. 44, pp. 3381–3387, 2003.
23. M. Paajanen, M. Karttunen, S. Kortet and O. Härkki, "Thermally more durable electromechanical films by POSS nanomodification," *Key Engineering Materials*, vol. 538, pp. 65–68, 2013.Key Engineering Materials.
24. W. Cao, Z. Li, G. Sheng, W. K Cao, Z. Li, G. H. Sheng and X. C. Jiang, "Insulating property of polypropylene nanocomposites filled with nano-MgO of different concentration," *IEEE Trans. Dielect Elect. Insul.*, vol. 24, no. 3, pp. 1430–1437, 2017.

25. Z. L. Li, Z. Y. Zhong and B. X. Du, "Dielectric relaxation and trap-modulated DC breakdown of polypropylene blend insulation," *Polymer*, vol. 185, 2019, Art. no. 121935.
26. Y. H. Wu, J. W. Zha, W. K. Li, S. J. Wang and, Z. M. Dang, "A remarkable suppression on space charge in isotatic polypropylene by inducing the β-crystal formation," *Appl. Phys. Lett.*, vol. 107, no. 11, 2015, Art. no. 112901.
27. A. Gopanna, R. N. Mandapati, S. P. Thomas, K. Rajan and M. Chavali, "Fourier transform infrared spectroscopy (FTIR), Raman spectroscopy and wide-angle X-ray scattering (WAXS) of polypropylene (PP)/cyclic olefin copolymer (COC) blends for qualitative and quantitative analysis," *Polym. Bull.*, vol. 76, no. 8, pp. 4259–4274, 2019.
28. T. C. K. Yang, S. S. Y. Lin and T. H. Chuang, "Kinetic analysis of the thermal oxidation of metallocene cyclic olefin copolymer (mCOC)/TiO2 composites by FTIR microscopy and thermogravimetry (TG)," *Polym. Degrad. Stabil.*, vol. 78, no. 3, pp. 525–532, 2002.
29. Z. G. Wang, R. A. Phillips and B. S. Hsiao, "Morphology development during isothermal crystallization. I. Isotactic and atactic polypropylene blends," *J. Polym. Sci.*, vol. 38, no. 19, pp. 2580–2590, 2000.
30. T. Teshome and A. Datta, "Two-dimensional graphene–gold interfaces serve as robust templates for dielectric capacitors," *ACS Appl. Mater. Interfaces*, vol. 9, no. 39, pp. 34213–34220, 2017.

Part IV
Electrode-Film Interface Modification

Chapter 12
Surface Molecular Grafting

12.1 Introduction

Metallized film capacitors (MFCs) are widely used in the conditions of power conversion, such as high voltage direct current (HVDC), renewable-energy generating and hybrid electric vehicles driving [1]. PP possesses the characteristics of facile processability, high withstand voltage strength, and outstanding charge–discharge efficiency, which makes it the most popular dielectric materials in the field of advanced power and electronics systems [2]. With the continuous improvement of voltage level and the thirsty demand for miniaturization of power facilities, the approach to manufacturing advanced metallized polypropylene film capacitors with higher electrical strength has been attracting increasing attention from scholars [3]. Despite the fact that PP enjoys the reputation as state-of-art dielectric materials, the dielectric material suffers from relatively low working temperatures in harsh environments [4]. The main reason why PP suffers from the limited operating temperature is the film loss under long-term high charge–discharge current [5]. The dielectric loss of PP caused by energy dissipation leads to an undesirable temperature rise [6]. The loss of PP films increases exponentially with the increase of temperature and the withstand strength decreases sharply [7]. Elevated temperature can accelerate the dielectric aging process and contribute to a destructive breakdown of insulation in turn [8].

Charge-injection under higher temperature and electric field is one of the determinants of polymers conductive loss [9]. High temperature excites carriers to overcome electrodes/dielectric interface barriers, resulting in a substantially drop of withstandvoltage and charge-efficiency [10]. The traditional method of regulating carrier migration behaviors is primarily based on body modification, which is found to be difficult in limiting the carrier injection process in the initial stage. Once the charge injection has been accomplished, the problem can be hardly avoided that the field distortion will lead to insulation breakdown, considering that the thickness of advanced membranes is usually less than 10 μm. Therefore, it is of significance to improve the electrical properties of dielectric capacitors by means of interfacial

B. Du and M. Xiao, *Polypropylene Film for HVDC Capacitors*,
https://doi.org/10.1007/978-981-96-3029-5_12

modification of PP films. The surface modification of dielectric films can effectively modify the carriers transport behavior and improve the dielectric performance of the membranes [11]. However, there are few reports on the methods of suppressing carrier injection by controlling the surface structure, so it is urgent to develop a method to manipulate the surface molecular structure based on chemical reactions. At present, the modification methods for the surface of PP films mainly focus on the construction of an inorganic charge barrier layer on the surface. However, this method is time-consuming, labor-intensive, and complex. In addition, due to the different expansion coefficients of inorganic and organic matter, the barrier layer is easy to fall off.

12.2 Surface Molecular Grafting Method

Polypropylene (PP) film possesses a stable molecular chain structure that resists undergoing chemical reactions readily. However, by selectively functionalizing it without compromising its inherent properties, enhancements can be achieved in its dielectric and energy storage performance. At present, the common modification methods are as follows: The PP surface is pretreated by surface treatment methods such as plasma and glow discharge. The active particles in the plasma will bombard and etch the surface of PP, oxidize it in the presence of air/O_2, and introduce polar groups such as carbonyl, carboxylic acid and ether bonds. This increases the trap energy level of the PP surface, which in turn improves its dielectric properties. No organic solvents are used in the treatment process, which is more environmentally friendly. However, the type, quantity and thickness of the groups produced by this method are uncontrollable and cannot be studied in depth, and the higher the reaction temperature, the greater the damage to the film body.

Through the grafting modification of monomers, polar functional groups are introduced into the PP molecular chain, and then the grafted modified PP is directly used in the preparation of composite films. This method can effectively improve the dielectric properties while minimizing the impact on the properties of the PP body. At present, the applicable modification methods for surface grafting mainly include radiation treatment methods. The irradiation grafting method uses high-energy rays to irradiate the polymer to induce the generation of free radicals and other active sites in its surface space, and then initiates the grafting reaction through these active sites. Since irradiation mainly acts on the surface of PP, it can avoid the problems of PP degradation and monomer residue in the process of melt/solution grafting, and at the same time, it has attracted much attention from researchers in recent years, such as high initiation efficiency, fast reaction rate, milder reaction conditions and fewer side reactions. Commonly used irradiation methods include accelerated electron beam, Co_{60}-γ rays, plasma, ultraviolet radiation, microwave radiation, etc. Among these, electron beam and γ-ray radiation are characterized by low linear energy transfer, providing significant penetration capabilities capable of generating macromolecular free radicals within the millimeter-scale region of the polymer surface. However, since capacitor PP films are only at the micron-level thickness, such radiation can potentially induce substantial damage to the microstructure. Nevertheless, ultraviolet radiation and other penetrating mechanisms are limited, leading to the generation of active centers solely within the nanoscale region beneath the polymer surface. This condition adequately fulfills the criteria for surface modification of capacitor polypropylene film.

Utilizing the facile dehydrogenation and bond cleavage properties of tertiary carbon atoms within the PP molecular chain under ultraviolet irradiation, hydrogen atoms in the molecule generate macromolecular free radicals. These radicals can directly graft with specified groups like sulfate, maleic anhydride (MAH), and others. In this process, the presence of initiators lowers the activation energy needed for the reaction, thereby accelerating the grafting reaction rate. Notably, sulfate groups exhibit lower stability and can be substituted by more stable hydroxyl (OH) groups following hydrolysis. Moreover, hydroxyl groups demonstrate high reactivity, facilitating grafting reactions with diverse groups such as amino and epoxy groups through chemical processes.

12.3 Surface Molecular Grafted Polypropylene

12.3.1 Effect on the Microstructure

In order to determine whether the grafting on the surface molecule is successful, a number of characterization methods are required to verify it, including water contact angle, X-ray diffraction, scanning electron microscopy (SEM), and atomic force microscopy. Figure 12.1a, b shows the change in the water contact angle and SEM morphology before and after grafting hydroxyl groups, and the compatibility between polar groups is better, so the surface grafted PP film has a smaller contact angle. Figure 12.1c also compares the static water contact angles of PP films under different processing times. With the increase of treatment time, the water contact angle decreased from 97.8° to 77.0°, which indicated that the hydrophilicity of the PP film was enhanced after grafting treatment, which verified the success of surface grafting treatment to a certain extent. Before surface grafting, the film section was homogeneous, and no multi-layer structure was observed. After surface grafting, an obvious dense structural layer can be observed on the surface of PP film. The thickness of the surface dense layer under different oxidation treatment times is shown in the figure, and the thickness of the surface treatment layer of the film is about 300 nm after 120s treatment.

In order to further verify the success of surface grafting, the XPS method was used to characterize the elemental content and chemical bond composition of PP films before and after modification, as shown in Fig. 12.1d. From the spectrum of C1s and O1s, it can be seen that the increased O content is mainly due to the introduction of C–O bonds, which indicates that the surface grafting treatment successfully introduced

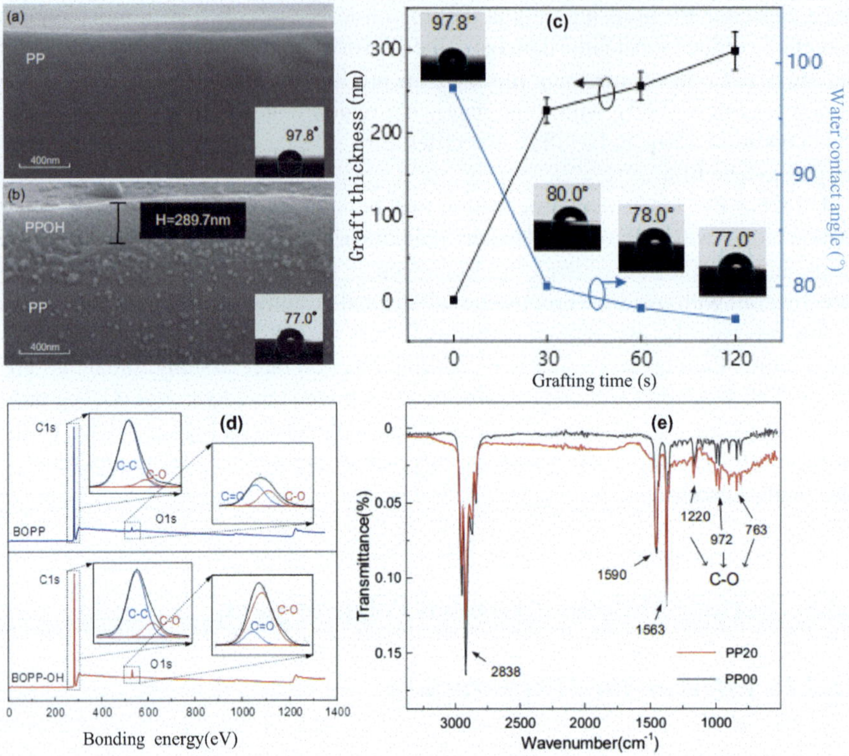

Fig. 12.1 **a**, **b** Cross-sectional morphology, **c** surface hydrophilicity contrast, **d** surface elements and chemical bonds and **e** FTIR spectra of the PP film with the hydroxyl groups surface grafted [12]. Copyright Clearance Center's RightsLink© License Number: 5811200243928

hydroxyl groups on the surface of PP films. At the same time, it can be seen that there is also a certain amount of C–O and C=O chemical bonds on the surface of the ungrafted PP film. Oxidants cause a certain degree of surface oxidation, and ultraviolet light can significantly accelerate this process, which can be verified by the cross-sectional structure of the film and the peak value of the XPS energy spectrum. The FTIR spectra are shown in Fig. 12.1e. A slight peak growth up took place in the lower wavenumber range (750–1250 cm^{-1}), because there were considerable sites on the thin film surface for the grafting reaction. The results show that the surface modification did not produce unexpected functional groups, while the change of the interfacial properties of the membranes was attributed to the C–OH groups. Similar characterization methods can also be used to study other groups in grafting.

Figures 12.2 and 12.3 are some of the methods for characterizing the butted maleic anhydride, hydroxyl group, amino group and epoxy group, respectively. These include the X-ray photoelectron spectroscopy, scanning electron microscopy (SEM), and atomic force microscopy. Through the XPS results, it can be seen that after grafting, specific elements and groups appear on the surface of the PP film. Elemental distribution test images in SEM can show the change in the elemental content on the surface of the film before and after grafting. AFM is employed to quantitatively assess the surface topography, investigating the change in surface roughness before and after grafting. As illustrated in Fig. 12.3c, d. The PP film has a smooth and uniform surface structure, and the arithmetical mean deviation value of the roughness (Ra) is 2.82 nm. While the surface of PP-3 film becomes much rougher, the fluctuation between atoms changes larger, with Ra measuring approximately 4.15 nm. In addition, the secondary treatment would further improve the roughness. The maximum mean square value (Rq) of roughness increased from 2.82 nm of PP to 6.87 nm of PP-A. This phenomenon can be attributed to the uneven molecular weight resulting from the grafting process.

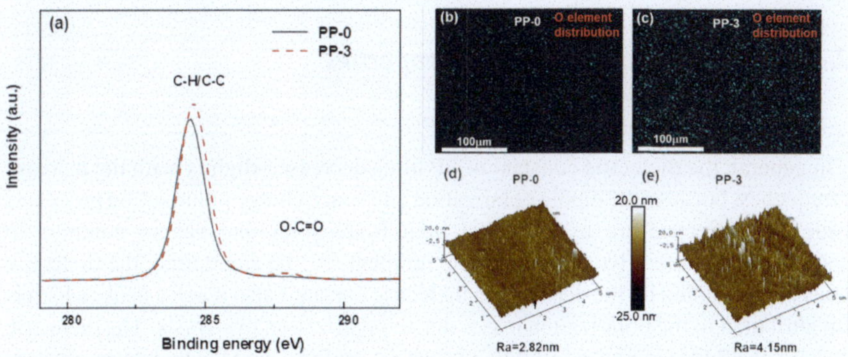

Fig. 12.2 **a** XPS spectra, **b**, **c** surface element distribution and **d**, **e** surface roughness of the PP film with maleic anhydride surface grafted [12]. Copyright Clearance Center's RightsLink© License Number: 5811200243928

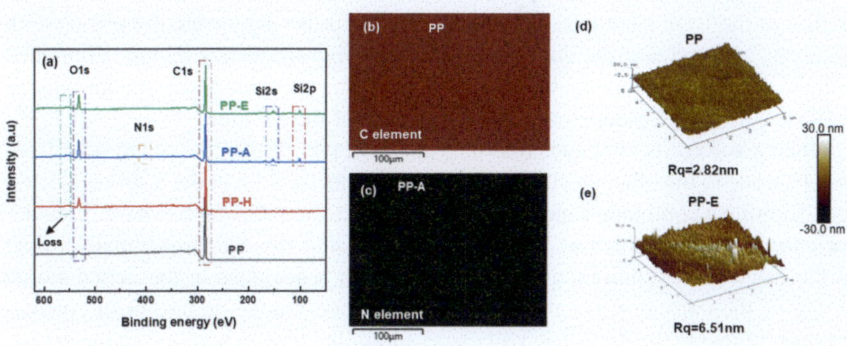

Fig. 12.3 **a** XPS spectra, **b**, **c** surface element distribution and **d**, **e** surface roughness of the PP film with hydroxyl (PP-H), amino (PP-A) and epoxy (PP-E) groups surface grafted [12]. Copyright Clearance Center's RightsLink© License Number: 5811200243928

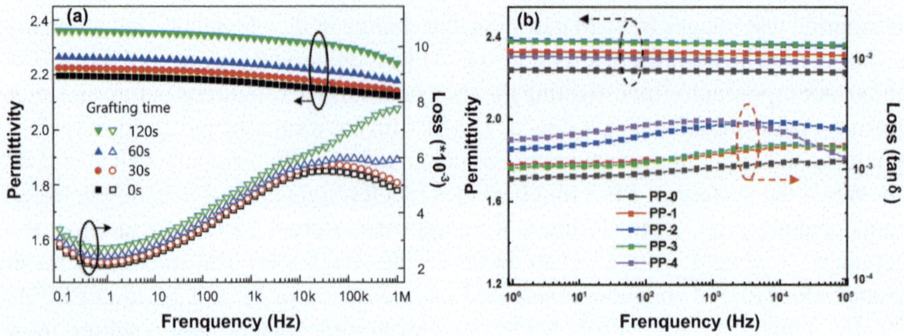

Fig. 12.4 Relationship between the permittivity, dielectric loss and the frequency with **a** hydroxyl groups and **b** maleic anhydride surface grafted

12.3.2 Effect on the Dielectric Properties

See Fig. 12.4.

In general, the dielectric constant of PP films decreases slightly with the increase of frequency, because the dipole polarization and space charge polarization processes inside the dielectric are hysteresis at high frequencies and cannot continue to contribute to the effective electric dipole moment. At the same time, the dielectric losses accompanied by the decrease in dielectric constant also reach a peak, which is also attributed to the polarization losses due to polarization hysteresis. The dielectric constant of PP film increased slightly by surface grafting, and the dielectric constant of PP film increased gradually with the increase of grafting time. In addition, PP-OH films show a significant increase in dielectric loss in the high-frequency region. Based on the trend of dielectric constant and dielectric loss, it can be concluded that the surface grafting introduces polar groups that increase the dielectric constant of PP film in the form of dipoles, and these dipoles further aggravate the polarization hysteresis and increase the dielectric loss in the high-frequency region. Combined with the characterization results of the PP film in the previous paper, the introduced dipole is a hydroxyl group grafted on the surface of the PP film.

When a voltage is applied across the capacitor, part of the charge migrates from one electrode to the other electrode through the dielectric film under a strong electric field, forming a continuous and stable conductive current. On the one hand, conductive current is also known as leakage current because the charge originally stored on the electrode is continuously lost, resulting in a decrease in the actual stored energy of the capacitor. On the other hand, the continuous migration of conductive current inside the dielectric will cause additional heat generation in the energy storage medium, which will exacerbate the vicious cycle of temperature rise and insulation deterioration in high temperature environments. Therefore, reducing the conductivity of PP film can improve the performance of the capacitor in terms of conserving electrical energy and suppressing temperature rise.

Figure 12.5 shows the conductivity of PP film before and after surface grafting. With the increase of the test temperature, the conductivity of each group of PP films generally increases. When the temperature is increased from 30 to 85 °C, the conductivity of the unmodified PP film increases. At different temperatures, the conductivity of the surface grafted modified film decreases to varying degrees. With the extension of grafting treatment time, the suppression effect of conductivity becomes more and more significant. At 30 °C, the conductivity of PP films treated with surface grafting for 120 s was reduced to 55.1%. The conductivity of PP-E is as low as 4.88×10^{-14} S/m at 120 °C, which is 82.7% lower compared to that of PP (2.83×10^{-13} S/m). The high ash content is the main reason for the sharp increase in the leakage conductivity of PP films at high temperatures. The ash content mainly comes from the catalyst system introduced during the synthesis of the polypropylene binder and the dust impurities introduced during the production of the PP film. The ionic groups provided by these ash impurities provide a conductive pathway for carrier migration inside the PP film, and the dissociation of charge at high temperatures further promotes carrier migration. Therefore, the conductance of PP films before and after grafting increased significantly at high temperatures.

At the same temperature, the conductivity of the surface-grafted PP films decreased to varying degrees. This suggests that the surface groups can play a role in limiting charge migration. It is reasonable to speculate that the inhibition effect of surface grafting on conductance mainly comes from the inhibition of electrode injection charge, and the specific mechanism of this mechanism needs to be discussed in more detail later.

In addition to the dielectric constant, the breakdown field strength of the energy storage medium is also the core parameter that determines its maximum energy storage density. Since the electric field strength is the quadratic term of the formula, it has a greater impact on the energy storage density than the dielectric constant. In addition, the capacitor film breakdown releases a large amount of energy instantaneously, and the metal electrodes near the breakdown point evaporate to form a

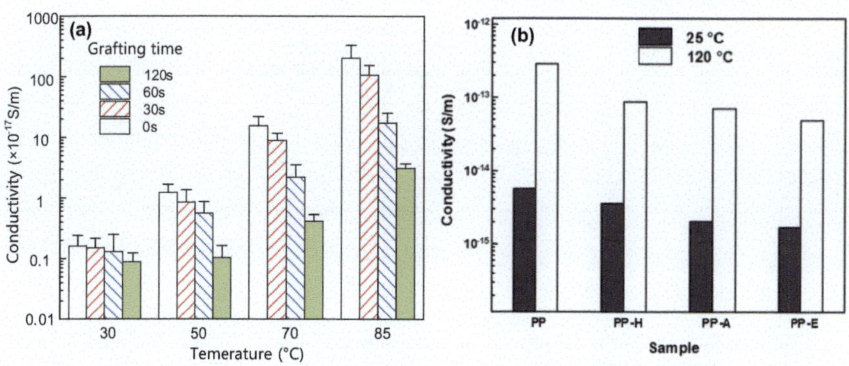

Fig. 12.5 Relationship between the conductivity and the temperature with **a** hydroxyl groups, **b** hydroxyl (PP-H), amino (PP-A) and epoxy (PP-E) groups surface grafted

self-healing zone to cut off the continuous short-circuit process. But this process is also accompanied by a loss of active electrode area and a decrease in capacitance. Therefore, in addition to increasing the maximum energy density, increasing the breakdown strength of PP film at high temperature can also delay the performance degradation rate of the capacitor at high temperature and prolong the service life of the high-temperature capacitor.

In this section, the effect of surface grafting on the dielectric strength of PP films under a DC electric field is tested. Figure 12.6 shows the breakdown field strength (E_b) and breakdown probability density distribution of PP films before and after grafting at different ambient temperatures. With the increase of temperature, the breakdown field strength of the original PP film decreased from 553.1 kV/mm at 30 °C to 386.4 kV/mm at 85 °C, a decrease of 30.1%. Compared with the original PP, the PP-OH film has better insulation performance at different temperatures. The introduction of MAH groups on the surface can increase substantially the E_b. The E_b changes from 662.8 kV/mm for PP-0 to 750.3 kV/mm for PP-3 at 25 °C, representing a noteworthy 21% enhancement. However, as the irradiation time increases further, the E_b of PP-4 begins to decrease to 715.9 kV/mm.

The energy storage density of PP films decreases rapidly at higher temperature. The drop of energy density forces the users of capacitor to reserve enough margin in selecting capacitors. The performance improvement at higher temperature can alleviate the concern of capacitor temperature rise in a general operating environment. It can be found that the U_e of the modified films has increased, and the PP-3 attains the peak value of 5.91 J/cm^3, which is 36.2% higher than that of PP-0 (4.34 J/cm^3) at 25 °C. The cause for the increase can be attributed to the improvement of the breakdown strength and permittivity. However, when the temperature was raised to 120 °C, a decline in energy density was observed. For example, that of PP-0 dropped from 4.34 to 2.49 J/cm^3. This reduction mainly comes from the diminished breakdown strength. Fortunately, the theoretical maximum U_e of the modified films still maintained a high value, reaching 4.02 J/cm^3 (1.61 times the PP-0). It is feasible

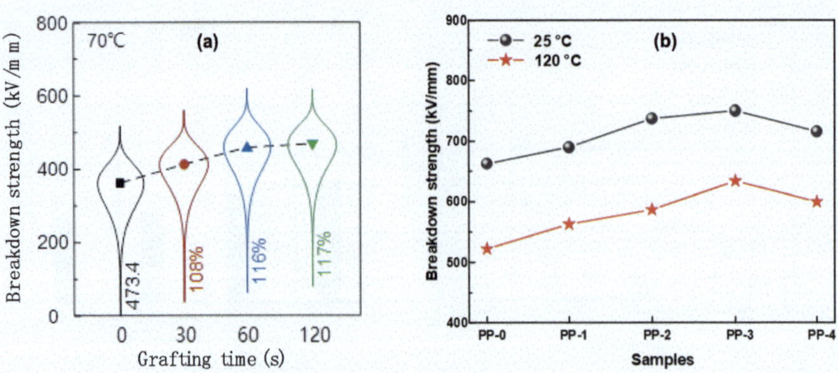

Fig. 12.6 Relationship between the breakdown strength and the temperature with **a** hydroxyl groups and **b** maleic anhydride surface grafted

to tune the surface structure of polymer by grafting MAH group to increase energy storage performance. As shown in Fig. 12.7a, as the electric field rises, the energy density increases and the efficiency decreases. The highest values of U_e of 5.91 J/cm^3 and η of 88.36% are achieved at 750 kV/mm for the PP-E film 25 °C, compared with U_e of 4.10 J/cm^3 and η of 82.25% at 650 kV/mm for the pristine PP film, which can be attributed to the higher ε_r and E_b. Excitingly, the efficiency of the modified films is higher than that of pristine PP. Higher efficiency means lower losses and less heat production, which is conducive to long-term safe and stable operation.

Since film capacitors need long-term work, heat accumulation and temperature increase will inevitably occur in the process, so it is of great significance to explore the changes in their energy storage performance at high temperatures. Figure 12.7b shows the energy density and charge–discharge efficiency of all sample films at high temperatures. PP-E film still has the highest U_e of 4.50 J/cm^3 at an electric field of 650 kV/mm, which is 68.5% higher than pristine PP (2.67 J/cm^3 at 650 kV/mm). The modified film also exhibits a high efficiency value. The η of the PP-E film is 73.0% at 120 °C, while the η of the PP film is only 71.4% under the same external electric field of 550 kV/mm. Hence, the PP-E film not only has improved energy storage density, but also maintains high charge–discharge efficiency.

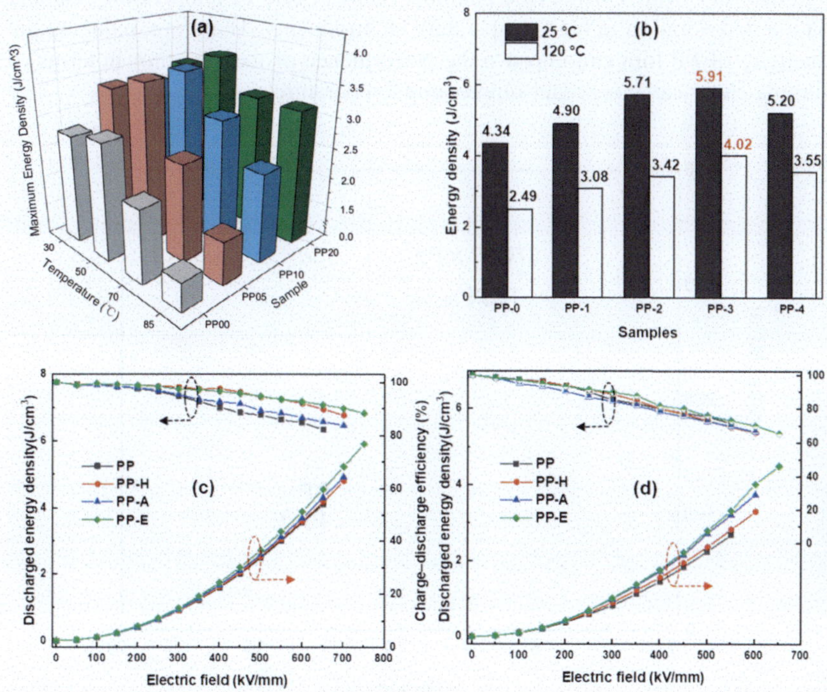

Fig. 12.7 Relationship between the energy density and the temperature with **a** hydroxyl groups and **b** maleic anhydride, hydroxyl (PP-H), amino (PP-A) and epoxy (PP-E) groups surface grafted

At present, most of the research tends to improve the energy storage density, while ignoring the high insulation performance, and the real sense of high energy storage film should be considered from the comprehensive performance. As shown in Fig. 12.8, it can be seen that the maximum stress and strain of PP film is 370.67 MPa and 16.25%, while the PP-E film rises to 359.80 MPa and 16.07%. The reduce in mechanical properties can be attributed to the network structure formed by the hydrolysis of silane coupling agents. After modification, the films still maintain excellent process-ability, and has great research value in practical application. It should be emphasized that since the surface functionalization method used in this article is processed at the molecular level, there is no need to worry about the occurrence of interfacial detachment and other conditions that affect the performance when the film is wound to prepare the capacitor.

When a voltage is applied across the capacitor, part of the charge migrates from one electrode to the other electrode through the dielectric film under a strong electric field, forming a continuous and stable conductive current. On the one hand, conductive current is also known as leakage current because the charge originally stored on the electrode is continuously lost, resulting in a decrease in the actual stored energy of the capacitor. On the other hand, the continuous migration of conductive current inside the dielectric will cause additional heat generation in the energy storage medium, which will exacerbate the vicious cycle of temperature rise and insulation deterioration in high temperature environments. Therefore, reducing the conductivity of PP film can improve the performance of the capacitor in terms of conserving electrical energy and suppressing temperature rise.

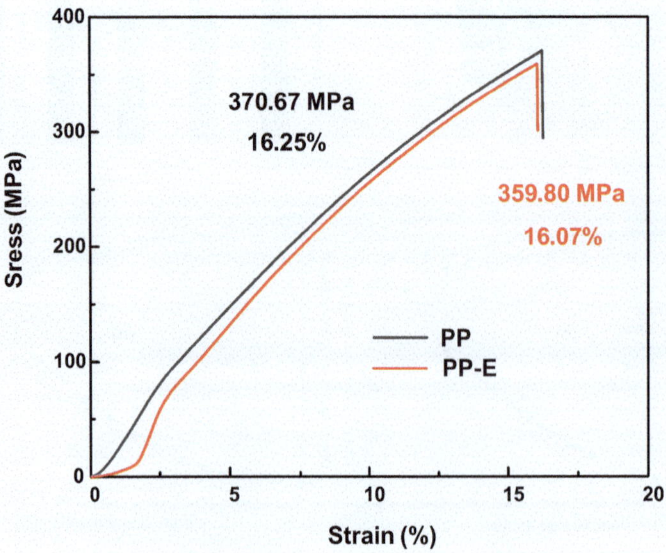

Fig. 12.8 Relationship between the stress and the strain

In addition to the dielectric constant, the breakdown field strength of the energy storage medium is also the core parameter that determines its maximum energy storage density. Since the electric field strength is the quadratic term of the formula, it has a greater impact on the energy storage density than the dielectric constant. In addition, the capacitor film breakdown releases a large amount of energy instantaneously, and the metal electrodes near the breakdown point evaporate to form a self-healing zone to cut off the continuous short-circuit process. But this process is also accompanied by a loss of active electrode area and a decrease in capacitance. Therefore, in addition to increasing the maximum energy storage density, increasing the breakdown field strength of PP film can also delay the performance degradation rate of the capacitor at high temperature and prolong the service life of the capacitor.

Comparing the increase of surface grafting treatment on breakdown field strength at different temperatures, it can be found that the performance advantages of surface grafted films are mainly reflected in high temperatures. According to the Schottky injection formula, the electrode injection current density (j) can be expressed as:

$$j(0,t) = AT^2 \exp\left(-\frac{ew}{kT}\right) \exp\left(\frac{e}{kT}\sqrt{\frac{eE(0,t)}{4\pi\varepsilon}}\right) \tag{12.1}$$

where A is the Richardson constant, T is the ambient temperature, e is the charge charge, w is the interface barrier height, k is the Boltzmann constant, E is the interface electric field strength, and ε is the dielectric constant of the dielectric. As the temperature increases, the charge injection rate increases significantly. At high temperatures, a large amount of charge accumulates at the defect of the film, causing partial discharge. At the same time, the high-temperature high-field reflux accelerated the impact of the molecular chain, resulting in the breaking of the molecular chain of the film and the deterioration of the material. The combination of these two factors causes the breakdown field strength of the film to decrease rapidly at high temperatures. The surface grafting treatment can inhibit the interfacial injection of charge, thereby reducing the deterioration rate of the film, so the breakdown field strength of the modified film is significantly improved at room temperature and high temperature, as shown in Fig. 12.9. Due to the decrease in conductivity and the increase in breakdown strength of the modified PP film, its energy storage performance has been improved, and the energy storage density and charge–discharge efficiency have been improved.

The experimental results show that surface grafting can effectively inhibit interfacial charge injection, reduce the leakage conductivity of PP films, and improve the breakdown field strength of the films. However, the influence mechanism of surface grafting on the interfacial charge injection process needs to be further analyzed.

Taking unmodified polypropylene (PP) and hydroxyl grafted polypropylene (PP-OH) as examples, quantum chemical calculations are used to analyze their molecular energy levels and distribution characteristics, and the simulation model is shown in Fig. 12.9. In polymer molecules, because different protons have different ability to attract electrons, the positive and negative charges may be unevenly distributed

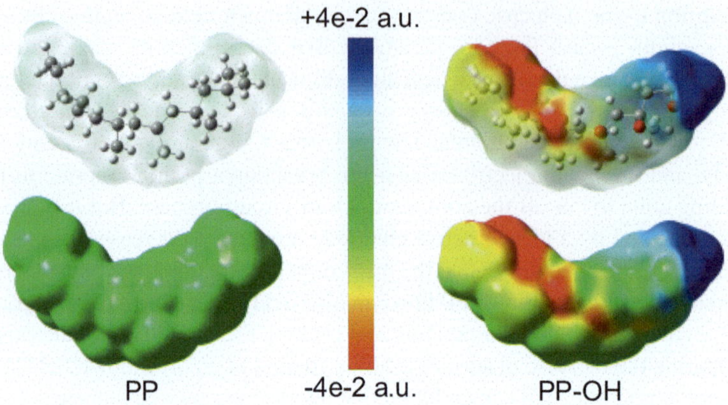

Fig. 12.9 Three-dimensional distribution of electrostatic potential of PP and PP-OH

at the molecular scale, resulting in a three-dimensional distribution of electrostatic potential on the molecular chain. The electrostatic potential of the molecule acts as a coulomb trapping center for positive and negative charges, which can help trap the charge and play a role in limiting carrier migration.

Figure 12.10a shows the electronic density of states of PP and PP-OH molecules. The highest occupied molecular orbital (HOMO) of PP molecule was -7.46 eV, the lowest unoccupied molecular orbital (LUMO) was 2.23 eV, and its band gap was 9.69 eV. After grafting, PP-OH introduced a higher valence band energy level (HOMO, − 5.72 eV) and a lower conduction band energy level (LUMO, 0.59 eV) into the PP band gap. These PP-OH orbitals have lower energy than PP molecular orbitals and are easier to trap moving carriers as charge traps. Figure 12.10b shows the three-dimensional spatial distribution of PP and PP-OH molecules LUMO and HOMO molecular orbitals and their adjacent molecular orbitals. It can be found that the molecular orbitals of PP are uniformly distributed throughout the molecular chain, which will lead to the formation of continuous migration of carriers on the PP molecular chain. However, the orbitals of PP-OH molecules are only distributed in the local positions of the molecular chain, and the charges at this energy level cannot move and accelerate freely along the PP-OH molecules. On the PP-OH molecule, the charge must first absorb enough energy from the outside world to jump to a higher molecular orbital to complete the movement, and the energy of these energy levels is the same as the LUMO or HOMO energy of the PP molecule. That is, the molecular orbital energy introduced by PP-OH is lower and dismittently distributed, which can effectively capture the charge and limit the charge movement.

The schematic diagram of the charge injection behavior before and after grafting MAH is shown in the Fig. 12.11a. Charge is injected from the electrode and aggregates in dielectric, which creates a high local electric field. As a result, the film has deteriorated insulation, reducing the breakdown strength. After grafting MAH, the injected charge density in the PP film decreases. The presence of MAH effectively

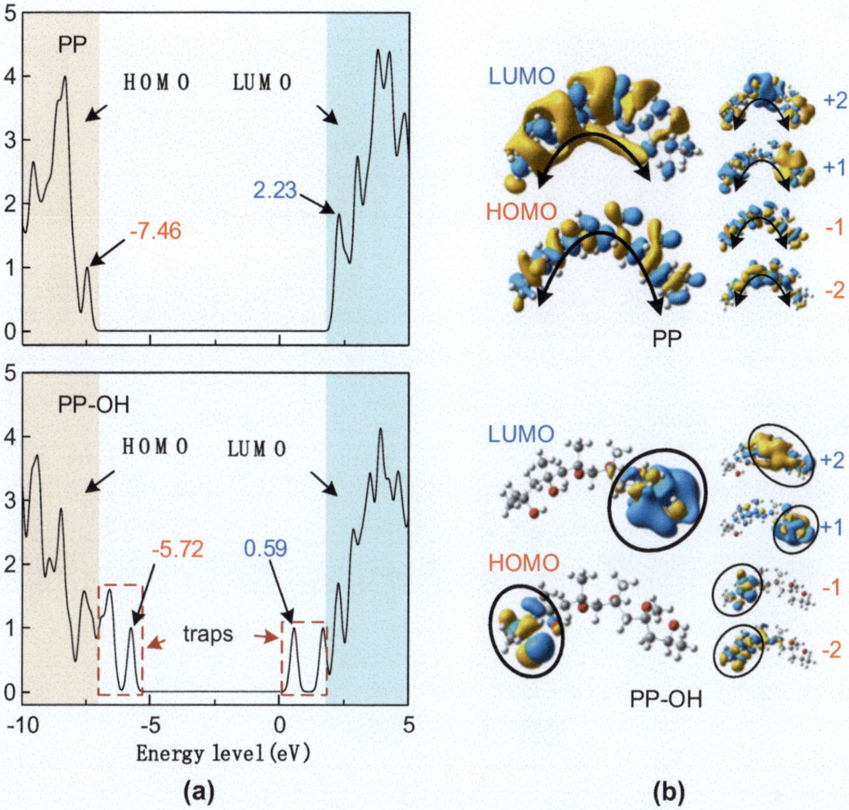

Fig. 12.10 Three-dimensional distribution of **a** electronic density states and **b** molecular orbitals of PP and PP-OH

confines carriers from permeating into the film, thereby elevating the electric breakdown strength. The trapped carriers have the capability to generate a reverse electric field through the Coulomb repulsion effect, thereby attenuating the external electric field. Consequently, this reduces the surface barrier height and impedes subsequent charge injection. Furthermore, the traps introduced via grafting reactions manifest as chemical traps, remaining stable even with increasing temperatures.

Finite element simulations were employed to conduct an in-depth analysis of the impact of electric field distribution on breakdown strength both prior to and following the grafting of maleic anhydride (MAH). The electric field distribution is given in Fig. 12.11b, where an external electric field of 200 kV/mm is applied. It can be observed that the maximum local electric field before grafting is 223.0 kV/mm, which is 271.0% higher than the minimum electric field of 60.1 kV/mm.

This great difference in the local electric field value might serve as a critical factor for the breakdown. As a consequence of the constraints associated with carrier injection, an excess of charge accumulates within the MAH layer rather than penetrating

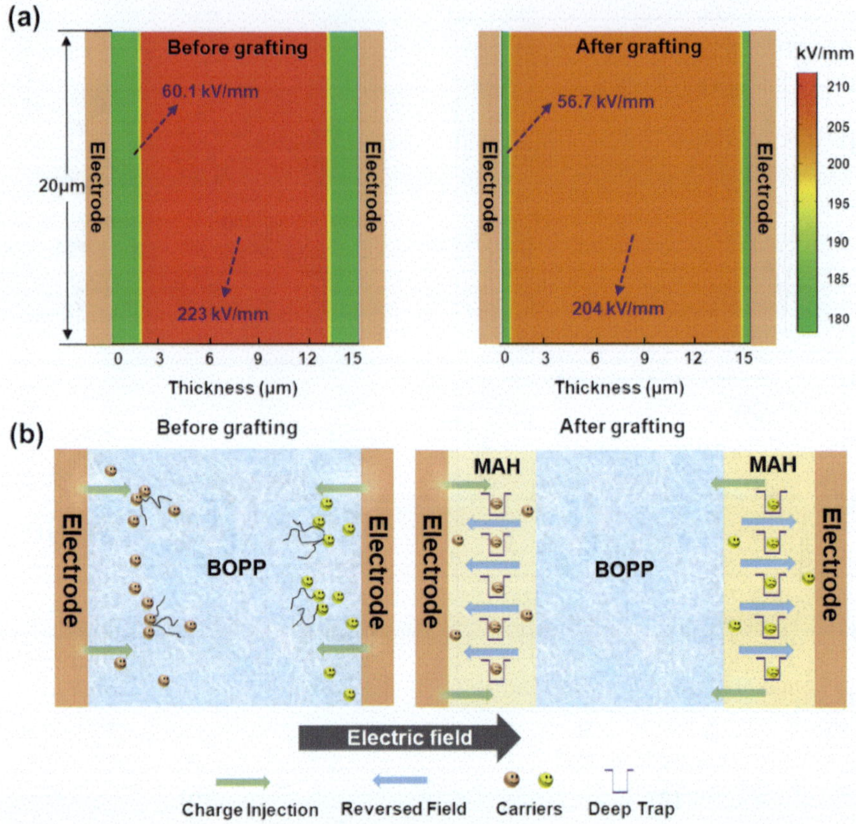

Fig. 12.11 Schematic diagram of **a** the charge injection and **b** the electric field distribution

into the dielectric material. A stronger reverse electric field is formed at the interface, which reduces the electric field strength at the modified film interface to 3.4 kV/mm. At the same time, the charge entering the dielectric is reduced, so that the electric field strength inside the dielectric is also decreased consequently. Eventually, the PP film after MAH grafting has the maximum and minimum electric field values of 204 kV/mm, 56.7 kV/mm, respectively. The distortion electric field strength is reduced and the breakdown strength is increased.

As shown in Fig. 12.12, the peaks on the left and right represent shallow traps and deep traps, respectively. Compared with the original PP, the density and energy level of the deep trap of the modified film have been improved, and it can be used as a trap trapping layer. PP-E films have the highest deep trap energy level and the largest deep trap density. For example, the deep trap density increased from 0.502 $\times$ 10^{21} eV^{-1} m^{-3} of PP to 1.277 $\times$ 10^{21} eV^{-1} m^{-3} of PP-E, the deep trap energy level increased from 0.899 eV of PP-H to 0.915 eV of PP-E. The deep trap layer captures more carriers, while also helping to inhibit carrier shedding and migration in the dielectric. In addition, carriers may generate an internal electric field opposite

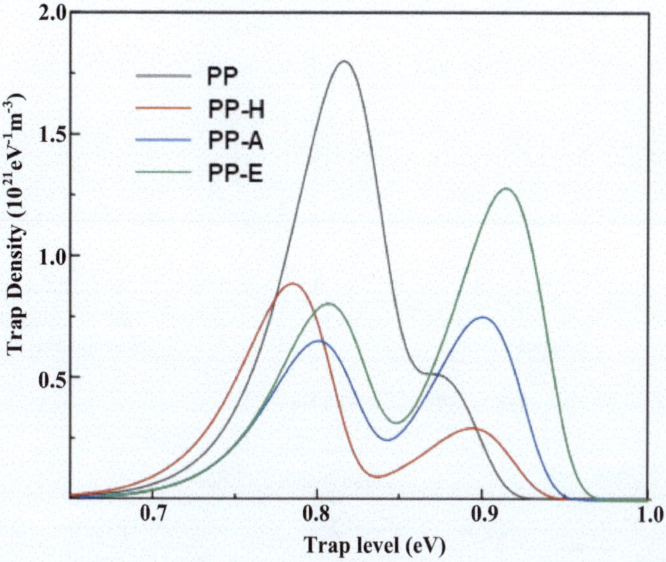

Fig. 12.12 Relationship between the trap density and the trap level

to the electric field after being trapped on the surface, raising the charge injection barrier and enhancing the electric field required for charge injection.

Furthermore, the effect of surface grafting on carrier migration behavior was further analyzed by the photoluminescence (PL) test as depicted in Fig. 12.13. PP-3 with the best insulation performance is chosen for comparison against PP-0. It becomes apparent that the PL strength of PP-0 is lower, on the contrary, PP-3 has the higher PL strength. Higher PL intensity shows weaker PL quenching, indicating lower electron mobility. This demonstrates that the film after grafting MAH can restrict the movement of carriers and enhance trap-assistant recombination to reduce electron mobility. PP-E has the highest PL intensity, indicating that it has the lowest electron mobility. The electron mobility and breakdown strength of the sample have the same trend.

It can be seen that the positive and negative electrostatic potentials of polypropylene are significantly increased after the introduction of hydroxyl groups, and the high electrostatic potential is mainly distributed near the hydroxyl group, which is mainly due to the stronger electronegativity of oxygen atoms than carbon atoms. After surface grafting, the injected charge from the electrode is trapped in the region of high electrostatic potential, making it more difficult to inject into the dielectric, resulting in charge surface trapping. Charge injection is thus suppressed.

Before surface grafting, because there is no interfacial layer limitation, the charge is injected into the PP film and migrates within the dielectric. The injected charge accumulates at the internal defects of the dielectric, causing distortion of the local electric field, which in turn causes partial discharge or even insulation breakdown. After surface grafting, the charge is first injected into the graft layer and trapped by

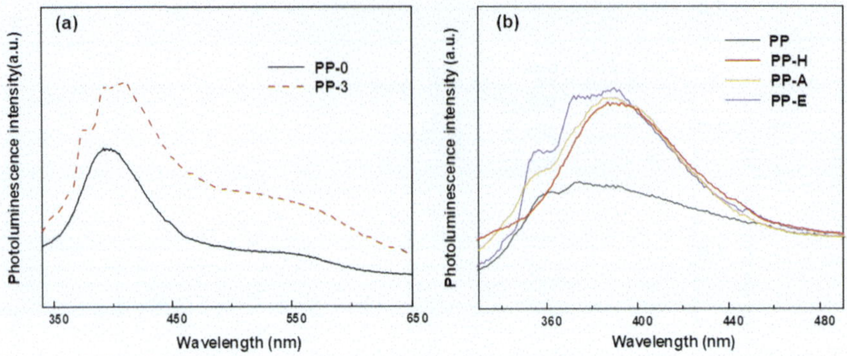

Fig. 12.13 Photoluminescence spectra of the samples

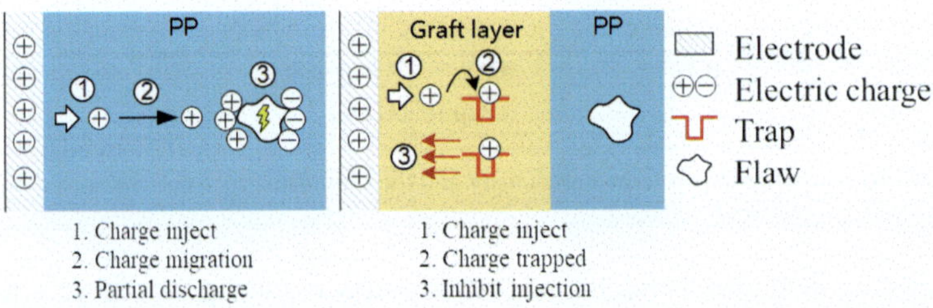

Fig. 12.14 Charge injection migration process of PP film before and after surface grafting

the electrostatic potential and the surface trap formed by the molecular orbitals, and is confined to the electrode/dielectric interface region. These trapped interfacial charges form a localized reverse electric field, which reduces the electric field strength at the interface, and the decrease in the electric field strength greatly reduces the charge injection rate. Therefore, the surface grafting layer inhibits charge injection, and the amount of charge accumulated in the internal defects of PP is reduced compared with that of ungrafted film, and the insulation deterioration rate of the film is reduced and the insulation performance is improved. The effect of surface grafting on the charge injection process is shown in Fig. 12.14.

12.4 Summary

In order to improve the performance of polypropylene capacitor film under high temperature and strong electric field, this section proposes a surface grafting method to optimize the interface structure and bulk structure of polypropylene capacitor film, so as to improve the insulation performance of polypropylene film. Hydroxyl groups grafted on the surface of polypropylene films can introduce charge deep traps on the surface of polypropylene films to capture the electrode injection charge. Copolar charges accumulate at the electrode/dielectric interface, weakening the interfacial electric field and inhibiting subsequent charge injection. After surface modification, the amount of charge accumulated in the internal defects of the polypropylene film is reduced, and the impact of high-energy carriers on the polypropylene molecular chain is weakened, and the insulation performance of the polypropylene film is improved.

References

1. M. Xiao, Y. S. Zhao, Y. N Song and B. X. Du, "Polypropylene films for HVDC capacitors with enhanced dielectric performance by irradiation crosslinking modification," *IEEE Trans. Dielect Elect. Insul.*, vol. 31, no. 2, pp. 666–673, 2024.
2. M. Xiao, Y. N. Song, H. L. Liu, M. D. Zhang, B. X. Du and Z. L. Xing, "Dielectric properties of cross-linked polypropylene by Gamma-ray irradiation for film capacitor," *IEEE Trans. Dielect Elect. Insul.*, vol. 31, no. 2, pp. 625–632, 2024.
3. M. Ritamki et al, "Performance metrics for a modern PP capacitor film," *IEEE Trans. Dielect Elect. Insul.*, vol. 26, no. 4, pp. 1229–1237, 2019.
4. M. Xiao et al, "Breakdown strength improvement of polypropylene by introducing long-chain branched structures for film capacitors," *J. Phys. D: Appl. Phys.*, vol. 55, no. 24, pp. 5501–5508, 2022.
5. M. Xiao, K. X. Wang, K. Chen and B. X. Du, "Effect of photoinitiator on dielectric properties of polypropylene films modified by UV irradiation," *IEEE Trans. Dielect Elect. Insul.*, vol. 31, no. 1, pp. 168–175, 2024.
6. Z. P. Li et al, "High dielectric constant polycarbonate/nylon multilayer films capacitors with self-healing capability," *ACS Appl. Polym. Mater.*, vol. 1, no. 4, pp. 867–875, 2019.
7. J. Ho and T. R. Jow, "High field conduction in biaxially oriented polypropylene at elevated temperature," *IEEE Trans. Dielect Elect. Insul.*, vol. 19, no. 3, pp. 990–995, 2012.
8. F. Zheng, Y. Miao, J. Dong, Z. An, Q. Lei and Y. Zhang, "Space charge characterization in biaxially oriented polypropylene films," *IEEE Trans. Dielect Elect. Insul.*, vol. 23, no. 5, pp. 3102–3107, 2016.
9. F. C. Chiu, "A review on conduction mechanisms in dielectric films," *Adv. Mater. Sci. Eng.*, vol. 2014, no. 7, pp. 1–18, 2014.
10. H. L. Liu, B. X. Du, M. Xiao and T. Tanaka, "High-temperature performance of dielectric breakdown in BOPP capacitor film based on surface grafting," *IEEE Trans. Dielect Elect. Insul.*, vol. 28, no. 4, pp. 1264–1272, 2021.
11. B. X. Du et al, "Improved carrier mobility dependent surface charge am flashover voltage of polypropylene film under DC and pulse voltages," *IEEE Trans. Dielect Elect. Insul.*, vol. 25, no. 3, pp. 1014–1021, 2018.
12. H. L. Liu, B. X. Du and M. Xiao, "Improved energy density and charge discharge efficiency of polypropylene capacitor film based on surface grafting," *IEEE Trans. Dielect Elect. Insul.*, vol. 28, no. 5, pp. 1539–1546, 2021.

Chapter 13
Parylene Doping

13.1 Introduction

Chapter 11 introduces a surface grafting method of polypropylene film, which could effectively suppress the charge injection from the electrode so as to improve the insulation performance of polypropylene film under different voltage conditions [1, 2]. However, there are some limitations in this method.

Through the structural characterization of the modified film, it can be found that the depth of interface modification is limited, and the total amount of interface traps introduced is small, which leads to the unsatisfactory effect of inhibiting charge injection. The marginal benefit of increasing the thickness of the modified layer by prolonging the UV irradiation treatment time is gradually reduced. The further improvement of surface grafting rate and grafting depth is facing challenges.

This method is based on grafting chemical reaction, and the film modification efficiency is low. In order to be applied to large-scale capacitor film production, the processing flow and equipment must be optimized.

Therefore, based on the principle of surface deep trap inhibiting charge injection, this chapter proposes a polypropylene film modification method by parylene doping and the doping method includes surface deposition and surface micro filling [3, 4]. This modification method has the advantages of good interface compatibility and high modification stability by depositing an organic modification layer on the surface of polypropylene film. It can not only realize the rapid treatment of capacitor film, but also accurately control the thickness of interface modification layer, which effectively makes up for the deficiency of surface grafting modification method. Furthermore, the dielectric properties of polypropylene films treated with different doping methods are compared, and the relevant modification mechanisms are also proposed.

B. Du and M. Xiao, *Polypropylene Film for HVDC Capacitors*,
https://doi.org/10.1007/978-981-96-3029-5_13

13.2 Parylene Doping Method

Parylene is a kind of aromatic compounds with extremely low dielectric loss and outstanding high-temperature resistance. The types and basic properties of the parylene used in this chapter are shown in Table 13.1.

The chemical vapor deposition was first proposed by John M. and Blocher Jr. et al. It is a process technology in which reactants undergo spatial gas-phase chemical reactions under gaseous conditions, directly transforming into solid substances on the surface of the solid substrate, and then forming coatings on the surface of the substrate. Essentially, this process belongs to the atomic category of gaseous mass transfer. The film-forming mechanism of chemical vapor deposition is shown as follows. The preparation process of film with parylene doped involves the cleavage of two connected methyl carbon bonds of parylene at high temperature, resulting in the formation of active para dimethylbenzene monomers. When the decomposition products enter the deposition chamber at room temperature from a high-temperature environment, unstable monomers will aggregate into a film. This polymerization process conforms to the mechanism of free radical chain polymerization. The free radical chain polymerization process is composed of a combination of elementary reactions such as chain initiation, chain growth, chain transfer, and chain termination. The rate and activation energy of these elementary reactions vary greatly [3, 5].

The polypropylene/parylene composite film prepared in the previous section is further hot pressed to obtain multilayer composite films. Stack four layers of polypropylene or polypropylene/parylene films with the same thickness and heat

Table 13.1 Properties of different types of parylene dimer

Category	Abbreviation	Molecular structure	Molecular formula	Dielectric constant	Dielectric loss
Parylene N Dimer	PN	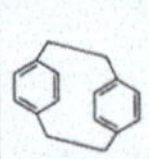	$C_{16}H_{16}$	2.6	0.0002
Parylene C Dimer	PC	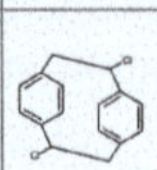	$C_{16}H_{14}Cl_2$	3.1	0.019
Parylene D Dimer	PD	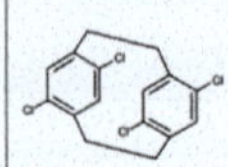	$C_{16}H_{12}Cl_4$	2.8	0.003
Parylene F Dimer	PF	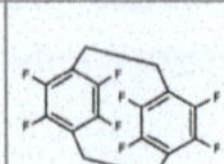	$C_{16}H_8F_8$	2.1	0.002
Parylene HT Dimer	PHT	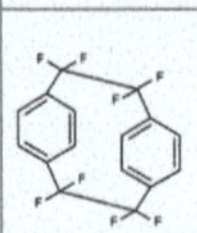	$C_{16}H_8F_8$	2.2	0.002

press them into one layer under the same conditions. According to the relative position of the films, the four-layer films could be divided into two outer films and two inner films. The outer layer is in direct contact with the electrode, so PN doping is introduced to capture charges. The inner layer is located in the middle of the composite film, so pure PP film is used to suppress charge excitation [4, 6].

13.3 Surface Deposition

13.3.1 Effect on the Microstructure

The cross-sectional structure of polypropylene/parylene composite films with different deposition layer thicknesses was observed by the scanning electron microscope (SEM), as shown in Fig. 13.1a. It can be seen that the film is obviously divided into two layers, indicating that chemical vapor deposition could form parylene deposition layers with different thicknesses on the surface of polypropylene film. The effectiveness of the chemical vapor deposition method and the thickness of the deposited layer could be further verified by using EDS (Energy-dispersive X-ray spectroscopy) to scan the intensity of the F atom characteristic peak along the thickness direction of the composite film, which characterizes the thickness of surface CVD layer, as shown in Fig. 13.1b. The thickness of the PP layer remains unchanged because the chemical reaction of bond breaking and reconnection occurs before the combination of PF and PP. In particular, the volume fraction of parylene in the polypropylene film in the four groups of samples is about 0, 5.6, 14 and 29 vol%, respectively, while the volume fraction of PF layer in the composite film is about 0, 5.4, 12 and 23 vol%.

13.3.2 Effect on the Dielectric Properties

(1) **Dielectric Constant and Dielectric Loss**

Figure 13.2 shows the dielectric constant and dielectric loss characteristics of surface deposited polypropylene film at different frequencies. With the increase of the deposition thickness, the dielectric constant of the composite film increases from 2.2 to 2.65. The dielectric constant generally shows an accelerating downward trend with the increase of frequency. Meanwhile, the dielectric loss of the composite film also increases with the increase of the deposition layer thickness and the peak position of different films are also different. The films with 0 and 5 vol% PF volume fraction have a dielectric loss characteristic peak. However, two loss characteristic peaks could be observed in the films deposited at 15 and 25 vol%, and the corresponding frequency of the dielectric loss peak gradually decreases with the increase of the deposition thickness.

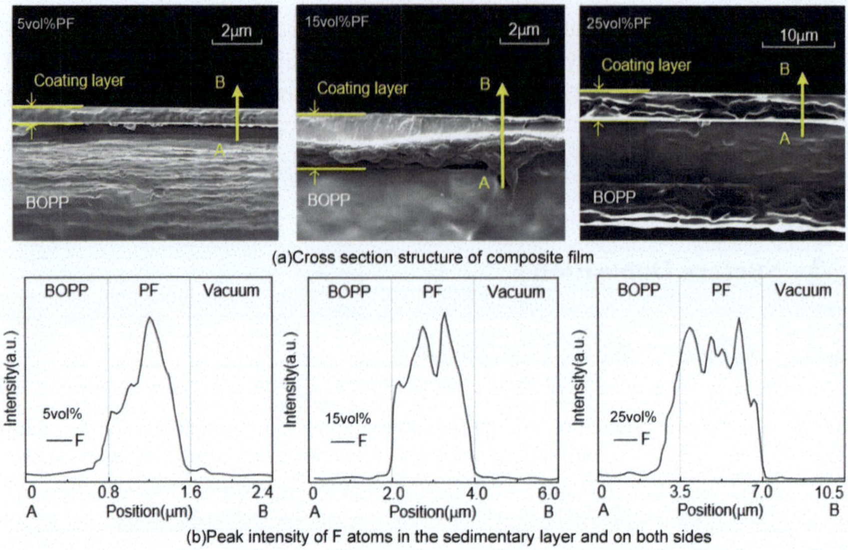

Fig. 13.1 **a** Cross section structure and **b** peak intensity of F atoms of surface deposited film [3]. Copyright Clearance Center's RightsLink© License Number: 5812530221446

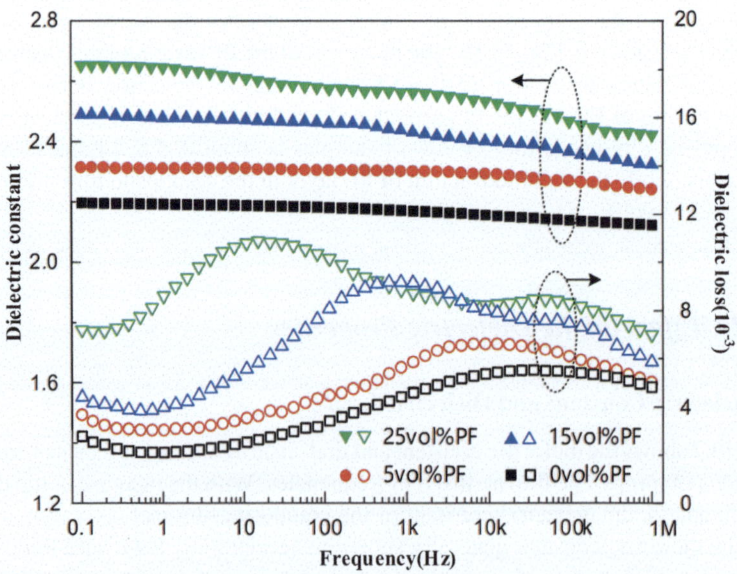

Fig. 13.2 Relationship between the dielectric constant, dissipation factor and frequency [3]. Copyright Clearance Center's RightsLink© License Number: 5812530221446

According to the theory of dielectric loss under alternating electric field, the peak value of dielectric loss corresponds to the frequency and polarization form. Therefore, it is inferred that the dielectric loss peak at a frequency of 100 kHz is attributed to the turning hysteresis of molecular chains and aggregated state level electric dipoles. The ideal polypropylene and PF molecular structures have high symmetry, so this loss may mainly come from the irregular arrangement of molecular chains caused by ash impurities, resulting in the formation of local electric dipole moments [7–9].

Furthermore, it could be clearly observed that a lower frequency loss characteristic peak shifts towards the lower frequency range as the deposition thickness increases. This is due to the interface polarization introduced by the BOPP/PF multilayer structure. As the thickness of the sedimentary layer increases, the accumulation of interface charges gradually increases. Therefore, the extreme loss caused by interface polarization gradually increases, and its characteristic frequency gradually decreases.

(2) **Conductivity**

The leakage conductivity of the surface deposited thin films at different temperatures is shown in Fig. 13.3. As the temperature increases, the conductivity of PP film increases by three orders of magnitude, from 5.10×10^{-15} S/m increased to 2.33×10^{-12} S/m. After depositing a 5 vol% thick PF interface layer, the conductivity of the film was effectively suppressed at all temperatures, decreasing to 22.9%, 17.8%, 7.57%, and 3.62% at 30, 50, 70, and 85 °C, respectively. After depositing a 25 vol% thick PF layer, the conductivity of the film increased by 31–1300%. When the temperature is below 50 °C, a 15% thick PF interface layer can also play a certain role in reducing conductivity. However, when the temperature is between 70 and 85 °C, the conductivity of the composite film does not decrease but instead increases.

The above phenomenon indicates that depositing a certain thickness of PF modified layer on the surface could indeed reduce the leakage conductivity of the thin film. However, it can also be observed that the conductivity of the composite film does not monotonically decrease with the increase of PF deposition layer thickness. When the surface deposition thickness exceeds 15 vol%, the structural modification of the composite film may have a negative optimization effect. This indicates that the leakage conductivity of thin films is also related to other carrier transport processes in addition to the mechanism of surface traps suppressing charge injection by capturing carriers as discussed earlier [10]. The relevant mechanisms will be discussed in detail later.

(3) **Breakdown Strength**

The breakdown strengths of the surface deposited thin films at different temperatures are shown in Fig. 13.4. As the temperature increases from 30 to 85 °C, the breakdown field strength of the original BOPP film decreases from 553.7 to 397.3 kV/mm. According to the Schottky injection formula, the intensification of electrode charge injection at high temperatures increases the rate of charge accumulation at internal defects in the film, and more charge carriers impact the molecular chains, which

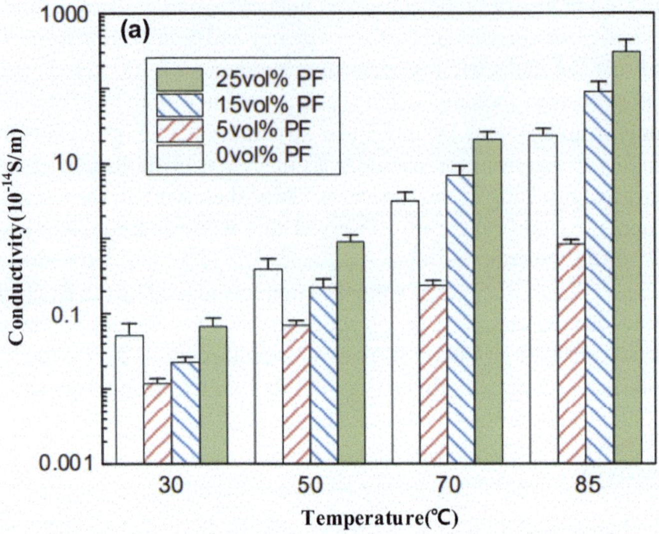

Fig. 13.3 Relationship between the conductivity and the temperature [3]. Copyright Clearance Center's RightsLink© License Number: 5812530221446

is one of the main reasons for the significant decrease in the insulation strength of the film [11–13]. After depositing a 5 vol% thick PF modified layer on the surface, the interface barrier effect of charge injection is formed. The decrease in carrier concentration inside the film slows down the rate of performance degradation at high temperatures. When the deposition thickness reaches 15%, the breakdown field strength of the composite film is basically equivalent to that of the original BOPP film. When the deposition thickness reaches 25 vol%, the insulation strength of the composite film decreases by 20–30%.

Considering that different dielectric materials have different resistance and capacitance characteristics, a dielectric parallel equivalent circuit could be constructed based on the material's resistance and capacitance parameters, as shown in Fig. 13.5a [14]. For the DC steady-state electric field studied in this chapter, the field strength borne by the material is determined by its resistance value, expressed by the following formula.

$$E_1 d_1 = \frac{R_1}{R_1 + R_2} E(d_1 + d_2) \tag{13.1}$$

where E is the average field strength of the composite film. If it is believed that the material has a uniform electrical resistivity, then the resistance could be measured by electricity resistance, expressed as $R = \rho d/S$. S is the area of the material perpendicular to the thickness direction. Assuming that the opposing areas of the two materials are the same, the formula can be derived as follows.

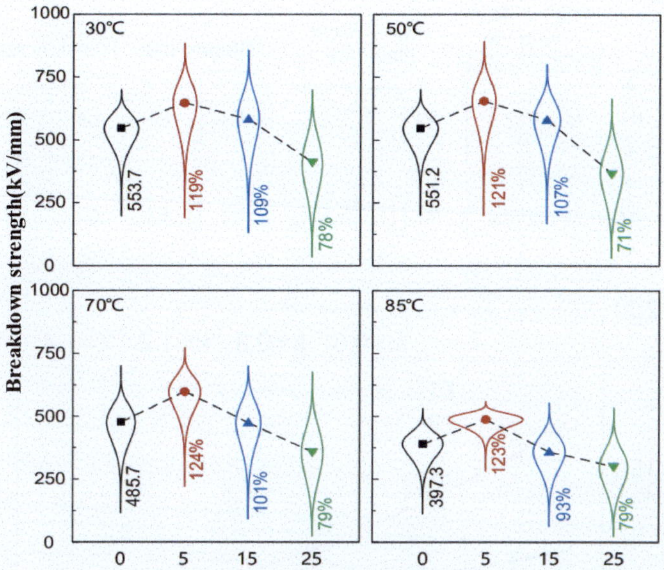

Fig. 13.4 Relationship between the breakdown strength and the temperature [3]. Copyright Clearance Center's RightsLink© License Number: 5812530221446

$$E_1 = \frac{\rho_1(d_1 + d_2)}{\rho_1 d_1 + \rho_2 d_2} E \tag{13.2}$$

Represent the resistivity of material 1 as the resistivity of material 2, i.e. $\rho_2 = k\rho_1$. The formula can be derived as follows.

$$E_1 = \frac{d_1 + d_2}{d_1 + kd_2} E \tag{13.3}$$

For the case of $0 < k < 1$ or $k > 1$, the variation of E_1 and E_2 electric field strength with the proportion of d1 thickness is shown in Fig. 13.5b. Assuming that the material 1 is PF and the material 2 is PP, the resistivity of PP is generally higher than that of aromatic compounds represented by PF $\rho_2 > \rho_1$, that is, $k > 1$. According to the curves of $k = 1.2$ and $k = 1.4$, it can be seen that as the thickness of PF deposition increases, the average electric field strength of both PF and PP layers gradually increases, and the double-layer thin film bears more severe insulation stress. Therefore, when the thickness of the PF deposition layer exceeds 15%, the breakdown field strength of the composite film rapidly decreases.

The above mentioned double-layer dielectric equivalent model discusses the effect of increasing the thickness of the PF deposition layer on the insulation performance of composite films, but the interface molecular structure introduced during the deposition process and the mechanism of the influence of the interface structure of composite films on the insulation performance of the films have not been involved.

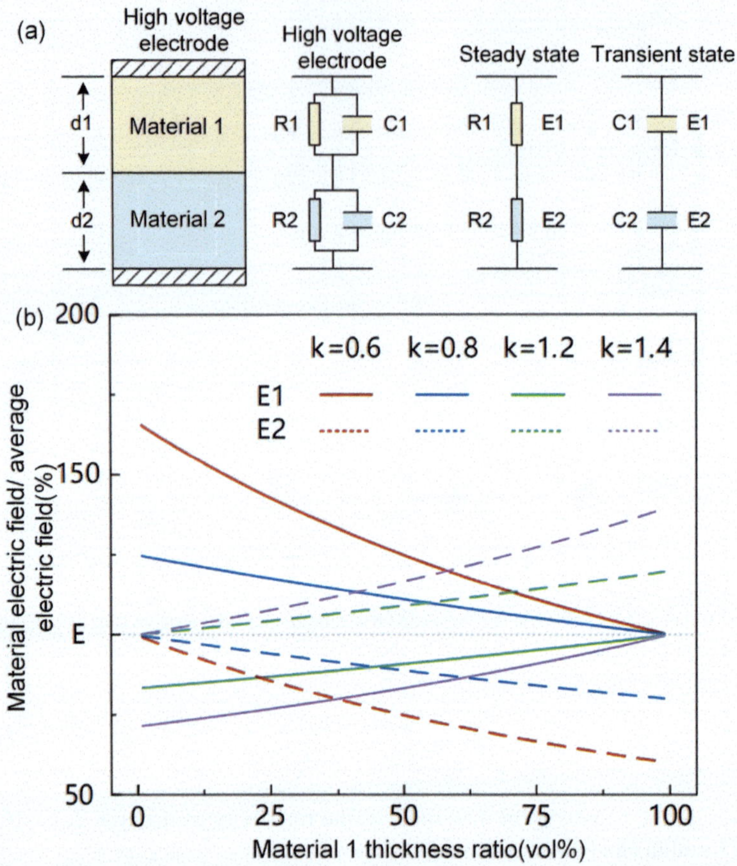

Fig. 13.5 **a** Equivalent circuit for parallel connection of dielectric materials, **b** electric field distribution of composite dielectrics

The energy level distribution characteristics of PF and PP molecules were analyzed and compared using quantum chemical calculation methods. Figure 13.6 show the electronic density of states and typical molecular orbital spatial distribution of PP and PF molecules. The HOMO energy level of PP molecule is − 7.46 eV, and the LUMO energy level is 2.23 eV. The bandgap width of PP molecule could be calculated to be 9.69 eV [15, 16]. The surface deposition of PF molecules introduces a higher valence band energy level (HOMO, − 5.72 eV) and a lower conduction band energy level (LUMO, 0.59 eV) in the PP band gap. Compared with PP molecular orbitals, PF orbitals have lower energy and are easier to capture moving carriers as charge traps [17, 18]. Figure 13.6b shows the three-dimensional spatial distribution of LUMO, HOMO molecular orbitals and their adjacent molecular orbitals (± 1, ± 2) of PP and PF molecules. It can be found that the LUMO and HOMO molecular orbitals of PP are evenly distributed throughout the molecular chain, which will lead to the continuous migration of carriers on the PP molecular chain [19]. However, the LUMO and HOMO orbitals of PF molecules are only distributed in the local position of the molecular chain, and the charge on this energy level cannot move and accelerate freely along the molecular chain. In the PF molecule, the charge must first absorb enough energy from the outside to transition to a higher molecular orbital (> LUMO + 2 or < HOMO-2) to complete the movement, and the energy of these energy levels is the same as the LUMO or HOMO energy of the PP molecule. It can be concluded that the molecular orbital energy introduced by PF is lower and discontinuously distributed, which can effectively capture the charge and limit the charge movement. It can be seen by observing the molecular structure that the two trap molecular orbitals are mainly contributed by the conjugated large π bond of the benzene ring. In addition, the F atom of the side chain of the PF molecule has strong electronegativity and can also play a role in capturing charge.

For the original PP, the charge migrates inside the PP after injection. For the BOPP/PF film, the trap level of the PF first captures the injected charge and limits the carrier movement when the charge is just injected from the electrode. Due to

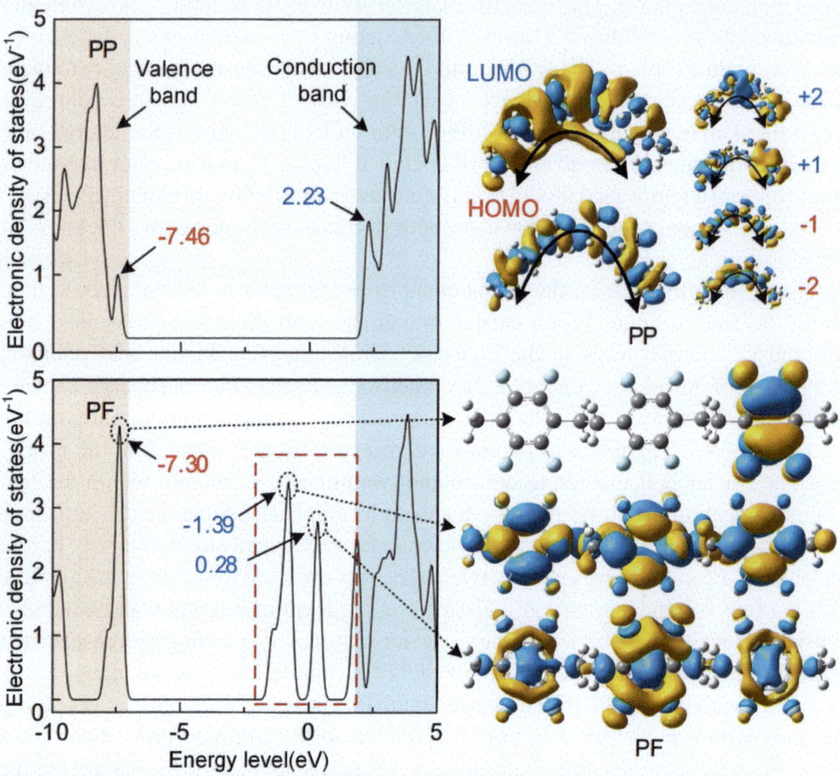

(a)The electronic density of states of PP and PF (b)The molecular orbital of PP and PF

Fig. 13.6 **a** Electronic density and **b** molecular orbital states of PP and PF the effect of the potential barrier at the PF/PP interface, more charges accumulate within the PF layer, thereby inhibiting subsequent charge injection. Therefore, the leakage current of BOPP film was suppressed to a certain extent after surface deposition of PF. Due to the lower internal charge of PP, the discharge generated by its internal defects or weak insulation points is suppressed, and the breakdown field strength of the thin film also increases consequently. Considering the charge capture and injection inhibition effects of PF molecules mentioned above, it is reasonable to predict that the improvement effect of surface deposition of PF layer on insulation performance should be positively correlated with the deposition thickness. However, the experimental results mentioned earlier clearly do not meet this expectation, indicating that a monotonic increase in the thickness of the deposition layer will have other adverse effects on the insulation strength of the thin film.

The proportion of free volume inside PF molecules is 3.6 times that of PP molecules. Therefore, charges within PF materials can accelerate more fully along the electric field to obtain kinetic energy. These high-energy carriers collide with molecular chains, causing PF materials to crack and produce small molecule free radicals. Active free radicals further promote material activation and lead to further degradation of molecular chains. The dielectric material gradually deteriorates and eventually forms insulation breakdown. However, the existing free volume theory faces difficult to explain problems. The free volume scale inside the material is extremely small, and the acceleration distance of charge carriers within the free volume is very limited. When charge carriers collide with molecular chains, their energy may be much lower than the bond energy of chemical bonds. Therefore, relevant studies have proposed an improved model to solve the above problems and explain the experimental phenomenon of breakdown strength decrease with increasing thickness of PF deposition layer [20].

As shown in Fig. 13.7a, the initial charge A accelerates along the electric field within the free volume. When carrier A collides with dielectric molecules, it is captured by internal traps in the dielectric, while releasing kinetic and potential energy. Part of the released energy is dissipated in the form of thermal energy, causing local temperature rise, while the other part radiates to the surrounding area in the form of photons. The released photons are absorbed by the carrier B in the nearby trap. The carrier B is excited to detach and continues to accelerate within the free volume, repeating the above process. It should be emphasized that the carrier B may absorb photons generated when multiple charges are trapped simultaneously before detachment. From an energy perspective, the energy obtained by accelerating a charge within a free volume can not only dissipate in the form of thermal energy, but also transfer and accumulate between charge carriers, ultimately reaching a level sufficient to break chemical bonds. The process of energy accumulation is shown in Fig. 13.7b. Based on the above model, the energy accumulation process of charge carriers during transport within a dielectric is not only related to the intrinsic properties of the dielectric, such as molecular chains and aggregated state structures, but also to the thickness of the dielectric material along the electric field direction. Assuming that the basic characteristics of the dielectric material remain unchanged. By increasing the thickness of the material alone, there is a higher probability that charge carriers will accumulate enough energy to cause damage to the dielectric material, leading to insulation material breakdown at lower electric field strengths.

For the double-layer composite film in this section, the leakage current decreases and the breakdown field strength increases when the PF deposition layer is thin. However, when the PF deposition layer is thick, the insulation performance of the composite film is not as good as that of the pure BOPP film. The transport process of charge carriers in BOPP and BOPP/PF films is shown in Fig. 13.7c. When the thickness of the PF layer is thin, the injected charge cannot accelerate the transport within the PF to obtain enough energy to cross the interface potential barrier of PF/PP.

Therefore, a large amount of charge accumulates in the PF layer, forming a weakening effect on the interface electric field and a suppression effect on subsequent

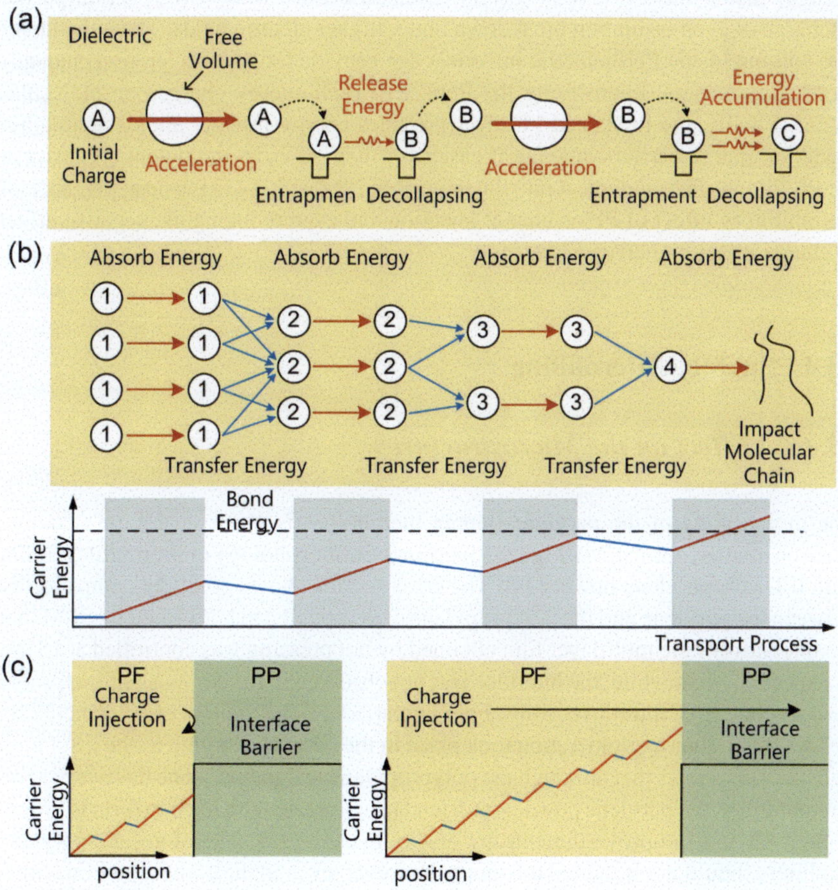

Fig. 13.7 Schematic diagram of **a** carrier acceleration and energy transfer process, **b** energy accumulation process and **c** carrier energy accumulation with PF deposition layer [3]. Copyright Clearance Center's RightsLink© License Number: 5812530221446

charge injection. When the PF layer is thicker, the acceleration distance of charge carriers is longer, and their energy accumulation is more complete. Since the potential barrier at the PF/PP interface remains unchanged, charge carriers are more likely to cross the interface barrier and migrate into the interior of PP. The accumulated charge within the PF layer is greatly reduced, and its inhibitory effect on subsequent injection is correspondingly weakened. Furthermore, due to the lower LUMO energy level of the PF layer, charges are continuously injected into the PF and subsequently into the PP, resulting in a rapid decrease in the resistivity and breakdown field strength of the composite film.

Based on the above analysis, the reasons for the decrease in insulation performance of thin films caused by excessively thick interface deposition layers are as follows. The relative size of the conductivity of PF and PP determines that increasing the proportion of PF layer thickness will simultaneously enhance the electric field strength inside the PF and PP layers. Dielectric thin films will generate partial discharge or even insulation breakdown under higher electric fields. Then, the larger free volume inside PF dielectric materials can provide longer acceleration channels for charge carriers, intensifying the impact of high-energy charges on molecular chains. Finally, the thickened PF layer provides energy transfer and accumulation space for charge carriers, making it easier for them to cross the potential barrier at the PF/PP interface and inject into the interior of PP, resulting in the disappearance of the inhibitory effect of PF on charge injection and even reducing the negative effect of charge injection barrier [20, 21].

13.4 Surface Microfilling

13.4.1 Effect on the Microstructure

The morphology of the parylene doping membranes is observed by the scanning electron microscope (SEM). Figure 13.8 shows the morphology of the multilayer film with 0.1 wt% parylene powder [4]. The cross section shows excellent compatibility between the parylene and the polypropylene dielectric. It can be verified by SEM that the thickness of the multilayer film obtained by hot pressing was controlled at 15 μm as expected. Meanwhile, the interface was not observed in the composite film formed by hot pressing of multilayer films. For comparison, if the content of pn is increased to 5 wt%, obvious agglomeration took place in the outer film acting as the insulation defects. This shows that although the polypropylene and the parylene have very good compatibility, the parylene powders still tend to aggregate with itself to form isotropic groups, so as to improve the entropy of the mixture [22]. Therefore, the parylene modifiers should not be added too much. However, even in the composite film with 5 wt% parylene addition, there is still no interface, which indicates that the hot-pressing method is feasible for manufacturing composite films. In addition, the agglomeration is only observed in the upper 1/4 part. It illustrates that the molecular chains in

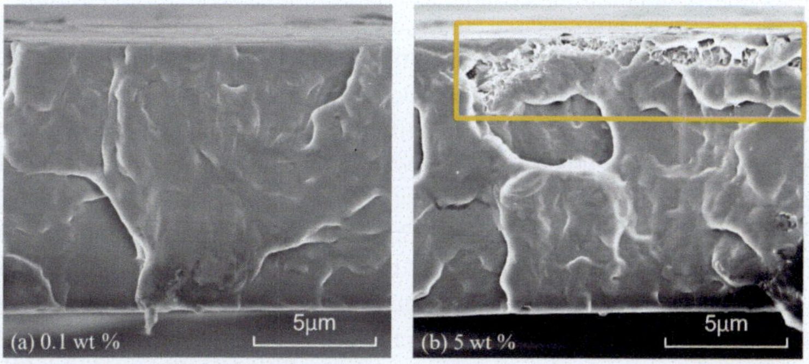

Fig. 13.8 Fractured morphology of the film with surface microfilling [4]. Copyright Clearance Center's RightsLink© License Number: 5811340086314

different layers does not flow and mix due to hot pressing at 190 °C. Following is the analysis. The contents of PN are very small, so that the polypropylene at the interface between the layers is isotropic. The thickness of different layers is basically the equal, and consequently, each layer is subjected to the same longitudinal extrusion force. As a result, it is the uniform transverse diffusion rather than the mixing between layers that occurs in the hot-pressing process. Moreover, in the experiment, it is observed that the fluidity of polypropylene is still poor at 190 °C, and the construction of multilayer composite films was considered to be successful [23].

13.4.2 Effect on the Dielectric Properties

For surface micro filling with parylene, the content of additives applied for modification is extremely small compared with PP substrate, the relative dielectric constant (ε_r) and dielectric loss characteristics is almost identical to that of pure polypropylene film. Other dielectric properties will be further discussed.

(1) **Conductivity**

The surface deposition modification of PP film suppresses charge injection by constructing an interface modification layer, reducing the conductivity of BOPP film and improving its breakdown field strength. However, an excessively thick interface modification layer provides an acceleration channel for the continuous energy accumulation of charge carriers, which could actually lead to a decrease in the insulation performance of the thin film. Therefore, if the material used for interface deposition is doped into the polypropylene matrix at an extremely low content and uniformly dispersed inside the polypropylene without forming a percolation network, it can introduce deep traps to capture charges while cutting off carrier transport channels, thereby suppressing charge transfer and improving the insulation strength of the film. Related studies have obtained PP/PN composite films with different contents through combined mixing and hot pressing. In order to avoid the formation of a conductive percolation network, the content of PN in the composite film is strictly controlled at 0.005, 0.010, 0.050, and 0.100 wt%, respectively.

The conductivity of the micro filling thin films doped with different PN contents is discussed. When the temperature is between 30 and 85 °C, doping PN molecules reduces the conductivity of PP films, indicating that the bulk doping of PN could also suppress the transport of charge carriers. Unlike the mechanism of surface modification in capturing charges at the electrode/dielectric interface mentioned earlier, the charge trap formed by PN doping is located inside the PP film, which reduces the leakage current of the film by suppressing the transport of charge carriers inside the PP film [24, 25]. At the same temperature, the PN doped PP film with a mass fraction of 0.010 wt% exhibits the best insulation performance. Although the composite film has higher insulation resistance compared to pure PP film when the PN doping content is greater than or equal to 0.050 wt%, the optimization effect of the film is gradually weakened under high doping content. Assuming that all PN molecules are uniformly dispersed within the polypropylene matrix as a single dimer unit. When the doping amount of PN exceeds 0.010 wt%, the average molecular distance rapidly decreases from 20 to 7 nm, and the greatly shortened molecular distance forms a conductive network, which may be the reason for the increase in conductivity. However, the average molecular spacing of 7 nm is much higher than the electron tunneling width of 1 nm, and this data is calculated based on the ideal situation where all molecules exist alone. If a certain amount of PN aggregates to form micro clusters, their average distance will further increase. Therefore, the main reasons for the increase in conductivity under high addition levels need to be further investigated.

Figure 13.9a shows the electrical conductivity of PN surface filled and uniformly filled films at different temperatures [26]. Compared with uniform filling, surface filling can further improve the insulation performance of the films. Uniformly filling PN could reduce the conductivity of polypropylene film to 27% at 85 °C. In contrast, surface filling with PN reduces the conductivity to 2%. In addition, the optimal content of PN under uniform filling is 0.010 wt%, while the optimal content of surface filling is 0.050 wt%. The surface filling with higher content of PN introduces more charge traps in the outer layer of the film, suppressing carrier migration and charge injection. However, by uniformly filling 0.050 wt%, too many narrow bandgap molecules were introduced inside the film, which excited more charges under high temperature and strong field conditions, resulting in a decrease in insulation performance.

Compared with the improvement in the insulation performance of PP films by PN filling, the effect of parylene side groups on the insulation performance of PP is relatively small, and the conductivity of each group of films has the same order of magnitude, as shown in Fig. 13.9b. At room temperature, the influence of various types of parylene is similar, and the insulation optimization effect of parylene with F-side groups is most significant at high temperatures, which is 0.6% of pure PP film and 30% of PN filled with the same content. The influence of PN side groups on the insulation performance of thin films will be analyzed in the following text.

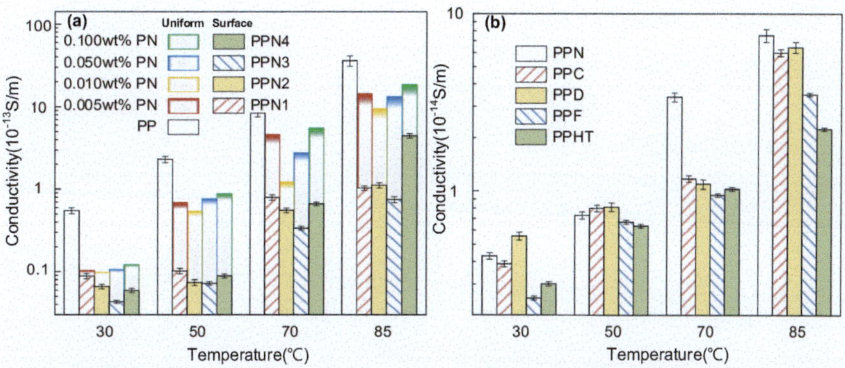

Fig. 13.9 Relationship between the conductivity and the parylene **a** content and **b** type [4]. Copyright Clearance Center's RightsLink© License Number: 5811340086314

(2) Breakdown Strength

The overall trend of the effect of PN doping on breakdown strength is similar to its effect on conductivity. The PP/PN composite film showed better breakdown field strength, and the modified film with 0.010 wt% PN doping had the highest breakdown field strength. However, the optimization effect of breakdown field strength of composite films under high PN doping content rapidly decreases, even slightly lower than that of unmodified PP films. In contrast, the experimental results of conductivity indicate that even PN with a content of 0.010 wt% still has an inhibitory effect on charge migration. It can be seen that the breakdown field strength and conductivity exhibit different variation characteristics, and the PP/PN film without forming a continuous charge migration network shows a trend of decreasing the effect of improving breakdown strength. The relevant mechanisms will be discussed in detail in the following text.

The breakdown strength and probability density distribution of the capacitor film under PN uniform filling and surface filling at different temperatures are shown in Fig. 13.10a. The breakdown field strength of the surface PN filled film and the uniformly filled PN film are basically the same at lower content and lower temperature. However, the electric heating field leads to an increase in the intrinsic excitation charge of PN at high temperatures, and the negative effect generated by the inner layer PN dominates, resulting in the improvement of the insulation performance of the thin film by surface filling compared to uniform filling. When the PN content is high, the breakdown field strength of the surface filled film is generally higher than that of the uniformly filled film. This is also due to the intensification of the excitation of PN molecular charges by the high-temperature strong field.

Compared with the two structures of surface filling and uniform filling, the maximum breakdown strength of the surface filling film is slightly higher than that of the uniform filling film, showing a better overall modification effect. Specifically, the PN filling content that achieves the highest breakdown field strength under the

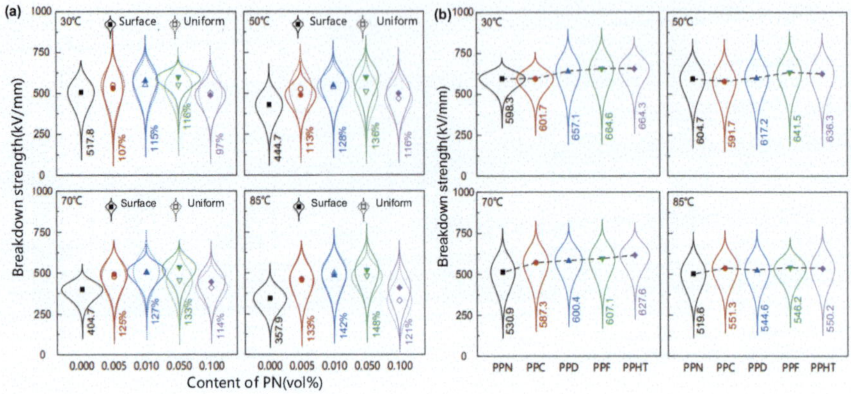

Fig. 13.10 Relationship between the breakdown strength and the parylene **a** content and **b** type [4]. Copyright Clearance Center's RightsLink© License Number: 5811340086314

two structures is different. Uniform filling occurs at 0.010 wt%, while surface filling occurs at 0.050 wt% (only the outer layer content). These two phenomena indicate that the outer PN traps could compensate for the insufficient excitation charge of the inner PN under lower content of uniform doping. However, the problem of excitation of charges by inner layer PN under surface filling no longer exists, and a higher content of outer layer PN promotes charge capturing.

The breakdown strength of micro filled films on the surface of different parylene molecules is shown in Fig. 13.10b. It can be seen that the side groups of parylene could further enhance the breakdown field strength of the film. Due to the fact that the PC, PD, PF, and PHT side groups are halogen elements located in different positions, further enhancement of breakdown may come from the Coulombic attraction of electrons by highly electronegative halogen functional groups. Owing to the stronger electronegativity of F atoms compared to Cl atoms, the performance of PPF and PPHT exhibits slightly better than that of PPC and PPD. However, the influence of PN structure on the insulation performance of thin films almost disappears at 85 °C. Unlike the stable charge trap introduced by molecular orbitals, attracting charges through Coulomb interactions may be overwhelmed by high carrier free energy at high temperatures.

The enhancement mechanism of polypropylene film doped with paralene is firstly investigated, where the PN is selected. This section establishes quantum chemical computational models for the molecular structures of PP and PN. The electronic density of states and the three-dimensional distribution of molecular orbitals are calculated. The calculation results indicate that PN molecules have higher HOMO and lower LUMO energy levels, thereby forming effective charge traps within the PP matrix. From the spatial part of molecular orbitals, it can be seen that the trap energy levels of positive and negative charges are mainly contributed by the benzene ring. The PN molecules doped increase the resistivity and breakdown field strength of the thin film by limiting carrier transport. It can be found from the previous introduction that the conductivity of circular materials is relatively high. However, the conductivity of the composite film made by blending this type of material with polypropylene is actually lower than that of pure polypropylene film, which is related to the content of the circular structure. When the insulation material is composed of multiple materials, its internal band structure is shown in Fig. 13.11a. The different bandgap widths of different molecular structures can create energy differences between molecules, making charges more easily captured by low-energy molecular orbitals. At this point, the narrow bandgap structure in the composite material plays a role in capturing charges. If the filling amount of narrow bandgap materials is strictly limited, that is, the average molecular spacing of narrow bandgap components is controlled to be greater than a specific threshold, it can limit the tunneling strength of charges between traps, thereby limiting charge movement. In addition to the charge trapping effect caused by the energy level difference between molecular orbitals, charges also absorb energy from the external environment and transition to higher orbitals. Obviously, the narrow bandgap reduces the difficulty of electronic transitions. However, due to the potential barrier between the narrow bandgap material and the wide bandgap materials, it is still difficult for the excited charges in the narrow bandgap to form continuous migration.

If composite materials are completely replaced with narrow bandgap molecules, the charge trapping effect caused by energy level differences will be weakened. In addition, the energy required to excite the charge to the conduction band is greatly reduced, resulting in a sharp increase in the material's conductivity, as shown in Fig. 13.11b. Therefore, the conductivity of PI thin films is significantly higher than that of wide bandgap materials such as PP and PE. As the PN content increases, the narrower bandgap will also cause potential insulation problems. According to quantum chemical calculations, the bandgap width of PN molecule (5.42 eV) is 44%

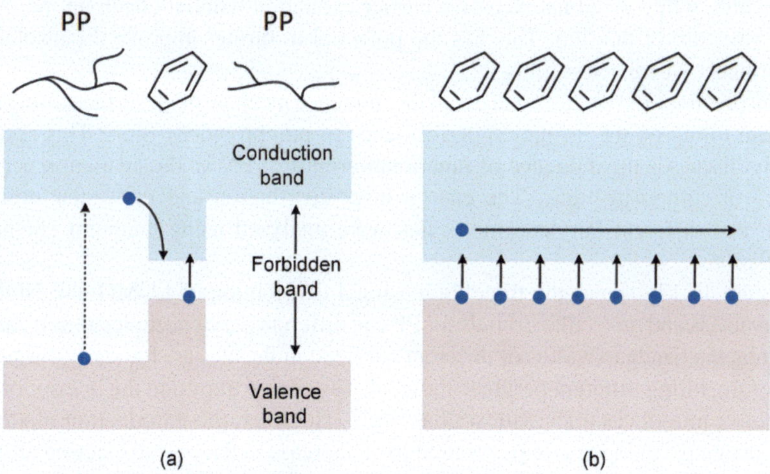

Fig. 13.11 **a** Microdoping of cyclic structured substances, **b** band structure of circular structure dielectric

narrower than that of PP molecule (9.69 eV). Under high temperature and strong field conditions, a narrower bandgap leads to more electrons transitioning from the valence band to the conduction band of PN molecules. The doped PN molecules themselves have evolved into sources of charge. As mentioned earlier, although the energy absorbed by charges is not sufficient to cross the PN/PP interface barrier, their accumulation at the two-phase interface will cause more severe electric field distortion, leading to partial discharge and breakdown. Therefore, the 0.100 wt% PN could reduce the conductivity of the thin film to some extent. However, the breakdown field strength of the thin film cannot be effectively improved.

The free charge in the dielectric can be mainly divided into excitation charge and injection charge according to their sources. If the anisotropy of the dielectric material is ignored, its excitation charge can be considered uniformly distributed at various positions of the dielectric material. The injected charge gradually migrates from the electrode to the inside of the thin film, and the carrier concentration gradually decreases with the capture process. Therefore, traps located at the electrode/dielectric interface make a greater contribution to suppressing charge carriers. At the same density, the charge capture efficiency of traps located deep in the dielectric is lower, which mainly exist as excitation centers for charges inside the dielectric under high temperature and strong fields. Therefore, it is necessary to further regulate the distribution of traps and reduce the impact of their excitation charges on insulation degradation while ensuring the efficiency of trap charge capture.

Based on the above conclusion, it can be concluded that introducing charge traps on the surface could suppress charge injection and improve the insulation performance of the capacitor film. Traps located inside the capacitor film have a lower efficiency in capturing charges, which can actually lead to a decrease in the performance of the capacitor film. Therefore, targeted structural design of the capacitor film can introduce charge traps to suppress interface injection on the outer layer of the film, while avoiding excessive carrier excitation without changing the inner layer structure of the film. This has the potential to further improve the insulation performance of capacitor films.

The previous text discussed in detail the influence mechanism of surface filling and uniform filling on the insulation performance of polypropylene films. This section mainly discusses the influence of functional groups of PN on the insulation performance of composite films. The energy level distribution and molecular orbitals of PP and different Parylene molecules were analyzed using quantum chemical calculations.

Figure 13.12a shows the three-dimensional distribution of LUMO and HOMO energy levels and molecular orbitals for PP and different parylene molecules. It can be seen that the bandgap widths of different parylene molecules are basically the same. Therefore, filling different parylene molecules as charge traps into the interior of the PP matrix introduces traps with similar depth. However, the introduction of strong electronegative halogen groups resulted in an increase in the overall electron affinity of PN molecules, as shown in Fig. 13.12b. The electron affinity of PN molecules containing side groups is significantly higher than that of PN molecules without side groups. According to the different side group elements and positions, PHT has the

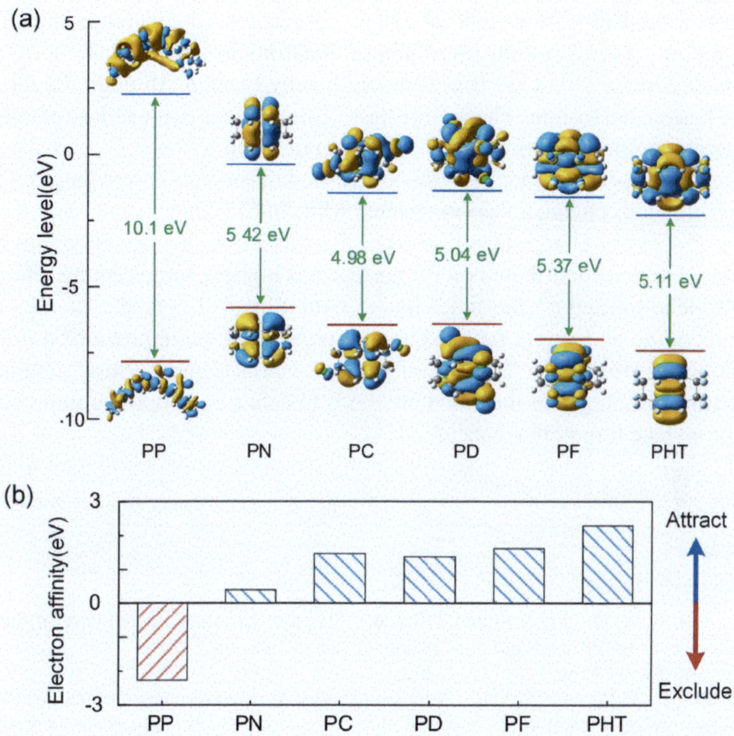

Fig. 13.12 **a** Spatial distribution and **b** electron affinity of PP and different parylene molecules

highest electron affinity energy. Therefore, PHT has the best optimization effect on the thin film at lower temperatures [27].

13.5 Summary

In order to improve the performance of polypropylene capacitor films under high temperature and strong electric fields, this chapter successively proposed the parylene surface deposition method and parylene surface micro filling method to optimize the interface structure and body structure, achieving the improvement of insulation performance of polypropylene films under strong electric and thermal fields. The main conclusions are as follows.

Continuously increasing the thickness of the surface deposition layer cannot infinitely improve the dielectric properties of the capacitor film. The increase in the volume fraction of the surface modified layer will simultaneously enhance the internal electric field of the modified layer and polypropylene layer, exacerbating the electric field distortion at internal defects. In addition, the increase in the thickness of the surface deposition layer allows electrons to continuously accumulate and transfer energy within the free volume, thereby breaking through the interface potential barrier and forming charge injection. This will lead to the failure of interface traps, thereby promoting insulation degradation of the thin film.

Uniformly dispersing trace amounts of parylene inside polypropylene can interrupt the continuous charge transport channel while introducing charge traps. Considering the process of charge injection from the surface of the capacitor film to the inner layer, parylene distributed on the surface has higher charge capture efficiency, while parylene located on the inner layer of the film will generate charges under high temperature and strong field excitation, weakening the improvement effect of its insulation performance. The use of parylene surface micro filling method can combine the advantages of the above methods to achieve the optimal improvement of polypropylene film performance.

References

1. H. L. Liu, B. X. Du, M. Xiao and T. Tanaka, "High-temperature performance of dielectric breakdown in BOPP capacitor film based on surface grafting," *IEEE Trans. Dielect Elect. Insul.*, vol. 28, no. 4, pp. 1264–1272, 2021.
2. H. L. Liu, B. X. Du and M. Xiao, "Improved energy density and charge discharge efficiency of polypropylene capacitor film based on durface grafting," *IEEE Trans. Dielect Elect. Insul.*, vol. 28, no. 5, pp. 1539–1546, 2021.
3. H. L. Liu, B. X. Du, M. Xiao, Z. Y. Ran and T. Tanaka, "Breakdown strength and energy density improvement of polypropylene by parylene deposition for film capacitor," *IEEE Trans. Dielect Elect. Insul.*, vol. 29, no. 3, pp. 981–989, 2022.
4. H. L. Liu, B. X. Du, M. Xiao and Z. Y. Ran, "Improved conductivity and breakdown performance of polypropylene film by parylene blending for power capacitor," *IEEE Trans. Dielect Elect. Insul.*, vol. 29, no. 3, pp. 997–1004, 2022.
5. N. Kawasaki et al, "Flexible picene thin film field-effect transistors with parylene gate dielectric and their physical properties," *Appl. Phys. Lett.*, vol. 96, no. 11, 2010, Art. no. 113305.
6. B. Han, P. Wang, H. C. Jin, Z. S. Hou, and X. Bai, "Wettability and surface energy of parylene F deposited on PDMS," *Phys. Lett. A*, vol. 384, no. 25, 2020, Art. no. 126628.
7. M. Xiao Z. Y. Zhang and B. X. Du, "Dielectric performance improvement of polypropylene film by hierarchical structure design for metallized film capacitors," *J. Phys. D: Appl. Phys.*, vol. 57, no. 38, pp. 5501–5509, 2024.
8. B. X. Du, J. H. Zhang, M. Xiao, H. L. Liu and Z. Y. Ran, "Dielectric breakdown performance of deashed polypropylene for HVDC film capacitors," *IEEE Trans. Dielect Elect. Insul.*, vol. 29, no. 6, pp. 2209–2217, 2022.
9. H. M. Umran, F. Wang, and Y. He, "Ageing: Causes and effects on the reliability of polypropylene film used for HVDC capacitor," *IEEE Access*, vol. 8, pp. 40413–40430, 2020.
10. H. L. Liu, B. X. Du, M. Xiao, Z. Y. Ran and Y. W. Ma, "Aromatic compound improving polypropylene film performance for power capacitor in magnetic field," *IEEE Trans. Dielect Elect. Insul.*, vol. 29, no. 5, pp. 1754–1762, 2022.
11. T. Umemura, K. Abe and K. Akiyama, "Thermal aging behavior of BOPP films," *IEEE Trans. Dielect Elect. Insul.*, vol. EI-22, no. 6, pp. 735–743, 1987.
12. P. Kreczanik, P. Venet, A. Hijazi and G. Clerc, "Study of supercapacitor aging and lifetime estimation according to voltage, temperature, and RMS current," *IEEE Trans. Dielect Elect. Insul.*, vol. 61, no. 9, pp. 4895–4902, 2014.

13. L. Y. Gao, D. M. Tu, S. C. Zhou and Z. L. Zhang, "The influence of morphology on the electrical breakdown strength of polypropylene film," *IEEE Trans. Dielect Elect. Insul.*, vol. 25, no. 3, pp. 535–540, 1990.
14. N. Hayakawa, J. Ishiguro, H. Kojima, K. Kato and H. Okubo, "Fabrication and simulation of permittivity graded materials for electric field grading of gas insulated power apparatus," *IEEE Trans. Dielect Elect. Insul.*, vol. 23, no. 1, pp. 547–554, 2016.
15. E. G. Batyev, "Pauli paramagnetism and Landau diamagnetism," *Phys. Usp.*, vol. 52, no. 12, pp. 1245–1246, 2009.
16. T. Takada, H. Kikuchi, H. Miyake, Y. Tanaka, M. Yoshida, and Y. Hayase, "Determination of charge-trapping sites in saturatedand aromatic polymers by quantum chemical calculation," *IEEE Trans. Dielect Elect. Insul.*, vol. 22, no. 2, pp. 1240–1249, 2015.
17. M. Meunier, N. Quirke and A. Aslanides, "Molecular modeling of electron traps in polymer insulators: Chemical defects and impurities," *J. Chem. Phys.*, vol. 115, no. 6, pp. 2876–2881, 2001.
18. J. G. Su et al, "Effects of antioxidant additives on electrical tree characteristics in EPDM under liquid nitrogen," *IEEE Trans. Appl. Supercond.*, vol. 29, no. 2, pp. 1–5, 2019.
19. F. Baudoin et al, "Bipolar charge transport model with trapping and recombination: an analysis of the current versus applied electric field characteristic in steady state conditions," *J. Phys. D: Appl. Phys*, vol. 41, no. 2, 2008, Art. no. 025306.
20. B. X. Du, J. G. Su and T. Han, "Effects of mechanical stretching on electrical treeing characteristics in EPDM," *IEEE Trans. Dielect Elect. Insul.*, vol. 25, no. 1, pp. 84–93, 2018.
21. C. Yuan et al., "Polymer/molecular semiconductor all-organic composites for high-temperature dielectric energy storage," *Nat. Commun.*, vol. 11, no. 1, pp. 1–8, 2020.
22. Y. C. Tao et al, "Direct measurement of the selective microwave-induced heating of agglomerates of dipolar molecules: The origin of and parameters controlling a microwave specific superheating effect," *J. Phys. Chem. B*, vol. 125, no. 8, pp. 2146–2156, 2021.
23. Z. Y. Ran, B. X. Du and M. Xiao, "High-temperature breakdown performance improvement of polypropylene films based on micromorphology control," *IEEE Trans. Dielect Elect. Insul.*, vol. 28, no. 5, pp. 1547–1554, 2021.
24. S. Kisin, J. D. Doelder, R. F. Eaton and P. J. Caronia, "Quantum mechanical criteria for choosing appropriate voltage stabilization additives for polyethylene," *Polym. Degrad. Stabil.*, vol. 94, no. 2, pp. 171–175, 2009.
25. Y. Yamano, "Roles of polycyclic compounds in increasing breakdown strength of LDPE film," *IEEE Trans. Dielect Elect. Insul.*, vol. 13, no. 4, pp. 773–781, 2006.
26. Y. F. Wang et al, "Ultrahigh energy density and greatly enhanced discharged efficiency of sandwich-structured polymer nanocomposites with optimized spatial organization," *Nano Energy*, vol. 44, pp. 364–370, 2018.
27. Y. Erdogdu and S. Erkoc, "Evolution of the electronic structure and properties of charged titanium doped aluminum nanoclusters," *Comp. Mater. Sci.*, vol. 79, pp. 599–610, 2013.

Chapter 14
Surface Fluoride Modification

14.1 Introduction

The surface modification of polypropylene film can optimize the interface between polypropylene film and electrode, which can not only regulate the insulation performance of polypropylene film, but also improve its surface insulation performance. There are many methods for surface modification of polypropylene film, including plasma technology, chemical vapor deposition method, surface molecular structure modification and so on. The low temperature plasma technology based on the principle of dielectric barrier corona discharge is used to modify the surface of polypropylene film, and the AC breakdown voltage is increased by about 12%. On the other hand, the surface morphology of polypropylene film becomes rough. If the electrode is metallized and evaporated, burrs may be generated between the dielectric and the metallized electrode, which will lead to the concentration of electric field and reduce the overall breakdown field strength. The principle of the composite film is to compound a thin layer of other materials on the surface of the substrate to improve the breakdown strength of the substrate. Studies have shown that multilayer ceramic-polymer composite dielectrics and multilayer all-polymer dielectric films can improve the energy storage density of the films [1, 2]. Some scholars have used plasma enhanced chemical vapor deposition to deposit a layer of silicon dioxide on the surface of polypropylene film, which increases the interface barrier between the electrode and the pattern, thereby reducing the space charge injected by the electrode into the polypropylene film [3]. The modification of polypropylene film by plasma technology and chemical vapor deposition will cause the surface of polypropylene film to be rough or the thickness of polypropylene film to increase, so it is not suitable for the modification of polypropylene film with micron thickness in DC support capacitor. The surface fluoride modification provides new ideas for improving the dielectric properties of polypropylene.

B. Du and M. Xiao, *Polypropylene Film for HVDC Capacitors*,
https://doi.org/10.1007/978-981-96-3029-5_14

14.2 Modification Theory

Fluorine atom is the atom with the smallest atomic radius except hydrogen atom. Its electronic polarization is low and its electronegativity is strong, which makes fluorine-containing compounds have low surface energy and strong chemical inertia. After the fluorine atom replaces the hydrogen atom, it can form a stable carbon–fluorine bond with the carbon atom. The bond energy is 485 kJ/mol, so its chemical stability and thermal stability are strong [4, 5]. Fluorinated polymers are extensively utilized in high-end industries such as automotive and aerospace engineering due to their exceptional electrical insulation, chemical resistance, and thermal properties. Their widespread application stems from these advantageous characteristics, which are crucial in demanding technological environments.

After the modification of the surface molecular structure of the polymer, the functional layer composed of carbon–fluorine bonds formed on the surface of the polymer improves the surface state of the polymer surface and retains the chemical structure inside the polymer matrix, thus having excellent electrical properties [6]. Studies have shown that the surface molecular structure modification technology is used to modify the surface of polyimide films, silicone rubber and other polymers. The functional layer formed on the surface of the polymer can capture charges. The electric field direction of the charge layer formed at the functional layer is opposite to the direction of the applied electric field, which inhibits the migration of charges to the deep traps inside the polymer. The surface charge migration of polypropylene film and the surface trap energy level of polypropylene film are very important for the development of surface discharge. The trap energy level of insulating material has a great influence on flashover, and the deep trap energy level can inhibit the development of flashover [7]. Surface molecular structure control technology can inhibit the injection and accumulation of surface charge [8, 9], improve the flashover voltage and electrical resistance of polymer materials [10, 11]. Controlling the molecular structure of the surface can also have a beneficial impact on inhibiting the accumulation of space charge, as referenced in [12, 13].

14.3 Surface Fluorinated Polypropylene

14.3.1 Effect on the Microstructure

The surface molecular structure modification technology uses the active chemical properties of fluorine gas to react with the polymer to form a functional layer composed of carbon–fluorine bonds on the surface of the polymer, which will neither destroy the microstructure of the polymer surface nor increase the thickness of the polymer. It provides a surface modification method for the optimization of the interface between the polypropylene film and the electrode of the DC supported capacitor [14]. The interface between polypropylene film and electrode is optimized without

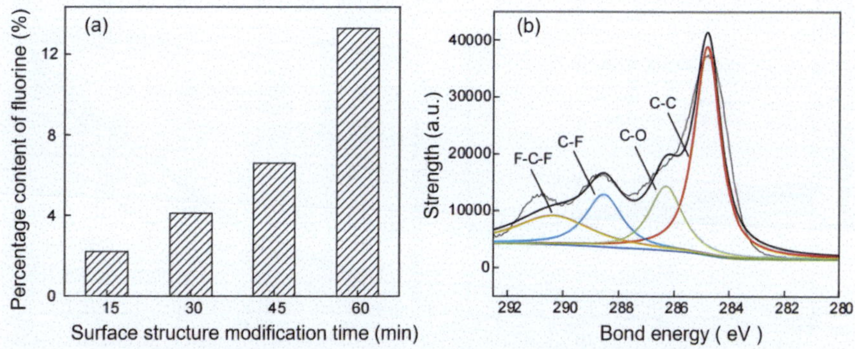

Fig. 14.1 **a** Fluorine content and **b** XPS spectra of surface molecular structure modified PP film. Copyright Clearance Center's RightsLink© License Number: 5825140990643

changing the internal insulation performance of polypropylene film, and the dielectric loss, breakdown and surface discharge characteristics of polypropylene film are regulated. Compared with the polypropylene film, the cross-section morphology of the polypropylene film regulated by the surface molecular structure has changed significantly. An extremely thin modified layer of nm level appears on the surface of the polypropylene film regulated by the surface molecular structure, which almost does not increase the thickness of the polymer. At the same time, the thickness of the fluorinated layer increases with the extension of the regulation time of the surface molecular structure.

In order to determine the successful introduction of fluorine in the modified layer of the surface molecular structure modified polypropylene film, EDAX was used to analyze the elements of the surface layer of the sample. As shown in Fig. 14.1a, fluorine is introduced into the surface layer of the surface molecular structure modified polypropylene film, indicating that fluorine reacts with the molecular chain of the surface layer of the polypropylene film. It can be further inferred that the modified layer of the surface layer of the polypropylene film is the fluorinated layer. At the same time, with the increase of surface molecular structure modification time, the percentage of fluorine element increases continuously. XPS was used to analyze the chemical bonds of polypropylene film and surface molecular structure controlled polypropylene film. The results are shown in Fig. 14.1b. Fluorine element exists in the form of C-F bond on the surface of surface molecular structure controlled polypropylene film.

14.3.2 Effect on the Dielectric Properties

(1) **Polarization and Loss Characteristics**

The permittivity of PP film by surface molecular structure regulation is shown in Fig. 14.2a. The surface molecular structure optimization modification slightly

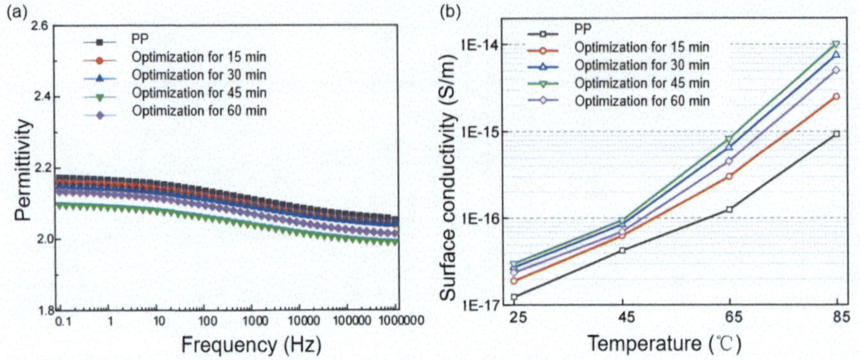

Fig. 14.2 **a** Relationship between the permittivity and the frequency with different optimization time. **b** Relationship between the surface conductivity and the temperature with different optimization time [15]

reduces the permittivity of the polypropylene film, and with the extension of the surface molecular structure modification time, it shows a trend of decreasing first and then increasing. In the short time optimization process, the tertiary hydrogen on the main chain of polypropylene is replaced by fluorine atoms, and the C–F bond movement on the main chain is limited. Therefore, the influence of the electronic polarization of fluorine atoms on the permittivity of polypropylene is mainly considered. As the optimization time of the surface molecular structure increases, it may cause the polypropylene molecular chain to break, and the formed small molecules lead to an increase in molecular steering polarization, which in turn leads to an increase in permittivity.

The surface conductivity of polypropylene film with different interface optimization time is shown in Fig. 14.2b. The interface optimization based on the regulation of surface molecular structure can significantly improve the surface conductivity of polypropylene film. When the interface optimization time increases to 45 min, the surface conductivity of polypropylene film shows a significant increase trend. When the interface optimization time is 60 min, the surface conductivity decreases slightly. Excessive surface molecular structure regulation may lead to the cleavage of polymer chemical bonds and chemical groups, destroy the regularity of the fluorinated layer containing-CF bonds, form deeper trap energy levels, and reduce the migration of carriers, resulting in a decrease in surface conductance [16, 17].

(2) **DC Breakdown Characteristics**

R. R. Xu found that when the interface optimization time was 45 min, the DC breakdown strength of the polypropylene film increased significantly. At high temperature, the breakdown strength of the interface-optimized polypropylene film increased by nearly 6% compared with pure polypropylene. From Fig. 14.3a, it can be seen that with the increase of the number of harmonics, the characteristic breakdown field strength of the interface optimized 45 min polypropylene film shows a

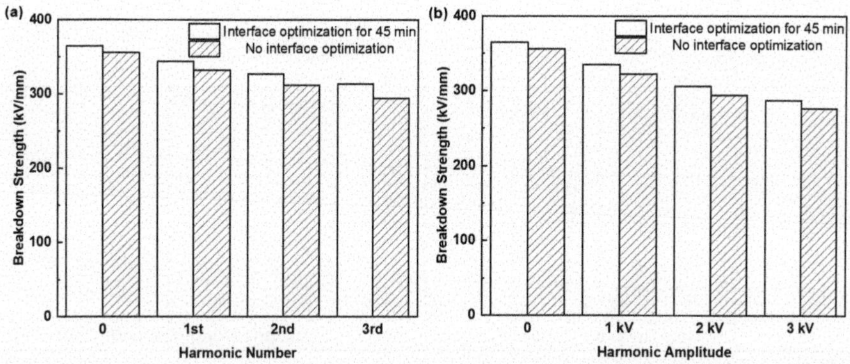

Fig. 14.3 Relationship between the breakdown strength and the harmonic voltages under **a** DC and **b** AC

downward trend. The breakdown field strength of polypropylene thin film after interface optimization for 45 min is slightly higher than that of polypropylene film, but the effect is very limited. As shown in Fig. 14.3b, when the temperature is 85 °C, the interface is optimized for 45 min of polypropylene film breakdown under DC superimposed harmonic voltages of different amplitudes. With the increase of the harmonic amplitude, the characteristic breakdown field strength of the interface optimized 45 min polypropylene film shows a downward trend. The breakdown field strength of polypropylene film with interface optimization for 45 min under DC superposition of different harmonic amplitude voltages is slightly higher than that of polypropylene film.

F atoms were introduced into the surface of BOPP film by direct fluorination method, and a charge shielding layer was formed at the electrode/film interface, which inhibited the charge injection. After fluorination modification, the breakdown field strength of BOPP film increases with the increase of fluoride concentration in a certain range. When the fluorination concentration is 10%, the breakdown field strength of the modified film is increased by 8.7%, and the energy storage density is increased by 18%. However, the grafted F atom will be eliminated to produce HF, which will corrode the electrode.

(3) **Process Analysis**

The trap energy level distribution of polypropylene film and interface optimized polypropylene film is shown in Fig. 14.4a. The deep trap and shallow trap energy levels of the interface optimized polypropylene film increase with the increase of temperature. By comparing the trap energy level distribution of polypropylene film and interface optimized polypropylene film, it can be seen that the trap energy level of interface optimized polypropylene film at different temperatures is slightly smaller than that of polypropylene film, and the deep trap density and shallow trap density are lower than those of polypropylene film.

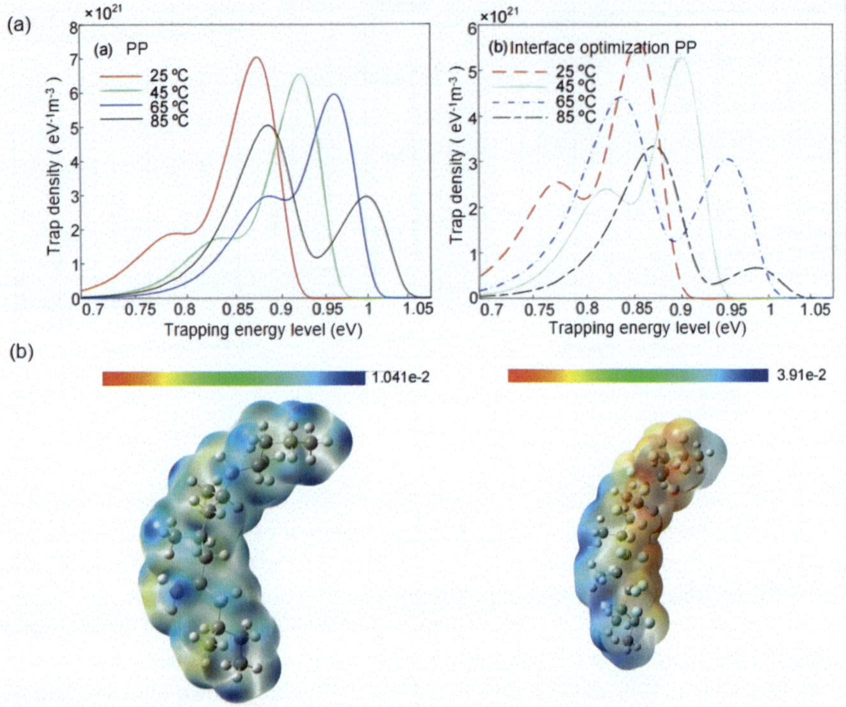

Fig. 14.4 **a** Relationship between the trap energy level distribution and the temperature. **b** Electrostatic potential and energy level distribution of interface optimized polypropylene molecular chain [15]. Copyright Clearance Center's RightsLink© License Number: 5825140990643

The electrostatic potential and energy level distribution of the polypropylene molecular chain and the interface-optimized polypropylene molecular chain are illustrated in Fig. 14.4b. The introduction of fluorine element into interface optimization can regulate the energy level distribution and charge transport of polypropylene film. The 3D electrostatic potential of the polypropylene molecular chain indicates that the polypropylene molecular chain is basically not polar. The electrostatic potential of the interface optimized polypropylene molecular chain near the fluorine atom is significantly enhanced, indicating that the local state center can be formed near the fluorine atom, that is, the trap energy level, which is easier to capture the charge.

The regulation mechanism of the presence of fluoride layer on charge transport is shown in Fig. 14.5. Compared with polypropylene film, interface optimized polypropylene film is easier to inject electron charge, while hole charge is more difficult to inject. The high electronegativity of fluorine results in a powerful affinity for electrons, exerting a pronounced influence on both electron adsorption and the mobility of bonded electrons. The high electron barrier at the interface results in electrons becoming firmly bound to the fluorinated layer upon injection, thereby posing challenges in their subsequent penetration into the polypropylene substrate.

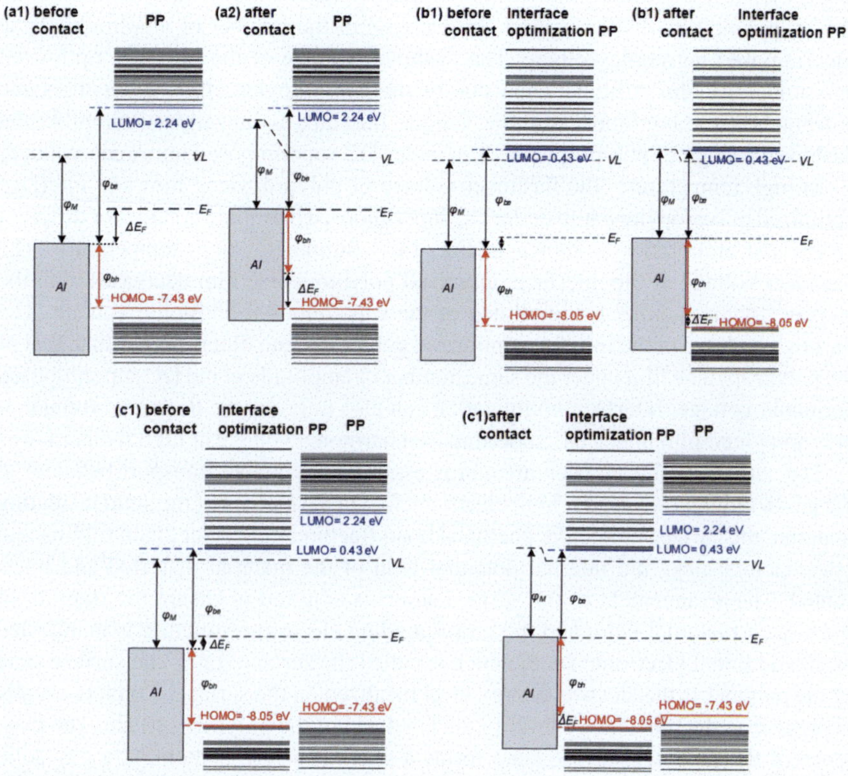

Fig. 14.5 Interface optimization on charge injection process

At the same time, the presence of electron charge accumulated on the fluorinated layer generates an electric field that attenuates the external electric field, thereby diminishing the injection of charges. Consequently, the breakdown field strength of the optimized interface in the polypropylene film increases.

14.3.3 Effect on the Surface Discharge

With the increase of interface optimization time, the flashover voltage scale parameters of PP film under DC voltage show a trend of increasing first and then decreasing, and the maximum can be increased by 10.58%. Using the method of surface molecular structure regulation to optimize the interface of polypropylene for a certain time can improve its flashover voltage and inhibit surface discharge.

The flashover voltage of polypropylene film and interface optimization polypropylene film under DC superimposed harmonic voltages of different orders and amplitudes is shown in Fig. 14.6. The flashover voltage of the interface optimized polypropylene film decreases with the increase of the number or amplitude of the superimposed harmonic voltage. The flashover voltage of the interface optimized polypropylene film is higher than that of the polypropylene film under the same order or the same amplitude harmonic voltage. Interface optimization can improve the flashover voltage of polypropylene film under DC superimposed harmonic voltage.

At high temperature, the flashover voltage of polypropylene film and interface optimized polypropylene film under DC superimposed harmonic voltages of different orders and amplitudes is shown in Fig. 14.7. Similar to room temperature. The flashover voltage of the interface optimized polypropylene film decreases with the increase of the number or amplitude of the superimposed harmonic voltage. The flashover voltage of the interface optimized polypropylene film is higher than that of the polypropylene film under the same number or amplitude of the DC superimposed harmonic voltage. Interface optimization can also improve the flashover voltage of polypropylene film under DC superimposed harmonic voltage at high temperature.

The surface flashover phenomenon is essentially a charge transport behavior at the gas–solid interface under high electric field. On the surface of the solid insulating material, the surface molecular chemical bond fracture, molecular chain folding and other factors make the internal potential field of the material distorted and interrupted. The arrangement of the surface atoms is adjusted to ensure the stability of the system potential field, and the corresponding electronic structure is manifested as an additional electronic state, which is called the surface state. The surface state of the polymer is the electron energy level localized on the surface, which is usually expressed as the trap energy level in the band gap of the polymer surface. The existence of the polymer surface state forms a special surface containing a gas–solid phase. The transport of charges on the complex surface is affected by the electric field and the surface trap energy level. Under the action of high electric field, the trap energy level on the surface of the polymer can capture the charge, which plays an important role in the migration of the charge. The surface flashover voltage of the

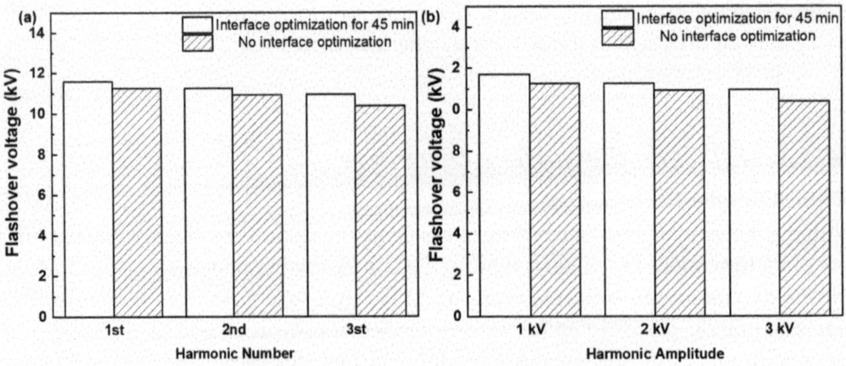

Fig. 14.6 Relationship between the flashover voltage and **a** the harmonic order and **b** the harmonic amplitude

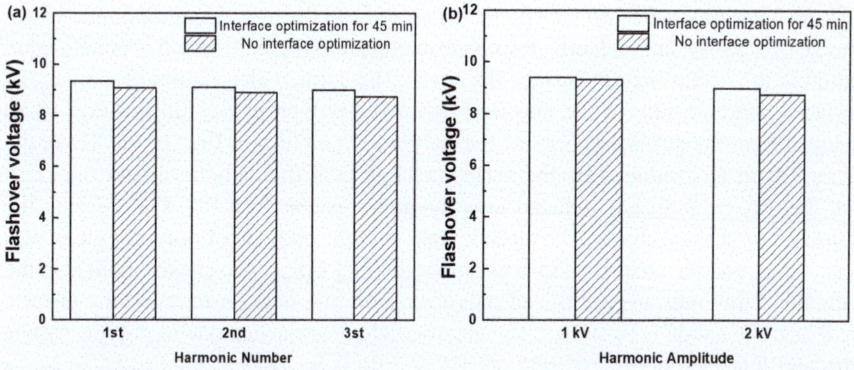

Fig. 14.7 Relationship between the flashover voltage and **a** the harmonic order and **b** the harmonic amplitude at 85 °C

solid insulating material is closely related to the generation, transport and interaction mechanism of the charge [18].

The surface flashover starts at the triple junction of air, electrode and polypropylene film surface. Under the action of electric field, field-induced electron emission occurs at the triple junction. The emitted electrons collide and ionize with gas molecules under the action of electric field, resulting in electron multiplication. Finally, electron avalanche is formed, and discharge triggers flashover, as shown in Fig. 14.8a.

As shown in Fig. 14.8b, the surface discharge of the interface-optimized polypropylene film starts from the fluorinated layer on the surface of the air, the electrode and the polypropylene film. On the one hand, because the bond energy of C-F bond is 485 kJ/mol, the chemical properties are more stable. On the other hand, the interface optimization reduces the trap energy level on the surface of the sample,

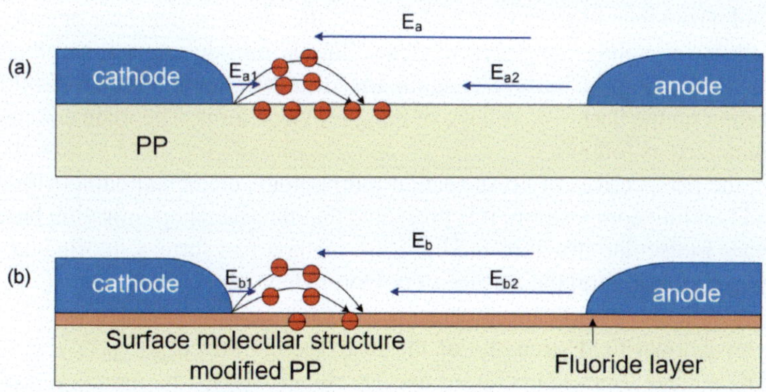

Fig. 14.8 Surface flashover model of interface optimization PP film [15]. Copyright Clearance Center's RightsLink© License Number: 5825140990643

making it easier for the surface charge to detrap from the trap, promoting the surface charge dissipation and reducing the electron impact ionization, which is beneficial to suppress the occurrence of surface flashover. The applied electric field $E_a = E_b$, the surface charge density of the interface optimized polypropylene film in Fig. 14.8b is lower than the surface charge of the polypropylene film in Fig. 14.8a. Then, the electric field E_{b1} induced by the surface charge near the cathode is less than E_{a1}, and the electric field E_{b2} induced near the anode is less than E_{a2}. Considering the influence of surface charge, the electric field near the cathode of polypropylene film is $E_a - E_{a1}$, and the electric field near the anode is $E_a + E_{a2}$. The electric field near the cathode of the interface optimized polypropylene film is $E_b - E_{b1}$, and the electric field near the anode is $E_b + E_{b2}$. The electric field near the cathode of polypropylene film and interface optimized polypropylene film is weakened. Due to $E_a + E_{a2} > E_b + E_{b2}$, the electric field on the anode side of polypropylene film is strengthened more seriously, so the charge is more likely to migrate to the anode and the discharge channel is easier to form. The electric field distortion on the anode side of interface optimized polypropylene film is lower than that of polypropylene film, so interface optimization is beneficial to suppress surface flashover.

14.4 Summary

In this chapter, the interface between polypropylene films and electrodes is optimized using surface layer molecular structure modification, and the influence of interface optimization on the loss, breakdown and along-surface discharge of polypropylene films is summarized. Firstly, the methods of surface layer molecular structure modification and the means of characterizing the interface are introduced. Then the influence of interface optimization on the loss and breakdown characteristics of polypropylene films is described. Based on a combination of experimental and simulation methods, the mechanism of interfacial modulation on the loss and breakdown characteristics of polypropylene films is revealed. Finally, the effect of interface optimization on the flashover voltage of polypropylene films is introduced, and the flashover voltage model along the surface is summarized to illustrate the effect of interface optimization on the flashover voltage of polypropylene films. The main conclusions are summarized as follows:

From the perspectives of cross-section morphology, elemental composition and chemical bonding composition, it is presented that interfacial optimisation based on the surface molecular structure modification method can form a fluorinated layer containing C–F bonds on the surface of polypropylene films. As the time of surface molecular structure modification increases.

The breakdown field strengths of the interface-optimized polypropylene films under DC voltage, high temperature and DC superimposed harmonic voltage are summarized and found to increase slightly compared to the polypropylene films. The trap energy levels of the optimized polypropylene films are found to be smaller than those of the polypropylene films and the trap energy level density is found to be lower.

The effect of surface fluorination modification on the DC flashover voltage of PP films is summarized. Interfacial optimization increased the DC flashover voltage of polypropylene films. Meanwhile, interface optimization improves the flashover voltage of polypropylene films at high temperature with DC superimposed harmonic voltage. The mechanism of the effect of interfacial optimization on the flashover voltage along the surface, obtained based on the polypropylene film and interfacial optimization polypropylene film flashover voltage model, is described.

References

1. Z. X. Sun et al, "Ultrahigh energy storage performance of lead-free oxide multilayer film capacitors via interface engineering," *Adv. Mater.*, vol. 29, no. 5, 2017, Art. no. 1604427.
2. Y. F. Wang et al, "Ultrahigh energy density and greatly enhanced discharged efficiency of sandwich-structured polymer nanocomposites with optimized spatial organization," *Nano Energy*, vol. 44, pp. 364–370, 2018.
3. Y. Zhou et al, "A scalable, high-throughput, and environmentally benign approach to polymer dielectrics exhibiting significantly improved capacitive performance at high temperatures," *Adv. Mater.*, vol. 30, no. 49, 2018, Art. no. 1805672.
4. D. M. Lemal, "Perspective on fluorocarbon chemistry," *J. Org. Chem.*, vol. 69, no. 1, pp. 1–11, 2004.
5. S. J. Blanksby, G. B. Ellison, "Bond dissociation energies of organic molecules," *Acc. Chem. Res.*, vol. 36, pp. 255–263, 2003.
6. V. C. Yadira, R. Hauge, L. Norman and W. E. Billups, "Enhancement of the chemical resistance of nitrile rubber by direct fluorination," *J. Appl. Polym. Sci.*, vol. 89, no. 4, pp. 971–979, 2010.
7. S. T. Li, Q. F. Huang, J. Sun, T. Zhang and J. Y. Li, "Effect of traps on surface flashover of XPLE in vacuum," *IEEE Trans. Dielect Elect. Insul.*, vol. 17, no. 3, pp. 964–970, 2010.
8. Y. Q. Liu, Z. L. An, J. Cang, Y. W. Zhang and F. H. Zheng, "Significant suppression of surface charge accumulation on epoxy resin by direct fluorination," *IEEE Trans. Dielect Elect. Insul.*, vol. 19, no. 4, pp. 1143–1150, 2012.
9. Y. Liu et al, "Rapid potential decay on surface fluorinated epoxy resin samples," *J. Appl. Phys.*, vol. 113, no. 16, 2013, Art. no. 164105.
10. B. X. Du and J. Li, "Surface charge coupling behavior of fluorinated polyimide film under DC and pulse voltage," *IEEE Trans. Dielect Elect. Insul.*, vol. 24, no. 1, pp. 567–573, 2017.
11. B. X. Du, Z. L. Li and J. Li, "Effects of direct fluorination on space charge accumulation in HTV silicone rubber," *IEEE Trans. Dielect Elect. Insul.*, vol. 23, no. 4, pp. 2353–2360, 2016.
12. Z. L. An, J. Cang, X. Chen and Y. Q. Liu, "Method for investigating threshold field of charge injection at electrode/dielectric interfaces by space charge observation," *Appl. Phys. Lett.*, vol. 101, no. 17, 2012, Art. no. 172902.
13. Z. L. An et al, "Correlation between space charge accumulation in polyethylene and its fluorinated surface layer characteristics," *J. Phys. D: Appl. Phys.*, vol. 45, no. 3, pp. 35302–35309, 2012.
14. X. Wang, X. Zhang, and B. F. Zhu, "Influence of surface fluorination treatment on properties of polypropylene Film," *IEEE Trans. Dielect Elect. Insul.*, vol. 28, no.6, pp. 2058–2064, 2021.
15. B. X. Du et al, "Improved charging behaviors and flashover voltage of polypropylene film based on surface molecular structure modification," *Thin Solid Films,* vol. 680, no. 30, pp. 12–19, 2019.

16. Z. Lv et al, "Study on the influential mechanism of fluorination on the flashover voltage of polypropylene film under nanosecond pulse voltage," *IEEE Trans. Dielect Elect. Insul.*, vol. 29, no. 4, pp. 1433–1440, 2022.
17. C. Y. Li, J. L. He and J. Hu, "Surface morphology and electrical characteristics of direct fluorinated epoxy-resin/alumina composite," *IEEE Trans. Dielect Elect. Insul.*, vol. 23, no. 5, pp. 3071–3077, 2016.
18. Y. Huang et al, "Surface flashover performance of epoxy resin micro-composites improved by electron beam irradiation," *Appl. Surf. Sci.*, vol. 406, no. 1, pp. 39–45, 2017.

The manufacturer's authorised representative in the EU is Springer Nature Customer Service Centre GmbH, Europaplatz 3, 69115 Heidelberg, Germany. If you have any concerns regarding our products, please contact ProductSafety@springernature.com

Printed and bound by CPI Group (UK) Ltd, Croydon, CR0 4YY

30/09/2026

02225302-0012